Airplane Stability and Control, Second Edition

From the first machines that flew for just minutes at a time to today's sophisticated aircraft, stability and control have always been crucial considerations. Following up their successful first edition, Malcolm Abzug and Eugene Larrabee forge through to the present day with their informal history of the personalities and the events, the art and the science of airplane stability and control.

The authors, widely known for their contributions to airplane design and development, have captured both the technological progress and the excitement of this important facet of aviation. Much of the book's content captures never-before-available impressions of those active in the field, from pre–Wright brothers airplane and glider builders straight through to contemporary aircraft designers. The chapters are arranged thematically, dealing with single subjects over their entire history. These include early developments, research centers, the effects of power on stability and control, the discovery of inertial coupling, the challenge of stealth aerodynamics, a look toward the future, and much more. This updated edition includes new developments in propulsion-controlled aircraft, fly-by-wire technology, redundancy management, applications, and safety. It is profusely illustrated with photographs and figures and includes brief biographies of noted stability and control figures along with a core bibliography. Professionals, students, and aviation enthusiasts alike will appreciate the comprehensive yet readable approach to this history of airplane stability and control.

Malcolm J. Abzug is the former president of ACA Systems.

E. Eugene Larrabee is Professor Emeritus at the Massachusetts Institute of Technology.

T0211551

Cambridge Aerospace Series 14

Editors:
MICHAEL J. RYCROFT AND WEI SHYY

Frontispiece George Hartley Bryan (1864–1928). The originator, with W. E. Williams, of the equations of airplane motion. Bryan's equations are the basis for the analysis of airplane flight dynamics and closed-loop control and for the design of flight simulators. (From *Obit. Notices of Fellows of the Royal Soc.*, 1932–1935)

Airplane Stability and Control,
Second Edition

A History of the Technologies That Made Aviation Possible

MALCOLM J. ABZUG
ACA Systems

E. EUGENE LARRABEE
Professor Emeritus, Massachusetts Institute of Technology

CAMBRIDGE
UNIVERSITY PRESS

CAMBRIDGE UNIVERSITY PRESS
Cambridge, New York, Melbourne, Madrid, Cape Town, Singapore, São Paulo

Cambridge University Press
The Edinburgh Building, Cambridge CB2 2RU, UK

Published in the United States of America by Cambridge University Press, New York

www.cambridge.org
Information on this title: www.cambridge.org/9780521809924

First published 2002
This digitally printed first paperback version 2005

A catalogue record for this publication is available from the British Library

Library of Congress Cataloguing in Publication data

Abzug, Malcolm J.
  Airplane stability and control : a history of the technologies that made aviation possible /
  Malcolm Abzug, E. Eugene Larrabee. – 2nd ed.
    p.   cm. – (Cambridge aerospace series; 14)
  Includes bibliographical references and index.
  ISBN 0-521-80992-4
  1. Stability of airplanes.   2. Airplanes – Design and construction – History.
  3. Airplanes – Control systems – History.   I. Larrabee, E. Eugene.   II. Title.   III. Series.
  TL574.S7 A2   2002
  629.132′36 – dc21   2001052847

ISBN-13 978-0-521-80992-4 hardback
ISBN-10 0-521-80992-4 hardback

ISBN-13 978-0-521-02128-9 paperback
ISBN-10 0-521-02128-6 paperback

"From th'envious world with scorn I spring,
And cut with joy the wond'ring Skies."

> From Odes II xx, by Horace, translated by Samuel Johnson in 1726.
> St. Martin's Press, New York, 1971.

"When this one feature [balance and steering] has been worked out, the age of flying
machines will have arrived, for all other difficulties are of minor importance."

> From *The Papers of Wilbur and Orville Wright,* Vol. 1.
> McGraw-Hill Book Co., New York, 1953.

"... any pilot can successfully fly anything that looks like an airplane."

> From "Airplane Stability and Control from a Designer's Point of View."
> Otto C. Koppen, *Jour. of the Aero. Sci.*, Feb. 1940.

Contents

Preface

After raising student enthusiasm by a particularly inspiring airplane stability and control lecture, Professor Otto Koppen would restore perspective by saying, "Remember, airplanes are not built to demonstrate stability and control, but to carry things from one place to another." Perhaps Koppen went too far, because history has shown over and over again that neglect of stability and control fundamentals has brought otherwise excellent aircraft projects down, sometimes literally. Every aspiring airplane builder sees the need intuitively for sturdy structures and adequate propulsive power. But badly located centers of gravity and inadequate rudder area for spin recovery, for example, are subtleties that can be missed easily, and have been missed repeatedly.

Before the gas turbine age, much of the art of stability and control design was devoted to making airplanes that flew themselves for minutes at a time in calm air, and responded gracefully to the hands and feet of the pilot when changes in course or altitude were required. These virtues were called flying qualities. They were codified for the first time by the National Advisory Committee for Aeronautics, the NACA, in 1943. Military procurement specifications based on NACA's work followed two years later.

When gas turbine power arrived, considerations of fuel economy drove airplanes into the stratosphere and increased power made transonic flight possible. Satisfactory flying qualities no longer could be achieved by a combination of airplane geometry and restrictions on center-of-gravity location. Artificial stability augmenters such as pitch and yaw dampers were required, together with Mach trim compensators, all-moving tailplanes, and irreversible surface position actuators. At roughly the same time, the Boeing B-47 and the Northrop B-49 and their successful stability augmenters marked the beginning of a new age.

Since then much of the art and science that connected airplane geometry to good low-altitude flying qualities have begun to be lost to a new generation of airplane designers and builders. The time has come to record the lore of earlier airplane designers for the benefit of the kit-built airplane movement, to say nothing of the survivors of the general-aviation industry. Accordingly, this book is an informal, popular survey of the art and science of airplane stability and control. As history, the growth of understanding of the subject is traced from the pre–Wright brothers' days up to the present. But there is also the intention of preserving for future designers the hard-won experience of what works and what doesn't. The purpose is not only to honor the scientists and engineers who invented airplane stability and control, but also to help a few future airplane designers along the path to success.

If this work has any unifying theme, it is the lag of stability and control practice behind currently available theory. Repeatedly, airplanes have been built with undesirable or even fatal stability and control characteristics out of simple ignorance of the possibility of using better designs. In only a few periods, such as the time of the first flights near the speed of sound, theoreticians, researchers, and airplane designers were all in the same boat, all learning together.

The second edition of this book brings the subject up to date by including recent developments. We have also used the opportunity to react to the numerous reviews of the first edition and to the comments of readers. One theme found in many reviews was that the first edition had neglected important airplane stability and control work that took place outside of the United States. That was not intentional, but the second edition has given the authors a new opportunity to correct the problem. In that effort, we were greatly aided by the following correspondents and reviewers in Canada, Europe, and Asia: Michael V. Cook, Dr. Bernard Etkin, Dr. Peter G. Hamel, Dr. John C. Gibson, Bill Gunston, Dr. Norohito Goto, Dr. Gareth D. Padfield, Miss A. Jean Ross, the late Dr. H. H. B. M. Thomas, and Dr. Jean-Claude L. Wanner.

The interesting history of airplane stability and control has not lacked for attention in the past. A number of distinguished authors have presented short airplane stability and control histories, as distinct from histories of general aeronautics. We acknowledge particularly the following accounts:

> Progress in Dynamic Stability and Control Research, by William F. Milliken, Jr., in the September 1947 *Journal of the Aeronautical Sciences.*
>
> Development of Airplane Stability and Control Technology, by Courtland D. Perkins, in the July–August 1970 *Journal of Aircraft.*
>
> Eighty Years of Flight Control: Triumphs and Pitfalls of the Systems Approach, by Duane T. McRuer and F. Dunstan Graham, in the July–August 1981 *Journal of Guidance and Control.*
>
> Twenty-Five Years of Handling Qualities Research, by Irving L. Ashkenas, in the May 1984 *Journal of Aircraft.*
>
> Flying Qualities from Early Airplanes to the Space Shuttle, by William H. Phillips, in the July–August 1989 *Journal of Guidance, Control, and Dynamics.*
>
> Establishment of Design Requirements: Flying Qualities Specifications for American Aircraft, 1918–1943, by Walter C. Vincenti, Chapter 3 of *What Engineers Know and How They Know It,* Johns Hopkins University Press, 1990.
>
> Evolution of Airplane Stability and Control: A Designer's Viewpoint, by Jan Roskam, in the May–June 1991 *Journal of Guidance.*
>
> Recollections of Langley in the Forties, by W. Hewitt Phillips, in the Summer 1992 *Journal of the American Aviation Historical Society.*

Many active and retired contributors to the stability and control field were interviewed for this book; some provided valuable references and even more valuable advice to the authors. The authors wish to acknowledge particularly the generous help of a number of them. Perhaps foremost in this group was the late Charles B. Westbrook, a well-known stability and control figure. Westbrook helped with his broad knowledge of U.S. Air Force–sponsored research and came up with several obscure but useful documents. W. Hewitt Phillips, an important figure in the stability and control field, reviewed in detail several book chapters. His comments are quoted verbatim in a number of places. Phillips is now a Distinguished Research Associate at the NASA Langley Research Center.

We were fortunate to have detailed reviews from two additional experts, William H. Cook, formerly of the Boeing Company, and Duane T. McRuer, chairman of Systems Technology, Inc. Their insights into important issues are used and also quoted verbatim in several places in the book. Drs. John C. Gibson, formerly of English Electric/British Aerospace, and Peter G. Hamel, director of the DVL Institute of Flight Research, Braunschweig, were helpful with historical and recent European developments, as were several other European and Canadian engineers.

Jean Anderson, head librarian of the Guggenheim Aeronautical Laboratory at the California Institute of Technology (GALCIT) guided the authors through GALCIT's impressive aeronautical collections. All National Advisory Committee for Aeronautics (NACA) documents are there, in microfiche. The GALCIT collections are now located at the Institute's Fairchild Library, where the Technical Reference Librarian, Louisa C. Toot, has been most helpful. We were fortunate also to have free access to the extensive stability and control collections at Systems Technology, Inc., of Hawthorne, California. We thank STI's chairman and president, Duane T. McRuer and R. Wade Allen, for this and for very helpful advice.

The engineering libraries of the University of California, Los Angeles, and of the University of Southern California were useful in this project. We acknowledge also the help of George Kirkman, the volunteer curator of the library of the Museum of Flying, in Santa Monica, California, and the NASA Archivist Lee D. Saegesser.

In addition to the European and Asian engineers noted previously, the following people generously answered our questions and in many cases loaned us documents that added materially to this work: Paul H. Anderson, James G. Batterson, James S. Bowman, Jr., Robert W. Bratt, Daniel P. Byrnes, C. Richard Cantrell, William H. Cook, Dr. Eugene E. Covert, Dr. Fred E. C. Culick, Sean G. Day, Orville R. Dunn, Karl S. Forsstrom, Richard G. Fuller, Ervin R. Heald, Robert K. Heffley, Dr. Harry J. Heimer, R. Richard Heppe, Bruce E. Jackson, Henry R. Jex, Juri Kalviste, Charles H. King, Jr., William Koven, David A. Lednicer, Dr. Paul B. MacCready, Robert H. Maskrey, Dr. Charles McCutchen, Duane T. McRuer, Allen Y. Murakoshi, Albert F. Myers, Dr. Gawad Nagati, Stephen Osder, Robert O. Rahn, Dr. William P. Rodden, Dr. Jan Roskam, Edward S. Rutowski, George S. Schairer, Roger D. Schaufele, Arno E. Schelhorn, Lawrence J. Schilling, Dr. Irving C. Statler, and Dr. Terrence A. Weisshaar.

Only a few of these reviewers saw the entire book in draft form, so the authors are responsible for any uncorrected errors and omissions.

This book is arranged only roughly in chronological order. Most of the chapters are thematic, dealing with a single subject over its entire history. References are grouped by chapters at the end of the book. These have been expanded to form an abbreviated or core airplane stability and control bibliography. The rapid progress in computerized bibliographies makes anachronistic a really comprehensive airplane stability and control bibliography.

Malcolm J. Abzug
E. Eugene Larrabee

Airplane Stability and Control, Second Edition

CHAPTER 1

Early Developments in Stability and Control

While scientists and mathematicians in the United States and Europe built the foundations of future advances by developing fundamental aeronautical theory, practical aeronautical designers invented and improved the airplane empirically. As recognized by the Wright brothers, solutions to the stability and control problem had to be found. This chapter presents the largely empirical development of airplane stability and control from the precursors of the Wrights through the end of the first World War. It was only then that aeronautical theory started to have an impact on practical airplane design.

1.1   Inherent Stability and the Early Machines

Pioneer airplane and glider builders who came before the Wright brothers recognized the importance of airplane stability. They had discovered that some degree of inherent stability in flight could be obtained with an appropriate combination of aft-mounted tail surfaces (Cayley and Pénaud), wing dihedral angle or lateral area distribution (Langley and Lanchester), and center of gravity location (Lilienthal).

However, very little thought had been given to the problem of control except for the provision of horizontal and vertical rudders (Langley et al.). It was commonly held that an airplane should hold its course in the air while the pilot decided what to do next. Then the pilot would deflect the rudder to steer it, more or less in the manner of a boat. Only the Wrights recognized that (1) an airplane has to be banked to turn in a horizontal plane; (2) an interaction exists between the banking or roll control and the yawing motion of an airplane; (3) excessive dihedral effects hinder pilot control unless sideslip is suppressed and makes the machine unduly sensitive to atmospheric turbulence; (4) wings can be stalled, leading to loss in control; and (5) control can be regained after stalling by reducing the angle of attack.

After the Wright brothers, Blériot and Levavasseur, the constructor and designer of the Blériot and Antoinette machines, respectively, pioneered in developing tractor monoplanes with normal tail surfaces and wing dihedral angles (Figure 1.1). These two airplanes had a fair amount of inherent stability, unlike the Wright biplanes. They had superior speed, which helped establish the aft tail as the normal arrangement. In fact, the Blériot and Antoinette machines were the transitional forms that led from the Wright brothers' biplanes to the famous pursuit airplanes of World War I.

1.2   The Problem of Control

Otto Lilienthal (1848–1896), Sir Hiram Maxim (1840–1916), and Dr. Samuel Pierpont Langley (1834–1906) followed the empirical route, much as did the Wrights, but they failed to demonstrate man-carrying mechanical flight mainly because they underestimated the problem of control. Lilienthal died of a broken back after losing control of his hang glider. Langley's airplane flew stably in uncontrolled flight as a quarter-scale model but broke up twice in full-scale launches. Maxim's steam-driven airplane might have flown, but it broke free of the down-holding rails on its test track and was wrecked.

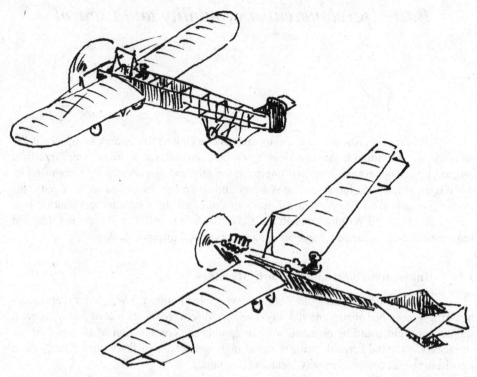

Figure 1.1   Two early flying machines with inherent longitudinal and lateral stability, the Blériot XI Cross-Channel airplane (*above*) and the Levavasseur Antoinette IV (*below*). Both used pronounced wing dihedral, unlike the Wright Flyers.

Maxim's well-engineered failure has had a continuing fascination for modern aeronautical engineers. Bernard Maggin, a noted stability and control engineer with a long career at NACA and the National Research Council, has done extensive research into Maxim's work for the National Air and Space Museum. Another stability and control expert, W. Hewitt Phillips, built and flew a rubber-powered, dynamically scaled, scale model of Maxim's large machine. In unpublished correspondence Phillips reports as follows:

> The model flies fine, despite the lack of vertical tail on the configuration that Maxim used when he ran it on tracks. It flies like a twin pusher, which is what it is. The big propellers aft of the center of gravity give it a marginal amount of directional stability.... Of course, the Reynolds number is far from the full-scale value, but this may not be very important since Maxim used thin airfoils....
>
> My conclusion is that Maxim's airplane would have flown, at least as a giant free-flight model ... I feel that Maxim should get more credit for his engineering contributions than has been given by historians.

The Wrights, on the other hand, addressed the control problem head-on. They taught themselves to fly with three experimental biplane gliders, each fitted with warpable wings for lateral control and all-moving foreplanes for pitch control. The third incorporated an all-moving vertical tail coupled to the wing warp for suppression of adverse yaw due to lateral control actuation, and they learned to fly it quite nicely by 1902. They applied for a patent, describing coupled lateral, or roll and yaw, controls.

In 1903 the Wrights built a powered machine based on the 1902 glider, with a four-cylinder gasoline engine geared to turn its two propellers, and they designed and built the engine and propellers too. They flew it first on 17 December 1903. Modern analysis by Professor Fred E. C. Culick and Henry R. Jex (1985) has demonstrated that the 1903 Wright Flyer was so unstable as to be almost unmanageable by anyone but the Wrights, who had trained themselves in the 1902 glider. In 1904 and 1905 the Wrights improved the lateral stability of their 1903 airplane by removing the downward arch of the wings as seen from the front (the so-called cathedral), reduced its longitudinal instability by ballasting it to be more nose-heavy, and improved its lateral control by removing the mechanical roll–yaw control interconnect.

Henceforth, appropriate roll–yaw control coupling would be provided by pilot skill. Finally, the Wrights learned to sense wing stall, especially in turning flight, and to avoid it by nosing down slightly. By practice they became masters of precision flight in their unstable machine. They also received a patent for their control innovations on 22 May 1905. Confident of their skill and achievements, they built two new machines and sent one to France in 1907.

1.3 Catching Up to the Wright Brothers

Two public demonstrations of perfectly controlled mechanical flight in 1908 by Wilbur Wright in France and by Orville Wright in the United States were clarion calls to the rest of the aeronautical community to catch up with and surpass their achievements. The airplane builders – Curtiss, Blériot, Levavasseur, the Voisins, Farman, Bechereau, Esnault-Pelterie, and others – responded; by 1910 they flew faster and almost as well; by 1911 they flew better. However, even after these momentous achievements, neither the Wrights nor their competitors still had any real understanding of aerodynamic theory.

1.4 The Invention of Flap-Type Control Surfaces and Tabs

Flap-type control surfaces, in which a portion of the wing or tail surface is hinged to modify the surface's overall lift, are at the heart of airplane control. Airplanes designed to fly at supersonic speeds often dispense with flap-type longitudinal controls, moving the entire horizontal surface. Also, some airplanes use spoiler-type lateral controls, in which a control element pops out of the wing's upper surface to reduce lift on that side. Aside from these exceptions, flap-type controls have been the bread-and-butter for airplane control since a few years after the Wright brothers.

It was in 1908 that the aviation pioneer Glenn Curtiss made the first flight of his June Bug airplane, which was equipped with flap-type lateral controls. This was an early, if not the first, advance in lateral control beyond the Wright brothers' wing warping. The Curtiss lateral controls were attached to the interplane struts between the biplane wings and were all-moving. Curtiss evidently saw them as lateral trim devices, since the wheel was connected to the rudder. The French called the flap-type lateral controls *ailerons* – little wings – and the name has persisted in the English language. The Germans call them *querrudern*, or lateral rudders.

The first true flap-type aileron control appears to have been on the French Farman biplane a year or two later. An aerodynamic theory for flap-type controls was needed, but it wasn't until 1927 that Hermann Glauert (Figure 1.2) supplied this need. Control surface tabs are small movable surfaces at the trailing edge, or rear, of a flap-type control. Tabs generate aerodynamic pressures that operate with a long moment arm about the control surface hinge line. Tabs provide an effective way to deflect main control surfaces in a direction opposite to the deflection of the tab itself relative to the main surface.

Figure 1.2 Hermann Glauert (1892–1934). In Glauert's short career he made important airplane stability and control contributions, in control surface, downwash, airfoil, wing, and propeller theory, and in the equations of motion. (From *Obit. Notices of Fellows of the Royal Soc., 1932–1935*)

The tab concept is due to the prolific inventor Anton Flettner, who first applied it to steamboat rudders. One may still find references in the literature to "Flettners," meaning tabs. Flettner received a basic German patent for the tab in 1922. This was for its application to aeronautics. Flettner's patent includes a description of a spring tab device (see Chapter 5), which was promptly forgotten. Glauert's aerodynamic theory for flap-type controls was extended to the tab case in 1928 by W. G. Perrin.

1.5   Handles, Wheels, and Pedals

Before the Wright brothers demonstrated their airmanship, little thought had been given to handles, wheels, and pedals for steering flying machines. Cayley provided his reluctant coachman-aviator with an oar having cruciform blades to "influence" the horizontal and vertical paths of his man-carrying glider. Langley provided Manley, his pilot and engine builder, with a cruciform tail that could be deflected vertically to control pitch attitude and horizontally to turn. Langley expected the dihedral angle of the tandem wings to keep them level, as they had done on his free-flying scale models.

Lilienthal shifted his weight sideways or fore-and-aft on his hang glider to control roll and pitch. This works, but it has limited effectiveness. A roll angle established by a hang glider pilot will make the machine turn if it has weathervane stability, that is, a fixed vertical tail. Hiram Maxim provided his steam-powered airplane with a gyroscopically controlled foreplane to regulate pitch attitude and thought of steering horizontally with differential power to its two independently driven pusher propellers. Fortunately he never had to try this arrangement in flight.

1.6   Wright Controls

In the Wright brothers' 1902 glider and their 1903 Flyer the pilot had a vertical lever for the left hand that was pulled back to increase foreplane incidence. The pilot lay on a cradle that shifted sideways on tracks to cause wing warp. To roll to the left the pilot decreased the incidence of the outer left wings and increased the incidence of the outer right wings. The rudder motion was mechanically connected to the wing warp mechanism to turn the nose left when the pilot wished to lower the left wing, and vice versa for lowering the right wing, thereby overcoming the adverse yaw due to wing warp.

When they began to fly sitting up in 1905, the Wrights retained the left-hand vertical lever for foreplane incidence but added a right-hand vertical lever for wing warp and rudder. They moved the new right-hand lever to the left for left wing down and forward for nose-left yaw. The right-hand lever was moved to the right for right wing down and aft for nose-right yaw. Turn coordination required the pilot to phase control motions, leading with yaw inputs. These unnatural control motions had to be learned and practiced on dual control machines or simple simulators. Bicyclists to the last, they never used their feet for control. They retained this scheme until 1909. Since wing warping involved considerable elastic deformation of the wing structure, they later changed the fore-and-aft motion of the right-hand lever to wing warp and mounted a new, short lever on its top for side-to-side movement to control the rudder. When the Wrights abandoned the all-moving foreplane array for an all-moving rear horizontal tail in 1911, the left-hand lever still controlled its incidence, but now reversed.

The Wrights' patent was for mechanically linked roll and yaw controls. Other airplane builders, notably Curtiss, built airplanes with ailerons, rudders, and elevators, providing independent three-axis control. Curtiss and others asserted that the Wright machine now had independent three-axis control, but U.S. courts upheld the Wright patent against them. The courts maintained that the coupling of roll and yaw controls in the Curtiss machines existed in the mind of the aviator and was essential to the art of flying. Therefore, the Curtiss independent three-axis control infringed on the Wright patent!

1.7   Blériot and Déperdussin Controls

Louis Blériot devised what has become the standard stick and rudder cockpit controls for small airplanes. A central stick between the pilot's legs is moved forward for nose down, aft for nose up, to the left for left wing down, and to the right for right wing down. The pilot's feet rest on a rudder bar from the ends of which a pair of cables run straight back to the rudder horns. Thus left foot forward deflects the rudder to the left and turns the machine to the left (Figure 1.3). Blériot fitted a nonrotatable wheel to the top of the control stick, perhaps to give the pilot a firmer grip for wing warping.

The Blériot rudder pedal convention, now quite standard, is just the opposite of bicycle or "Flexible Flyer" sled steering, where operators turn the handlebars or hand grips in the direction of the desired turn. Igor Sikorsky thought that the Blériot convention was backward. Sikorsky crossed the rudder wires on all of his airplanes, to make them steer like bicycles. He warned conventionally trained pilots not to try to fly these particular machines.

Before the war, the company Société pour Avions Déperdussin (SPAD) produced a series of military airplanes and racers that were designed by Bechereau. These streamlined airplanes were fitted with Blériot-style rudder bars and a vertical wheel that could be moved fore and aft for pitch and turned sideways for wing warp. The wheel's increased mechanical

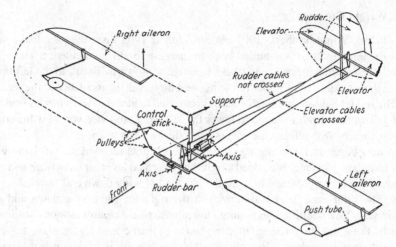

Figure 1.3 Diagrammatic sketch of a simple airplane control system. When the controls are moved as shown by arrows on the stick and rudder bar, the surfaces move as shown by the arrows. (From Chatfield, Taylor, and Ober, *The Airplane and Its Engine*, McGraw-Hill, 1936)

advantage as compared with levers was needed to warp wings of increased torsional rigidity. The Déperdussin wheel is the ancestor of modern control yokes.

1.8   Stability and Control of World War I Pursuit Airplanes

By 1917 trial and error during the first World War had established the wire braced biplane with aft-tailed surfaces as the normal configuration. Diagonal brace wires between the wing struts and fuselage and within the wing frames made a torsionally rigid structure that resisted twisting and instability failure in high-speed dives. The heavy engine in front and the generous tail surfaces behind tended to keep the fuselage and wings aligned with the velocity of flight. The pilot could apply roll control by aileron deflection, yaw control by rudder deflection, and pitch control by elevator deflection – all independently. Aerodynamic hinge moments tended to center the controls. By ground adjustment of wing, fin, and tailplane rigging the airplane could be made to maintain level flight with cruising power in calm air for a minute or so.

Violent maneuvering in combat was provided mainly by the elevator, which had sufficient authority to bring the airplane to a full stall. Horizontal turning flight required rolling the airplane about its longitudinal axis quickly, which was most often accomplished by combined rudder and aileron deflection. The rudder-induced sideslip produced an unsymmetric stall and a snap or flick roll that could be checked at the desired angle by relieving stick back pressure and centering the rudder and ailerons.

The ailerons were difficult to deflect at combat speeds but could be used to produce a slow or barrel roll. An important use of the ailerons was to produce a cross-controlled (e.g., right rudder and left stick) nonrolling sideslip for glide path control while landing. The glide angle could be steepened appreciably by sideslipping in a steep bank, incidentally giving the pilot a good view of the touchdown point.

A dangerous aspect of stability and control of the otherwise benign World War I airplanes was inadvertent stalling and spinning at low altitudes, the so-called arrival and departure stalls (and spins). Moderate sideslip at stall would provoke a snap roll, which rapidly developed into the dreaded tail spin, or spinning nose dive. Generally there was insufficient room for recovery before ground contact.

Arrival stalls are still produced in modern airplanes by attempting to rudder the airplane around to the proper heading on final approach at a low speed without banking. The inner wing stalls and drops. The pilot attempts to pick it up with aileron deflection, which aggravates the situation. The airplane stalls and spins into the intended turn. The pilot who survives complains that the ailerons did not work.

Departure stalls are more spectacular. The pilot takes off from a small field. As the obstacles at the end of the field get near, with the engine at full power, the pilot rolls with the ailerons to a steep bank angle and turns away. The airplane has insufficient power to climb in steeply turning flight, so the pilot applies top rudder to hold up the nose. The resulting sideslip stalls the top wing, and the airplane performs an over-the-top snap roll and spin entry, followed by a fiery crash at full power.

Because of the stall–spin propensity of World War I airplanes, student pilots were given flight instruction on spin entry and recovery in airplanes with generally docile behavior. However, some airplanes, notably the Sopwith Camel, killed many student pilots because of its particularly vicious stalling characteristics. The Camel's main fuel tank was behind the pilot, and the fully loaded center of gravity was so far aft that the airplane was unstable in pitch just after takeoff. Constant pilot attention was required to keep it from stalling.

Not only that, but, like many other World War I airplanes, the Camel's vertical tail was too small. Any stall automatically became a snap roll spin entry, even without intentional rudder deflection. Finally, once spinning, the Camel required vigorous rudder deflection against the spin to stop the motion. A well-behaved airplane, on the other hand, has to be held in a spin; letting the controls go free should result in automatic recovery. Directional instability was so common among World War I airplanes that the Royal Air Force (R.A.F.) resisted closed cockpits for years so that pilots could use wind on one cheek as a sideslip cue.

Another dangerous feature of World War I airplanes was the gyroscopic effect of rotary engines. According to Gibson (2000), engine gyroscopic effect in the Sopwith Camel required left rudder for both left and right turns and caused a departure if full power was used over the top of a loop at too low an airspeed. Pilots were warned to attempt their first hard right turns only above 1,000 feet.

1.9   Contrasting Design Philosophies

Comparison of the 1917 British (Royal Aircraft Factory) S.(scouting) E.(experimental)-5 and the Fokker D-VII shows an interesting contrast between the design philosophies of the Royal Aircraft Factory designers, who had been exposed to primitive airplane stability theory, and Anthony H. G. Fokker and his co-worker, Reinhold Platz, neither of whom had any formal technical training. Platz had been trained in the art of acetylene gas welding, which he applied to the construction of steel tube airplane fuselages, while Fokker was an experienced craftsman, pilot, and small boat sailor with an instinct for aerodynamics.

The strong dihedral (5 degrees) of the S.E.-5 wings (Figure 1.4) is evidence of an attempt to give the airplane inherent spiral stability. On spirally stable airplanes, if the pilot establishes a banked turn, the rudder and elevator have to be held in a deflected position to continue the turn. If the pilot centers the rudder bar and control stick, a correctly rigged airplane will automatically, but slowly, regain wings-level flight.

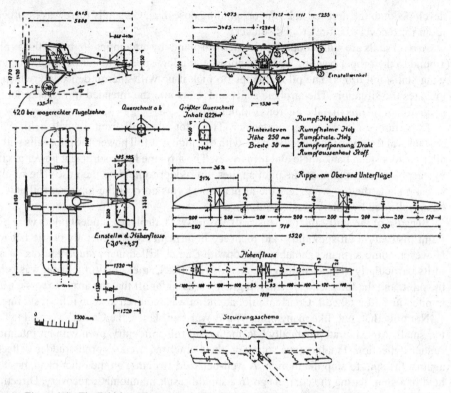

Figure 1.4   The British paid attention to inherent spiral stability during World War I days, building 5 degrees of dihedral into the S.E.-5. (From Jane's *All the World's Aircraft*, 1919. Jane's used a German source for these drawings since the S.E.-5 was still classified in Britain in 1919.)

The S.E.-5's control surfaces had no aerodynamic balance and were difficult to move at diving speeds. Thin wing sections were used. The designers also had embraced a whim for numerology; the wings had 250 square feet of area and 5-foot chords; they were set at 5 degrees with respect to the thrust line, and so on.

Modern flight tests of World War I fighters (using the Shuttleworth Collection) give the S.E.-5A high ratings. Ronald Beaumont says this airplane was

> perhaps the best handling fighter on either side, with excellent pitch and yaw control and inherent stability on both axes, and with light and responsive ailerons up to the quite high speed of 130 mph.

The Fokker D VII (Figure 1.5) had wooden-frame cantilever wings, almost without dihedral, with a thick airfoil section, an early result of Prandtl/Lanchester circulation theory. David Lednicer reported (2001) that the D VII wing airfoil was close to the Göttingen 418. The D VII had a steel-tube–welded fuselage and tail assembly. Horn balances (called elephant ears) were provided to lessen the pilot effort to deflect the ailerons, elevators, and rudder.

When Fokker flew the first version he realized he had created a dangerous airplane. Before the German Air Ministry officials could get a good look at it, he rebuilt it secretly in the hangar, moving the wings aft to make it less unstable, lengthening the fuselage, and modifying the vertical tail to incorporate a fixed fin. As a result of the D VII's long tail moment arm; blunt-nosed, cambered airfoil sections; and mechanically limited up elevator

Figure 1.5   The Fokker D-VII, built without wing dihedral, showing no concern for spiral stability.
This machine had horn aerodynamic balances at the tips of all control surfaces, to reduce control forces.
(From *Progress in Airplane Design Since 1903*, NASA Publication L-9866, 1974.)

deflection, stability and control at low speeds and climb rate were quite good. In its final form
it pleased everyone so much that it was mentioned in the Treaty of Versailles as a military
airplane that had to be surrendered to the Allied authorities, the only one so designated.

1.10   Frederick Lanchester

Airplane stability and control theory in the modern sense began with Frederick
William Lanchester. Lanchester was not really a theoretician but a mechanical engineer who
devoted most of his effort to the construction of very innovative motor cars. He performed
aeronautical experiments with free-flying gliders. He speculated correctly on the vortex
theory of lift and the nature of the vortex wake of a finite wing but was unable to give
these ideas a useful mathematical form. His free-flying gliders were inherently stable and
exhibited an undulating flight path, which he analyzed correctly in 1897. He misnamed
the motion the "phugoid," intending to call it the "flying" motion; actually he called it the
"fleeing" motion, having forgotten that the Greek root already existed in the English word
"fugitive."

Lanchester published two books, *Aerodynamics* in 1907 and *Aerodonetics* in 1908, which
expressed his views and the results of his experiments. He even talked with Wilbur Wright,
evidently to no avail, because Wilbur had no understanding of inherent stability in flight,
already demonstrated by Pénaud, Langley, and Lanchester on a small scale.

1.11   G. H. Bryan and the Equations of Motion

The mathematical theory of the motion of an airplane in flight, considered as a rigid
body with 6 degrees of freedom, was put into essentially its present form by Professor George
Hartley Bryan (frontispiece) in England in 1911. In an earlier (1903) collaboration with W. E.
Williams, Bryan had developed the longitudinal equations of airplane motion only. Bryan's
important contribution rested on fundamental theories of Sir Isaac Newton (1642–1727)
and Leonhard Euler (1707–1783). Today's stability and control engineers are generally

$$W\left(\frac{du}{gdt} + \frac{qw}{g} - \frac{rv}{g}\right) = W\sin\theta + H \quad - X \qquad . \quad .$$

$$W\left(\frac{dv}{gdt} + \frac{ru}{g} - \frac{pw}{g}\right) = W\cos\theta\cos\phi \quad - Y \qquad . \quad .$$

$$W\left(\frac{dw}{gdt} + \frac{pv}{g} - \frac{qu}{g}\right) = -W\cos\theta\sin\phi \; - Z \qquad . \quad .$$

$$A\frac{dp}{gdt} - F\frac{dq}{gdt} + (C-B)\frac{rq}{g} + F\frac{pr}{g} = \quad - L \qquad . \quad .$$

$$B\frac{dq}{gdt} - F\frac{dp}{gdt} + (A-C)\frac{pr}{g} - F\frac{qr}{g} = \quad - M \qquad . \quad .$$

$$C\frac{dr}{gdt} \qquad + (B-A)\frac{pq}{g} - F\left(\frac{p^2-q^2}{g}\right) = -Hh - N$$

$$p = \dot{\phi} + \dot{\psi}\sin\theta$$
$$q = \dot{\theta}\sin\phi + \dot{\psi}\cos\theta\cos\phi$$
$$r = \dot{\theta}\cos\phi - \dot{\psi}\cos\theta\sin\phi$$

Figure 1.6 G. H. Bryan's modern-looking 6-degree-of-freedom equations of airplane motion, supplemented by the Euler angular rate equations. For NASA symbols interchange Y and Z, M and N, q and r, and v and w. A, B, and C are moments of inertia about NASA's X-, Y-, and Z-axes, respectively. (From Bryan, *Stability in Aviation*, 1911)

$$W\frac{du}{gdt} \qquad\qquad = W\,\epsilon\cos\theta_o + \delta H - uX_u - vX_v - rX_r$$

$$W\left(\frac{dv}{gdt} + \frac{rU}{g}\right) = -W\epsilon\sin\theta_o \quad - uY_u - vY_v - rY_r$$

$$C\frac{dr}{gdt} \qquad\qquad = -h\delta H \qquad - uN_u - vN_v - rN_r$$

$$W\left(\frac{dw}{gdt} - \frac{qU}{g}\right) = -W\phi\cos\theta_o -- wZ_w - pZ_p - qZ_q$$

$$A\frac{dp}{gdt} - F\frac{dq}{gdt} = \qquad\qquad - wL_w - pL_p - qL_q$$

$$B\frac{dq}{gdt} - F\frac{dp}{gdt} = \qquad\qquad - wM_w - pM_p - q\,M_q$$

Figure 1.7   The perturbation form of Bryan's equations of airplane motion. The longitudinal equations are above, the lateral equations below. Note the absence of control derivatives. (From Bryan, *Stability in Aviation*, 1911)

astonished when they first see these equations (Bryan, 1911). As his book's (Bryan, 1911) title indicated, he focused on airplane stability, not control. Aside from minor notational differences, Bryan's equations are identical to those used in analysis and simulation for the most advanced of today's aircraft (Figures 1.6 and 1.7).

Not surprisingly, at this early date he does not cover in detail control force and moments, nor does he treat the airplane as an object of control. The perturbation equations in Fig. 1.7 include stability but not control derivatives. The influence of external disturbances such as gusts is also not addressed, although he recognizes this and other problems by presenting a summary of questions not covered in his book that set an agenda for years of research.

Bryan calculated stability derivatives based on the assumption that the force on an airfoil is perpendicular to the airfoil chord. W. Hewitt Phillips points out that while this theory is not the most accurate for subsonic aircraft, it is quite accurate for supersonic aircraft, particularly those with nearly unswept wings, such as the Lockheed F-104. Thus, Bryan might be considered even more ahead of his time than is usually acknowledged.

Bryan obtained solutions for his equations and arrived at correct modes of airplane longitudinal and lateral motion. At the end of *Stability in Aviation*, Bryan reviews earlier stability and control theories by Captain Ferber, Professor Marcel Brillouin, and MM. Soreau and Lecornu of France; Dr. Hans Reissner of Germany; and Lieutenant Luigi Crocco of Italy.

Little progress was made at first in the application of Bryan's equations because of the difficulties of performing the calculations and the uncertainties in estimating the airloads corresponding to airplane motions. The airloads associated with rolling, pitching, and yawing motions, the so-called rotary loads, were a particular problem. Early efforts were made at the National Physical Laboratory in England to measure these rotary airloads in a wind tunnel.

The evolution of Bryan's equations of airplane motion into an indispensable tool for stability and control researchers and designers is traced in Chapter 18 of this book.

1.12 Metacenter, Center of Pressure, Aerodynamic Center, and Neutral Point

Joukowski's theory for a monoplane wing shows that the pitching moment coefficient about a point one-quarter of the chord length behind the wing leading edge is a constant at all angles of attack. This point is the wing's aerodynamic center. The constant pitching moment coefficient at the aerodynamic center is negative, or nose-down, for positively cambered wings and positive for negatively cambered wings.

The combination of the lift coefficient, varying linearly with the angle of attack, and a constant pitching moment coefficient can be replaced with a single vector lift force. The point of action on the wing chord of the single vector force is called the wing's center of pressure. Early longitudinal trim and stability calculations use the center of pressure concept, rather than the concept of aerodynamic center.

Dr. Charles McCutchen speculates that center of pressure methods came to the aeronautical field through aeronautical engineers originally trained as naval architects, such as

Figure 1.8 Bennett Melvill Jones (1887–1975), an airplane pilot as well as an important stability and control figure. His stability and control section in Durand's *Aerodynamic Theory* instructed many engineers in airplane dynamics. Jones made contributions in stall research. (From *Biog. Memoirs of Fellows of the Royal Soc.*, 1977)

Dr. Jerome C. Hunsaker. Ship stability against roll is determined by metacentric height, the distance between the ship's center of gravity and the line of action of the buoyancy force in a roll. Ship buoyancy force acting through the metacenter is analogous to the wing lift force acting through the center of pressure, except that other forces are involved in airplane stability.

B. Melvill Jones (Figure 1.8) proposed the term "metacentric ratio" for what we now call static margin, the distance in wing chords between the center of gravity and the center of gravity position for neutral stability of the complete airplane (Jones, 1934). That center of gravity position is called the neutral point. According to J. C. Gibson (2000), an article published by S. B. Gates in a 1940 *Aircraft Engineering* magazine explained the neutral point concept in physical terms, providing a "stunning revelation" to engineers trained on center of pressure methods.

CHAPTER 2

Teachers and Texts

2.1   Stability and Control Educators

The gap between aeronautical theory and stability and control practice has never entirely closed. However, the number of aeronautical engineers trained in stability and control theory has grown greatly since the subject started to be taught in the aeronautical engineering schools that sprang up starting around 1920.

By 1922, there were already five U.S. universities with programs in aeronautical engineering: the Massachusetts Institute of Technology, the California Institute of Technology, the University of Michigan, the University of Washington, and Stanford University. In that same year, Drs. Alexander Klemin and Collins Bliss of New York University offered elementary aerodynamics as an option in mechanical engineering, launching the aeronautical program there. Other U.S. colleges and universities, too numerous to mention, followed in later years. By 1997, the American Institute of Aeronautics and Astronautics (AIAA) had no fewer than 145 student branches in colleges and universities around the world.

Otto C. Koppen was an aviation pioneer who taught airplane stability and control while continuing as a designer of new airplanes. William F. Milliken, Jr. (1947) had this to say about Koppen's contributions to his own field of airplane dynamics:

> Since about 1930 the course of airplane dynamics in this country has been widely and continuously influenced by the research and teachings of Otto C. Koppen. His theoretical investigations, wind tunnel work (oscillators), and achievements in airplane design are now well known. For years his course in stability and control at MIT was unique in its treatments of the complete dynamics, and many current trends in design and research may be traced directly to his work as an educator....

European educators were busy as well. At the time (1911) he wrote *Stability in Aviation*, G. H. Bryan was a mathematics professor at the University of North Wales. Twenty problems for further research may be found at the end of that book. Airplane stability and control was soon taught widely in Great Britain.

In 1920, the first edition of Leonard Bairstow's famous book *Applied Aerodynamics* appeared. That year he was appointed Professor of Aerodynamics at Imperial College. In 1945, Ernest F. Relf and William J. Duncan, both already identified with airplane stability and control, served on a group that established the Cranfield College of Aeronautics, for postgraduate education. Relf became principal, and Duncan became a legendary Cranfield professor.

In Japan, the late Professor Kyuichiro Washizu (1921–1981) played somewhat the same role as did Otto Koppen in the United States. Washizu introduced the concept of airplane stability and control to Japan, at the Department of Aeronautics, the University of Tokyo. His students spread all over Japan, eventually leading stability and control education and research in that country. Pioneer stability and control educators in continental Europe were the Belgian Professor Frederic Charles Haus, the Dutch Professor Otto H. Gerlach, and the German Professor Karl-H. Doetsch.

Airplane stability and control is still being taught in universities around the world, bringing fresh talent into the field. Stability and control research is carried on in many of these schools, usually as graduate programs. Governmental and commercial research institutions are important contributors, as well.

Photographs of some stability and control educators and engineers who are mentioned in the text appear in Figure 2.1.

2.2   Modern Stability and Control Teaching Methods

The digital computer has revolutionized the teaching of airplane stability and control, just as it has its practice. In precomputer times, flight dynamics students had to learn numerical techniques for factoring high-degree polynomials and producing linearized transient responses. Eigenvalues or roots of the equations of airplane motion were extracted by factorization, and the flight modes of motion were found.

Computer programs for root extraction and a great deal more are at the modern engineer's fingertips, and present-day teachers of flight dynamics have found ways to use the digital computer to improve their courses. A few instances follow. Stanford University Professor Arthur Bryson's book *Control of Aircraft and Spacecraft* uses Matlab$^\circ$ computer routines in many examples and problem assignments. For example, pages 199–201 show a student how to synthesize an optimal climb-rate/airspeed stability augmentation system using Matlab. As with other mathematics computer packages, Matlab is available in a low-cost student edition.

The State University of New York at Buffalo Professor William J. Rae assigns exercises that use a 6-degree-of-freedom computer program called SIXDOF to explore in detail the solutions of the nonlinear equations of airplane motion. This supplements normal instruction in the modes of motion and control theory using linearized equations. Still another approach has been pursued by University of Florida Professor Peter H. Zipfel. He makes available to his students a CADAC CD-ROM disk, with which to build modular aerodynamics, propulsion, and guidance and control computer models. As in the previous cases, students are able to solve realistic stability and control problems without getting lost in routine mathematical detail.

In France, Professor Jean-Claude Wanner, on the staffs of several universities, is developing an advanced flight mechanics teaching tool, in the form of a CD-ROM. A stability volume computes the time response of an airplane specified by the user to control and throttle inputs, presenting results in the form of conventional strip charts, but also in real time as viewed from the cockpit or the ground. There are preliminary interactive chapters, including text and exercises, on subjects such as phugoid motion and accelerometer instruments.

The only cautions that might be applied to these modern approaches are the same ones that must be observed in the practice of engineering, using powerful digital computers. Both student and working engineer must keep in mind the assumptions that lie behind flight dynamics computer programs, their limitations as well as their capabilities. Good practice also requires reasonableness checks on computer output using independent simple methods.

2.3   Stability and Control Research Institutions

Stability and control research is carried out worldwide in a variety of institutions. There are the governmental laboratories and privately owned research firms that are found in almost every country where airplanes are built. Universities are also major centers of

Figure 2.1   Stability and control engineers mentioned in the text.

Duane T. McRuer   Steven Osder   Courtland D. Perkins

William H. Phillips   Lloyd D. Reid   Ernest F. Relf

Jan Roskam   A. Jean Ross   Robert F. Stengal

Fred C. Weick   Roland J. White   Charles H. Zimmerman

Figure 2.1   (*continued*)

research. One can assume that any university where stability and control is in the curriculum has corresponding faculty or graduate research projects. Finally, aircraft manufacturers can be assumed to have stability and control research in progress at some level, depending on how advanced their products are.

As in many research areas, the lines of responsibility are blurred by cooperative projects among the three groups mentioned – laboratories, universities, and manufacturers. This is especially true when expensive or unique equipment is involved. A prime example is the regulated use of large governmental wind tunnels for aircraft that will enter production, rather than be used for research. Another example is the use of the highly specialized in-flight simulators operated by governmental and private laboratories to improve the flying qualities of future production aircraft.

The world-wide extent of stability and control institutions is evident in the sources named for reports and papers in the References and Core Bibliography section of this book.

2.4 Stability and Control Textbooks and Conferences

Sharing honors with the educators in bringing stability and control theory into practice are a number of textbooks. B. Melvill Jones' "Dynamics of the Aeroplane" section of W. F. Durand's *Aerodynamic Theory* (1934) is the earliest textbook with a widespread impact in the field. The second edition of Leonard Bairstow's *Applied Aerodynamics* (1939) was also widely used. A truly landmark textbook appeared in 1949: *Airplane Performance, Stability and Control*, by Courtland D. Perkins and Robert E. Hage. This book was and still is a favorite for undergraduate stability and control instruction. It is very well balanced, giving space to many important topics. Aside from the Jones, Bairstow, and Perkins texts, we list a number of other stability and control textbooks (in English), in the order published. The pace of publishing new books in the field seems to be accelerating in recent times, with about as many new titles since 1990 as in all previous years.

W. J. Duncan, *Control and Stability of Aircraft*. Cambridge, 1952

B. Etkin, *Dynamics of Flight: Stability and Control*. Wiley, 1959, 1982, 1985 (with L. D. Reid)

A. W. Babister, *Aircraft Stability and Control*. Pergamon, 1961

W. R. Kolk, *Modern Flight Dynamics*. Prentice-Hall, 1951

E. Seckel, *Stability and Control of Airplanes and Helicopters*. Academic, 1964

T. Hacker, *Flight Stability and Control*. Elsevier, 1970

J. Roskam, *Flight Dynamics of Rigid and Elastic Airplanes*. U. Kansas, 1972

B. Etkin, *Dynamics of Atmospheric Flight*. Wiley, 1972

D. McRuer, I. Ashkenas, and D. Graham, *Aircraft Dynamics and Automatic Control*. Princeton, 1973

H. Ashley, *Engineering Analysis of Flight Vehicles*. Dover, 1974

A. W. Babister, *Aircraft Dynamic Stability and Response*. Pergamon, 1980

F. O. Smetana, *Computer-Assisted Analysis of Aircraft Performance, Stability and Control*. McGraw-Hill, 1983

J. M. Rolfe and K. J. Staples, eds., *Flight Simulation*. Cambridge, 1986

R. C. Nelson, *Flight Stability and Automatic Control*. McGraw-Hill, 1989

J. H. Blakelock, *Automatic Control of Aircraft and Missiles*. Wiley, 1991

B. L. Stevens and F. L. Lewis, *Aircraft Control and Simulation*. Wiley, 1992

A. E. Bryson, Jr., *Control of Aircraft and Spacecraft*. Princeton, 1994

B. W. McCormick, *Aerodynamics, Aeronautics and Flight Mechanics*. Wiley, 2nd ed., 1995

G. J. Hancock, *An Introduction to the Flight Dynamics of Rigid Aeroplanes.* Horwood, 1995

M. B. Tischler, ed., *Advances in Aircraft Flight Control.* Taylor & Francis, 1996; AIAA, 2000

D. Stinton, *Flying Qualities and Flight Testing of the Airplane.* AIAA, 1996

J. Russell, *Performance and Stability of Aircraft.* Arnold, 1996

M. V. Cook, *Flight Dynamics Principles.* Arnold, 1997

J.-L. Boiffier, *The Dynamics of Flight: The Equations.* Wiley, 1998

L. V. Schmidt, *Introduction to Aircraft Flight Dynamics.* AIAA, 1998

M. J. Abzug, *Computational Flight Dynamics.* AIAA, 1998

B. Pamadi, *Performance, Stability, Dynamics, and Control of Airplanes.* AIAA, 1998

J. Hodgkinson, *Aircraft Handling Qualities.* AIAA, 1999

P. H. Zipfel, *Modeling and Simulation of Aerospace Vehicle Dynamics.* AIAA, 2000

R. W. Pratt, ed., *Flight Control Systems: Practical Issues in Design and Implementation.* AIAA, 2000

As one of the founders of NATO's Advisory Group for Aerospace Research and Development, or AGARD, Dr. Theodore von Kármán helped greatly in the advance and dissemination of stability and control knowledge in the years following the second World War. In 1997, AGARD was incorporated into the Research and Technology Organization (RTO) of the Defense Research Group of NATO, as a budgetary measure.

While AGARD was still active, it brought together stability and control experts from all of the NATO countries in a wide variety of periodic meetings. For example, there were meetings of flight mechanics and guidance and control panels, symposiums, lecture series, and consultant and exchange programs. The publications and meetings of AGARD and its successor RTO remain a useful source for stability and control research and development.

In the United States, the work of the AGARD and RTO groups is paralleled by that of the Atmospheric Flight Mechanics Committee of the American Institute of Aeronautics and Astronautics, or AIAA. Under the committee's direction, the AIAA holds valuable atmospheric flight mechanics and guidance and control conferences yearly. The Society of Automotive Engineers (SAE) has an active Aerospace Committee (formerly A-18) that briefs its members periodically and has a publication and meetings program.

In Europe, the Royal Aeronautical Society, the German Aerospace Society DGLR, and the French National Academy hold periodic conferences that address flight mechanics and control.

Flying Qualities Become a Science

The stability and controllability of airplanes as they appear to a pilot are called flying or handling qualities. It was many years after airplanes first flew that individual flying qualities were identified and ranked as either desirable or unsatisfactory. Even more time passed before engineers had design methods connected with specific flying qualities. A detailed and fascinating account of the early work in this area is given in Chapter 3 of Stanford University Professor Walter G. Vincenti's scholarly book *What Engineers Know and How They Know It*. We pick up the story in 1919, with the first important step in the process that made a science out of airplane flying qualities.

3.1   Warner, Norton, and Allen

Vincenti found that the first quantitative stability and control flight tests in the United States occurred in the summer of 1919. MIT Professor Edward P. Warner (Figure 3.1), working part time at the NACA Langley Laboratory, together with two NACA employees, Frederick H. Norton and Edmund T. Allen, made these tests using Curtiss JN-4H "Jennies" and a de Havilland DH-4. They made the most fundamental of all stability and control measurements: elevator angle (with respect to the fixed part of the tail, or stabilizer) and stick force required for equilibrium flight as a function of airspeed.

Warner and Norton made the key finding that the gradient of equilibrium elevator angle with respect to airspeed was in fact an index of static longitudinal stability, the tendency of an airplane to return to equilibrium angle of attack and airspeed when disturbed. The elevator angle–airspeed gradient thus could be correlated with the 1915–1916 MIT wind-tunnel measurements by Dr. Jerome C. Hunsaker of pitching moment versus angle of attack on the Curtiss JN-2, an airplane similar to the JN-4H. In the words of Warner and Norton (1920):

> If an airplane which is flying with the control locked at a speed corresponding to the negatively sloped portion of the elevator position curve is struck by a gust which decreases its angle of attack, the angle will continue to decrease without limit. If the speed is low enough to lie on the positively sloping portion of the curve, the airplane will return to its original speed and angle of trim as soon as the effect of the gust has passed. *A positive slope [of the elevator angle-airspeed gradient] therefore makes for longitudinal stability.* (Italics added)

A strange aspect of the Warner and Norton JN-4H test results was the effect of airspeed on static longitudinal stability. The JN-4H was stable at airspeeds below about 55 miles per hour and unstable above that speed (Figure 3.2). One would be tempted to look for an aeroelastic cause for this, except that wind-tunnel tests of a presumably rigid model showed the same trend. The cause remains a mystery. The 1915–1916 MIT wind-tunnel tests were supplemented in 1918 by the U.S. Air Service at McCook Field with JN-2 wind-tunnel tests, in which the model had an adjustable elevator angle.

The McCook Field group was active in stability and control flight tests at the same period. As part of an armed service procurement activity, McCook's primary interest was in airplane

Figure 3.1   Edward Pearson Warner (1894–1958). His DC-4E flying qualities requirements launched a new science. (From National Air and Space Museum)

suitability for military use, rather than in aeronautical research. Thus, it is understandable that there were no measurements at the level of sophistication of the Norton and Allen tests at the NACA. Captain R. W. (Shorty) Schroeder was one of the Air Service's top test pilots. His 1918 (classified Secret) report on the Packard-Le Pere LUSAC-11 fighter airplane's handling qualities was completely qualitative.

In the course of the pioneering stability and control flight tests at the NACA Langley Laboratory, instrumentation engineers including Henry J. E. Reid, a future Engineer-in-Charge at Langley, came up with specialized devices that could record airplane motions automatically, freeing pilots from having to jot down data while running stability and control flight tests. Langley Laboratory individual recording instruments developed in the 1920s measure control positions, linear accelerations, airspeed, and angular velocities.

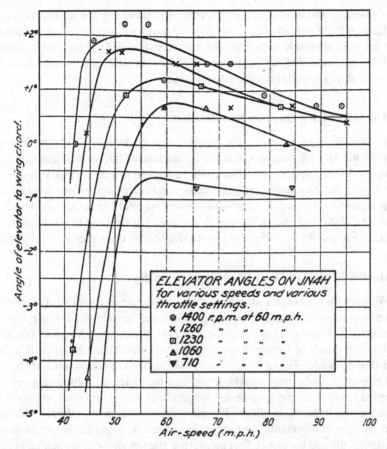

Figure 3.2 Warner and Norton's measurements of elevator angles required to trim as a function of airspeed and power for the Curtiss JN4H (Jenny) airplane. They correctly interpreted the data to show static longitudinal instability at airspeeds above the peaks of the curves. (From NACA Rept. 70, 1920)

In each recording instrument, a galvanometer-type mirror on a torsion member reflects light onto a photographic film on a drum. A synchronizing device keys together the recordings of individual instruments, putting timing marks on each drum. Frederick Norton said in later years that the work at Langley in which he took the most pride was the development of these specialized flight recording instruments (Hansen, 1987).

The instrument developments put NACA far in front of other groups in the United States who were working on airplane stability and control. The photorecorder was typical technology at other groups running stability and control tests, such as the U.S. Army Air Corps Aircraft Laboratory at Wright Field. In the photorecorder, stability measurement transducers, ordinary flight instruments, and a stopwatch are mounted in a bulky closed box and photographed by a movie camera. Data are then plotted point by point by unfortunate technicians or engineers reading the film.

As another indication of NACA's advanced flying qualities measurement technology, one of this book's authors (Abzug) who served in the U.S. Navy during World War II remembers having to borrow a stick force measuring grip from NACA to run an aileron roll test on a North American SNJ trainer.

NACA flying qualities research in the 1920s and early 1930s also trained a group of test pilots, including Melvin N. Gough, William H. McAvoy, Edmund Allen, and Thomas

Carroll, in stability and control research techniques, including the ability to reach and hold equilibrium flight conditions with accuracy. As with all good research test pilots, the NACA group worked closely with flight test engineers and in fact took part in discussing NACA's flying qualities work with outsiders. All of this helped lay the groundwork for the comprehensive flying qualities research that followed.

3.2   The First Flying Qualities Specification

·   Edward P. Warner, acting as a consultant to the Douglas Aircraft Company in the design of the DC-4E transport, has the distinction of having first embodied flying qualities research into a specification that could be applied to a new airplane design, much as characteristics such as strength and performance had been specified previously. Warner's 1935 requirements were based on interviews with airline pilots, industrial and research test pilots, and NACA staff engineers. Warner also recognized the need to put flying qualities requirements on a sounder basis by instrumented flight tests correlated with pilot opinions.

3.3   Hartley Soulé and Floyd Thompson at Langley

Warner's ideas were picked up by the NACA (Warner was, after all, a member of the main committee; his ideas counted), and the grand comprehensive attack on airplane flying qualities started. The authorizing document was NACA Research Authorization number 509, "Preliminary Study of Control Requirements for Large Transport Aircraft" (Hansen, 1987). Hartley A. Soulé (Figure 3.3), a portly, worldly-wise staff member at the NACA Langley Aeronautical Laboratories in Hampton, Virginia, ran tests the following year (1936) that attempted to correlate the long-period longitudinal or phugoid mode of motion with pilots' opinions on handling qualities. The phugoid motion involves large pitch attitude and height changes at essentially constant angles of attack. Eight single-engine airplanes were tested by Soulé and his group. This pioneering attempt showed that neither period nor damping of the phugoid motion had any correlation with pilot opinion.

However, the NACA was fairly launched on the idea of correlating flying qualities measurements with pilots' opinions. Soulé and his associate, Floyd L. Thompson, outlined the practical steps needed to carry out Warner's ideas. Flying qualities had to be defined "in terms of factors known to be susceptible of measurement by existing NACA instruments or by instruments that could be readily designed or developed."

Thompson and Soulé started with what we would now call a set of "straw man" requirements based on Warner's work, but modified to be measurable by NACA's instruments. They used a Stinson Reliant SR-8E single-engined high-wing cabin airplane (Figure 3.4) for the tests. It turned out that the only instruments that needed to be specially developed for the Stinson tests were force-measuring control wheel and rudder pedals. These used hydraulic cells developed by the Bendix Corporation as automobile brake pedal force indicators.

The "straw man" NACA requirements seemed to ignore Soulé's previous findings of the unimportance of the longitudinal phugoid motion, and a reasonably well-damped oscillation of period not less than 40 seconds was specified. Even more curiously, F. W. Lanchester's research on the phugoid period was quite overlooked in the straw man requirements, although Lanchester's results were given in the well-known 1934 "Dynamics of the Airplane," by B. Melvill Jones, which was included in Volume V of W. F. Durand's *Aerodynamic Theory*. Lanchester had shown that the phugoid period for all aircraft was linearly proportional to airspeed and would invariably fall below the required 40 seconds at airspeeds under about 150 miles per hour.

Figure 3.3 Hartley A. Soulé (1905–), a pioneer in flying qualities research. (From Hansen, *Engineer in Charge*, NASA SP-4305, 1987)

Aside from this cavil, Soulé's research followed reasonable lines. Each straw man requirement was stated, test procedures to check each requirement were spelled out, and the test results were presented and discussed. Some of Soulé's 1940 test procedures have come down to our day virtually unchanged except for the increased sophistication of data recording. For example, there were measurements of elevator angle and stick force for equilibrium flight at various airspeeds, measurements of time to bank to a specified angle, and, most advanced of all, measurements of the period and damping of the phugoid oscillation

Figure 3.4 The Stinson SR-8E airplane used in Hartley Soulé's pioneering stability and control flight test measurements. (From NACA Rept. 700, 1940)

as a function of airspeed (Figure 3.5). The Lanchester approximation for phugoid period is shown as a dashed line in Figure 3.5(a).

In his published report Soulé (1940) provides the variations with airspeed for equilibrium flight of both the elevator angle and the control column position from the dashboard. These data would give exactly the same trends were it not for stretch of the control cables that connect the two, under load. Vincenti's book tells the interesting story of the discovery of the effects of cable stretch on the Stinson data.

Soulé's report was reviewed in preliminary form by engineers at the Chance Vought Aircraft plant in Connecticut, who noticed that different incidence settings of the horizontal tail affected the variations in elevator angle for equilibrium flight, an unexpected outcome. C. J. McCarthy of Chance Vought wrote to Soulé suggesting that the discrepancy might be explained by control cable stretch if the elevator angle had been deduced from the control column position, rather than having been measured directly at the surface itself. According to Vincenti:

> Robert R. Gilruth, a young engineer who had recently taken over the flying quality program when Soulé moved to wind tunnel duties, measured the stretch under applied loads and found that Chance Vought's supposition was in fact correct . . . In tests of later airplanes, elevator angles were measured directly at the elevator. Such matters seem obvious in retrospect, but they have to become known somehow.

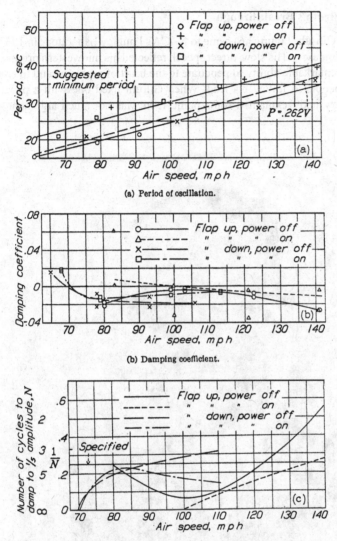

Figure 3.5 Dynamic stability measurements for the Stinson SR-8E, made around 1937 by Hartley Soulé. (From NACA Rept. 700, 1940)

Some Stinson measurements called for by the straw man requirements are definitely archaic and not a part of modern flying qualities. Very specific requirements were put on the time needed to change pitch attitude by 5 degrees; these were checked. Likewise, the need to limit adverse yaw in aileron rolls was dealt with by measuring maximum yawing acceleration and comparing it with rolling acceleration. The yaw value was supposed to be less than 20 percent of the roll value. However, all of the pieces were in place now and ready for the next major step.

After the Stinson tests the NACA had the opportunity to test a large airplane, the Martin B-10B bomber. Those results went to the Air Corps in a confidential report of 1938. According to Vincenti, Edward Warner was able to feed back both the Stinson SR-8E and Martin B-10B results to his flying qualities requirements for the Douglas DC-4E, which was just beginning flight tests.

3.4   Robert Gilruth's Breakthrough

Robert R. Gilruth (Figure 3.6) came to NACA's Langley Laboratory in 1937 from the University of Minnesota. His slow, direct speech reflected his midwestern origins. He is remembered for a remarkable ability to penetrate to the heart of problems and to convince and inspire other people to follow his lead. When Gilruth fixed one with a penetrating stare and, with a few nods, explained some point, there was not much argument. Many

Figure 3.6   Robert R. Gilruth (1913–2000). An early expert in airplane flying qualities and design methods. He played a leading role later on in NASA's space program. (From Hansen, *Engineer in Charge*, NASA SP-4305, 1987)

years later, when NACA became NASA, Gilruth was tapped by the government to head the NASA Manned Spacecraft Center.

Gilruth's seminal achievement was to rationalize flying qualities by separating airplanes into satisfactory and unsatisfactory categories for some characteristic, such as lateral control power, by pilot opinion. He then identified some numerical parameter that could make the separation. That is, for parameter values above some number, all aircraft were satisfactory, and vice versa. The final step was to develop simplified methods to evaluate this criterion parameter, methods that could be applied in preliminary design.

The great importance of this three-part method is that engineers now could design satisfactory flying qualities into their airplanes on the drawing board. Although proof of good flying qualities still required flight testing, engineers were much less in the dark. The old way of doing business is illustrated by an NACA report (W-81, ACR May 1942) on the development of satisfactory flying qualities on the Douglas SBD-1 dive bomber. Discussing a Phase III series of tests in September 1939, the report said, "The best configuration from this phase was submitted to a pilot representative from the [Navy] Bureau [of Aeronautics], who considered that insufficient improvements [in control force characteristics] had been made."

Two applications of this new method were published (1941) by Gilruth and co-authors Maurice D. White and W. N. Turner to static longitudinal stability and to lateral control power, respectively. White had joined Gilruth at Langley in 1938. Fifteen airplanes ranging in size from the Aeronca K to the Boeing B-15 were tested in the first series, on longitudinal stability. Gilruth and White suggested a design value of 0.5 for the gradient of elevator angle with angle of attack, for the propeller-idling condition, to ensure power-on stability and adequate stick movement in maneuvers.

In the lateral control application of the new method, 28 different wing–aileron combinations were tested, including alterations to the wings and ailerons of two of the airplanes tested (Figure 3.7). The famous lateral control criterion function pb/2V came into being as a result of this work. pb/2V is the helix angle described by a wing tip during a full-aileron

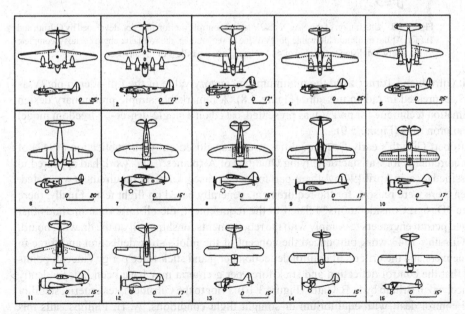

Figure 3.7 The 15 airplanes tested by Gilruth and White to get data for their longitudinal stability estimation method. (From NACA Rept. 711, 1941)

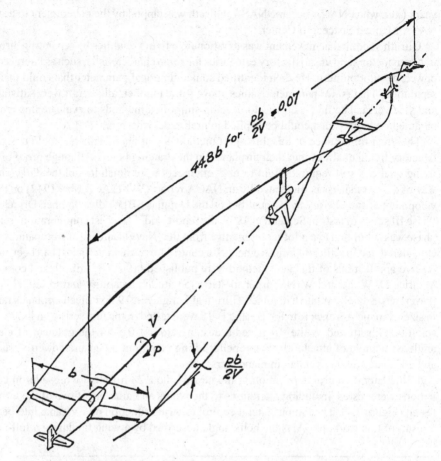

Figure 3.8   Illustration of the NACA pb/2V lateral control criterion function developed by Gilruth and Turner. At the minimum allowable pb/2V value of 0.07 radian, the roll helix angle creates a complete roll in a forward distance traveled of 44.8 wing spans, regardless of the airspeed.

roll. Gilruth and Turner fixed the minimum satisfactory value of the full-aileron pb/2V as 0.07, expressed in radian measure (Figure 3.8). A remarkably simple preliminary design estimation technique for pb/2V was presented, based on a single-degree-of-freedom model for aileron rolls (Figure 3.9).

Robert Gilruth's early flying qualities work was closed out with publication (1943) of "Requirements for Satisfactory Flying Qualities of Airplanes." This work had appeared in classified form in April 1941. A three-part format was used. First, the requirement was stated. Then there were reasons for the requirement, generally based on flight tests. Finally, there were "Design Considerations" related to the requirement, the all-important methods that would permit engineers to comply with the requirements for ships still on the drawing board.

Gilruth's 1943 work introduced the concept of the pilot's stick deflection and force in maneuvers and the criteria of control deflection per g and stick force per g. Vincenti points out that the control deflection and stick force per g criteria may have been independently conceived in Britain by S. B. Gates (Figure 3.10). Prior to the Gilruth/Gates criteria, stability and control dealt with equilibrium or straight flight conditions. W. H. Phillips calls this quantization of maneuverability one of Gilruth's most important contributions to airplane flying qualities.

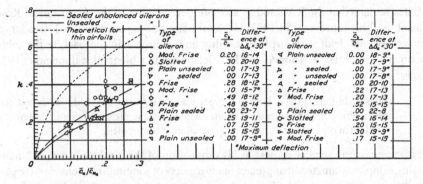

Figure 3.9   The control surface effectiveness derivative κ, back-figured from flight tests of 28 different airplane configurations. Wing twist, control system stretch, and nonlinearities at large control angles all account for the markedly lower values than the Glauert thin airfoil theory, shown dotted. (From Gilruth and White, NACA Rept. 715, 1941)

3.5   S. B. Gates in Britain

Sidney Barrington (Barry) Gates had a remarkable career as an airplane stability and control expert in Britain, spanning both World Wars. He left Cambridge University in 1914, in his words, "as an illegitimate member of the Public Schools Battalion of the Royal Fusiliers." Somebody with unusual perception for those days saw that the young mathematician belonged instead in the Royal Aircraft Factory, the predecessor of the Royal Aircraft Establishment or RAE, and he was transferred there.

Figure 3.10   Sidney B. Gates (1893–1973), contributor to understanding of airplane spins; originator of neutral and maneuver points, stick force per g, and many other flying qualities matters. (From Thomas and Küchemann, *Biographical Memoirs of Fellows of the Royal Society*, 1974)

Gates remained at the RAE and as a committee member of the Aeronautical Research Council until his retirement in 1972. His total output of papers came to 130, a large proportion of which dealt with airplane stability and control. His approach to the subject is described by H. H. B. M. Thomas and D. Küchemann in a biographical memoir (1974), as follows:

> Gates's life-long quest – how to carve a way through the inevitably harsh and complex mathematics of airplane motion to the shelter of some elegantly simple design criterion based often on some penetrating simplification of the problem.

This approach is seen at its best in his origination of the airplane static and maneuver margins, simple parameters that predict many aspects of longitudinal behavior. He gave the name "aerodynamic center" to the point on the wing chord, approximately 1/4 chord from the leading edge, where the wing pitching moment is independent of the angle of attack. The wing aerodynamic center concept led to current methods of longitudinal stability analysis, replacing wing center of pressure location.

Gates is also remembered for a long series of studies on airplane spinning, begun with "The Spinning of Aeroplanes" (1926), co-authored with L. W. Bryant. Gates' other airplane stability and control contributions over his career cover almost the entire field. They include work in parameter estimation, swept wings, VTOL transition, flying qualities requirements, handling characteristics below minimum drag speed, transonic effects, control surface distortion, control friction, spring tabs, lateral control, landing flap effects, and automatic control.

Together with Morien Morgan, in 1942 Gates made a two-month tour of the United States, "carrying a whole sackload of RAE reports on the handling characteristics of fighters and bombers." The pair met with the main U.S. aerodynamics researchers and designers at the time, including Hartley Soulé, Gus Crowley, Floyd Thompson, Eastman Jacobs, Robert Gilruth, Hugh Dryden, Courtland Perkins, Walter Diehl, Jack Northrop, Edgar Schmued, W. Bailey Oswald, George Schairer, Kelly Johnson, Theodore von Kármán, and Clark Millikan. As Morgan comments, the scope and scale of the 1942 "dash around America" showed what a towering reputation Gates had worldwide.

3.6   The U.S. Military Services Follow NACA's Lead

Following NACA's lead, both the U.S. Air Force and the U.S. Navy Bureau of Aeronautics issued flying qualities specifications for their airplanes. Indeed, after the war, the allies discovered that the Germans had established military flying qualities requirements at about the same time. In April 1945 the U.S. Air Force and Navy coordinated their requirements, recognizing that some manufacturers supplied airplanes to both services. The coordinated Air Force specification got the number R-1815-A; the corresponding Navy document was SR-119A. In 1948 the final step was taken and the services put out a joint military flying qualities specification, MIL-F-8785. This document went through many subsequent revisions, the most significant of which was the 1969 MIL-F-8785B(ASG).

The main difference between Gilruth's NACA requirements and the military versions was in the detailed distinctions made by the military among different types of aircraft. For example, in the military version, maneuvering control force, the so-called stick force per g, was generalized to apply to airplanes with any design limit load factor. NACA had recognized only two force levels, for small airplanes with stick controls and for large ones with wheel controls. Special requirements were given in the joint specification for aircraft meant to fly from naval aircraft carriers. MIL-F-8785 and its revisions were incorporated into the procurement specifications of almost all military aircraft after 1948.

The transformation of one particular flying qualities requirement from the original NACA or Gilruth version through successive military specifications can be traced (Westbrook and McRuer, 1959). This requirement is for the longitudinal short-period oscillation. The longitudinal short period is a relatively rapid oscillation of angle of attack and pitch attitude at relatively constant airspeed. The original NACA requirement applies to the stick-free case only, as follows:

> When elevator control is deflected and released quickly, the subsequent variation of normal acceleration and elevator angle should have completely disappeared after one cycle.

Gilruth (1943) goes on to give reasons for this requirement, as follows:

> The requirement specifies the degree of damping required of the short-period oscillation with controls free. A high degree of damping is required because of the short period of the motion. With airplanes having less damping than that specified, the oscillation is excited by gusts, thereby accentuating their effect and producing unsatisfactory rough-air characteristics. The ratio of control friction to air forces is such that damping is generally reduced at high speeds. When the oscillation appears at high speeds as in dives and dive pull-outs, it is, of course, very objectionable because of the accelerations involved.

The first U.S. Air Force specification, C-1815, relaxed the NACA requirement, allowing complete damping in two cycles instead of one. This was done because opinions collected from Air Force pilots and engineers were that the response with the stick fixed was always satisfactory, and so the short-period oscillation was of no importance in design. However, by the time of a 1945 revised specification, R-1815-A, further experience led back to the original NACA requirement of complete damping for the stick-free case in one cycle. A refinement to an analytically more correct form, one better suited for design and flight testing, was made in the next revision, R-1815-B. This is damping to 1/10 amplitude in one cycle, corresponding to a dimensionless damping ratio of 0.367.

Modern design trends, especially higher operating altitudes and wing loadings, decreased the damping in the stick-fixed case, while with irreversible controls the stick-free case essentially disappeared. When the initial stick-fixed damping requirements were set, in MIL-F-8785, the level was set at a damping ratio of 0.110, or damping to a half-amplitude in one cycle. This relatively low damping requirement is based on NACA experience with research airplanes, whose pilots seem agreeable to low damping levels. As service experience was gained with high-altitude, dense airplanes the trend was reversed and damping requirements were increased again.

This uncertainty in the desirable level of longitudinal short-period damping was typical of what led to ambitious, reasonably well-funded Air Force and Navy research programs to rationalize the flying qualities data base. In the United States, flying qualities flight and ground simulator testing went on all over the country, especially at the NACA laboratories, the Cornell Aeronautical Laboratory, Systems Technology, Inc., NATC Patuxent River, Wright Field, and at Princeton University. British, German, Dutch, and French laboratories also became active in flying qualities research at this time.

Under U.S. Air Force sponsorship, the Cornell Aeronautical Laboratory used a variable-stability jet fighter to make a systematic attack on the longitudinal short-period damping question. Robert P. Harper and Charles R. Chalk ran experiments with variations in short-period damping and frequency at constant levels of stick force and displacement per unit normal acceleration. They found a "bull's-eye" of good damping and natural frequency combinations, surrounded by regions of acceptable to poor performance.

This and similar efforts went into successive revisions of MIL-F-8785, reaching at last the "C" revision of November 1980. All along, the specification writers were guided by peer reviews and conferences involving specification users in the industry. At one point, the U.S. Navy Bureau of Aeronautics requested Systems Technology, Inc. to search for weaknesses in the specification. The resultant report (Stapleford, 1970) was issued with the attention-getting title, "Outsmarting MIL-F-8785(ASG)." Good summaries of the revision work may be found in Chalk (1969) and Ashkenas (1973).

3.7   Civil Airworthiness Requirements

Military aircraft are procured under comprehensive flying qualities specifications. These specifications are contractual obligations by the manufacturer to a single military customer. On the other hand, the flying qualities of commercial aircraft are generally not governed by contracts with individual customers, but rather by governmental agencies, acting to protect the flying public.

Civil flying qualities requirements are found in airworthiness requirement documents. Compliance with airworthiness standards is proved in flight testing, leading to the award of airworthiness certificates and freedom to market the aircraft. Civil airworthiness requirements are the minimum set that will ensure safety. This is a different objective than military requirements, which not only address safety, but also the effectiveness of military airplanes in their missions. Thus, flying qualities requirements found in civil airworthiness requirements are much less detailed than specified requirements for military airplanes. This is a key distinction between the two sets of requirements.

3.8   World-Wide Flying Qualities Specifications

As mentioned earlier, the German air forces in World War II operated under a set of military flying qualities requirements related to the Gilruth set of 1943. The growth of civil aviation after the war led to a number of national and world-wide efforts to specify flying qualities requirements, in order to rationalize aircraft design and procurement in each country and the international licensing of civil aircraft. The goal of internationally agreed upon civil aircraft flying qualities standards is the responsibility of the International Civil Aviation Organization (ICAO), an arm of the United Nations. Annex 8 of the ICAO Standards deals with airworthiness, which includes adequate flying qualities (Stinton, 1996).

Standards have also been adopted by individual countries for both civil and military machines. An earlier section traced the evolution of U.S. flying qualities specifications for military aircraft. Similar evolutions took place all over the world. British military specifications are in the UK DEF STAN publications. In particular, DEF-STAN 00-970, issued in 1983, is similar in style to MIL-F-8785C and provides much the same information (Cook, 1997).

British civil flying qualities requirements were embodied initially in the BCARs, or British Civil Airworthiness Requirements. European standards now apply, as found in the European Joint Aviation Requirements, or JARs, issued by the Joint Aviation Administration. The U.S. versions are the Federal Air Regulations, or FARs, parts 21, 23, 25, and 103 of which deal with airplanes. The wording of the stability and control airworthiness requirements of the FARs is similar to the Gilruth requirements of 1943, which were also concerned with minimum rather than optimum requirements.

3.9 Equivalent System Models and Pilot Rating

The 1980 military flying qualities specification MIL-F-8785C represents the culmination of the representation of airplanes by classical transfer functions (see 20.4), the transfer functions of bare airframes augmented only by simple artificial damping and crossfeeds, where needed. In pitch, the bare airframe transfer function of pitching velocity as an output to control surface angle as an input has an inverse second-order denominator and a first-order numerator under the constant airspeed assumption. Three parameters define this function: natural frequency and damping ratio in the denominator and the numerator time constant. The classical bare-airframe transfer function models are called equivalent systems because they can only approximate the transfer functions of complex, augmented flight control systems, such as command augmentation systems and the newer superaugmented systems for highly unstable airframes. The uses to which equivalent system models are put in specifying longitudinal and lateral flying qualities are illustrated in Chapter 10, "Tactical Airplane Maneuverability."

The 1980 specification MIL-F-8785C represented another culmination in the development of airplane flying qualities as a science. This is assigning a numerical scale to pilot opinion. In the 1950s A. G. Barnes in the United Kingdom used the initials G, M, and B for good, medium, and bad, with + and − modifiers. The numerical scale, running from 1 to 10, was proposed by George E. Cooper in 1961. The MIL-F-8785C uses the Cooper-Harper rating scale (Figure 3.11), in which the experience of NASA and Calspan are combined (Cooper and Harper, 1969).

A successor to the Cooper-Harper rating scale originated at the College of Aeronautics, Cranfield University (Harris et al., 2000) to deal better with modern fly-by-wire aircraft. The proposed new scale, called the Cranfield Aircraft Handling Qualities Rating Scale, or CAHQRS, considers separately five parameters – longitudinal, lateral, directional, trim, and speed control – and rates behavior in subtasks according to a Cooper-Harper-type scale, and also a criticality scale. The CAHQRS has been tested initially on a flight simulator. Further

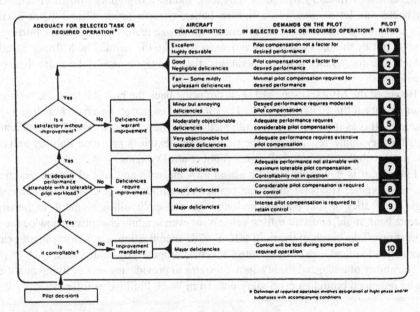

Figure 3.11 The Cooper-Harper pilot numerical rating scale, now a definitive standard. (From NASA TN D-5153, 1969)

experience with this new approach is needed to confirm its expected benefits relative to the Cooper-Harper standard.

The next phase in the unfolding history of the science of flying qualities involves a new level of sophistication, freeing the subject from the constraint of equivalent systems. Mathematical models of the human pilot as a sort of machine are combined with airplane and control system mathematical models and are treated as a combined system. Human physiology and psychology are now enlisted in the study of flying qualities requirements. These interesting developments are treated in Chapter 21, "Flying Qualities Research Moves With the Times."

3.10   The Counterrevolution

In the late 1980s a counterrevolution of sorts took place, a retreat from authoritative military flying qualities specifications. A new document (1987), called the Military Standard, Flying Qualities of Piloted Vehicles, MIL-STD-1797 (USAF), merely identifies a format for specified flying qualities. Actual required numbers are filled into blanks through negotiations between the airplane's designers and the procurement agency. As explained by Charles B. Westbrook, the idea was to let MIL spec users know that "we didn't have it all nailed down, and that industry must use some judgement in making applications."

A large handbook accompanies the Military Standard, giving guidance on blank filling and on application of the requirements. The handbook is limited in distribution because its "lessons learned" includes classified combat airplane characteristics. The Military Standard development for flying qualities is associated with Roger H. Hoh of Systems Technology, Inc., and with Westbrook, David J. Moorhouse, and the late Robert J. Woodcock, of Wright Field.

The demise of the authoritative MIL-F-8785 specification was part of a general trend away from rigid military specifications, with the intent of reducing extraneous and detailed management of industry by the government. Industry designers said in effect, "Get off our backs and let us give you a lighter, better, cheaper product" and "Quit asking for tons of reports demonstrating compliance with arcane requirements." Some horror stories brought out by the industry people did seem to make the point. The Military Standard is in fact ideal for "skunk works" operations; their managers don't like more than general directions.

However, the Military Standard seems to bring back the bad old days, the "straw man" requirements of the 1930s, established by pilots and engineers based on hunch and specific examples. It is as if the rational Gilruth method had never been invented. A justification of sorts for the counterrevolution is the tremendous flexibility provided stability and control designers with the new breed of digital flight control systems.

Literally, it is now possible to have an airplane with any sort of flying qualities that one can imagine. Tiny side sticks can replace conventional yoke or stick cockpit controls. Right or left stick or yoke controls no longer have to apply rolling moments to the airplane. Instead, bank angle, constant rolling velocity, or even heading change can now be the result. By casting off the bonds of the rigid MIL-F-8785 specification, a procuring agency can take advantage of radical, innovative control schemes proposed by contractors.

The ability of advanced flight control systems to provide any sort of flying qualities that can be imagined brought a cautionary note from W. H. Phillips, as follows:

> The laws of nature have been very favorable to the designers of control systems for old-fashioned subsonic, manually-controlled airplanes. These systems have many desirable

features that occur so readily that their importance was not realized until new types of electronic control systems were tried.

Don Berry, a senior engineer at the NASA Dryden Research Center, had similar views:

> We have systems capable of providing a wide variety of control responses, but we are not sure what responses or modes are desirable.

A further step in the dismantling of "rational" Gilruth flying qualities specifications is the recent appearance of independent assessment boards, charged with managing the flying qualities (and some performance) levels of individual airplanes. Such a board, called the "Independent Assessment Team," was formed for the Navy's new T-45A trainer. Team members for the T-45A included the very senior, experienced engineers William Koven, I. Grant Hedrick, Joseph R. Chambers, and Jack E. Linden.

3.11 Procurement Problems

In either case, whether airplane flying qualities are specified by a standardized specification such as MIL-F-8785 or by negotiations involving a Military Standard, there is still the matter of getting new airplanes to meet flying qualities requirements. In other words, the science of flying qualities is useless unless airplanes are held to the standards developed by that science.

In recent years, new airplanes are being bought by the U.S. armed services in a way that seems designed for poor flying qualities. Program officers are given sums of money sufficient to produce a fixed number of airplanes on a schedule. Military careers rest on meeting costs and schedules. These are customarily optimistic to begin with, having gotten that way in order to sell the program against competing concepts or airplanes.

The combination of military career pressures and optimistic cost and schedule goals usually leads to the dreaded (by engineers) "concurrency" program. Production tooling and some manufacturing proceed concurrently with airplane design and testing, rather than after these have been completed. When flying quality deficiencies crop up late in a concurrent program, requiring modifications to tooling and manufactured parts, it is natural for program officers and their counterparts in industry to resist.

Three notable recent concurrent programs were the Lockheed S-3 Viking anti-submarine airplane, the Northrop B-2 stealth bomber, and the U.S. naval version (T-45A) of the British Aerospace Hawk trainer, being built by McDonnell Douglas/Boeing. The Lockheed S-3 and McDonnell Douglas/Boeing T-45A concurrency stories are involved with the special flying qualities requirements of carrier-based airplanes and are discussed in Chapter 12, on that subject.

3.12 Variable-Stability Airplanes Play a Part

A variable-stability airplane is a research airplane that can be made to have artificially the stability and control characteristics of another airplane. Waldemar O. Breuhaus credits this invention to William M. Kauffman, at the NASA Ames Research Center, about the year 1946 (Breuhaus, 1990). The colorful story that Breuhaus tells is of Kauffman looking out of the window at the Ames flight ramp and seeing three Ryan FR-1 Fireball fighters sitting side by side. Each FR-1 had a different wing dihedral angle. The airplanes had been so modified to try to find in flight testing the minimum amount of effective dihedral angle that pilots would accept. Kauffman said, according to Steve Belsley and some others, "There has to be a better way."

Ames modified a Grumman F6F-3 Hellcat into the first variable-stability airplane by a mechanism that moved the ailerons in response to measured sideslip angles. An electric servo motor, adapted from a B-29 gun turret drive, moved the F6F's aileron push–pull rods in parallel to the pilot's stick input. With this parallel arrangement, the pilot's stick is carried along when the servo works in response to measured sideslip. This is suitable for automatic pilots, where it is often acceptable and even desired for the pilot's controls to reflect automatic pilot inputs. However, it does not serve the function of a variable-stability airplane, where the action of the variable-stability mechanism is supposed to be unnoticeable to the pilot.

In the case of the pioneering F6F-3 variable-stability airplane, pilot stick motions were suppressed approximately by an ingenious scheme that canceled the aerodynamic hinge moment corresponding to the commanded aileron deflection. This was done by driving the aileron tab through its own servo motor with a portion of the same signal that was used to drive the aileron push–pull rod.

The F6F-3 variable-stability airplane was followed in the next 30 years by at least 20 other airplanes of the same type. The majority were built by NACA/NASA; the Cornell Aeronautical Laboratories, later Calspan; the German Aerospace Center, or DLR; and the Royal Aircraft Establishment, later DERA. Princeton University, the Canadian National Research Council, Boeing, and research agencies in France and Japan also built them.

The crude compromises of the early machines have given way to ever more sophisticated ways of varying airplane stability and control as seen by the test pilot. Later models, such as the Calspan Total In-Flight Simulator, or TIFS, and the Princeton University Variable-Response Research Aircraft, or VRA, have special side-force generating surfaces.

3.13   Variable-Stability Airplanes as Trainers

The objectives of most of the variable-stability programs were either to apply the Gilruth method of obtaining flying qualities requirements by exposing pilots to different stability and control levels or to present the flying characteristics of a future machine for evaluation. However, quite by chance, a different use for variable-stability airplanes cropped up. Breuhaus reports that Gifford Bull, the project engineer and safety pilot of a Calspan variable-stability USAF B-26 airplane, was chatting with members of the Navy Test Pilot School at the Patuxent River Naval Air Test Center. The B-26 was at Patuxent to run Navy-sponsored tests on minimum flyable longitudinal handling qualities under emergency conditions. Test Pilot School staffers were struck by what looked like

> the unique suitability of the variable stability airplane to serve as a flying class room or laboratory to demonstrate to the school the effects of the myriad flying quality conditions that could be easily and rapidly set up.

A trial run in 1960 was such an instant success that the program was broadened to include the Air Force Test Pilot School at Edwards Air Force Base, and a second B-26 was added. The aging B-26s were eventually replaced by two variable-stability Learjet Model 24s. By the end of 1989 nearly 4,000 service, industrial, and FAA pilots and engineers had instruction or demonstrations using the variable-stability B-26s and Learjets.

In a more recent application of an airplane modified to fly like another airplane for training, NASA used a Grumman Gulfstream G-2 in a high drag configuration to train pilots to fly the Space Shuttle's steep, fast-landing approach profile, starting at an altitude of about 30,000 feet.

3.14   The Future of Variable-Stability Airplanes

The engineers at NASA, Calspan, DERA, the Canadian NRC, Princeton University, and other European and Asian laboratories who had so much to do with the development of variable-stability airplanes can point to impressive accomplishments using these devices. Variable-stability airplanes shed light on many critical issues, such as the role of roll-to-yaw ratios on required Dutch roll damping, permissible levels of spiral divergence, and the effect of longitudinal flying qualities on instrument landing system (ILS) landing approaches. Variable-stability airplanes have also provided a preliminary look at the flying qualities of radical new airplanes such as the Convair B-58 Hustler; the Rockwell X-15, XB-70, B-1, and Space Shuttle Orbiter; the Lockheed A-12 and F-117A; the Grumman X-29A; various lifting body projects; and the Anglo-French Concorde before those new airplanes flew.

The TIFS machine, based on a reengined Convair C-131B transport, has had a particularly productive career (Figure 3.12). Calspan engineers provided the TIFS with the ability to add aerodynamic forces and moments to all 6 degrees of freedom. Flight tests are carried out from an evaluation cockpit built into the airplane's nose, while a safety crew controls the airplane from the normal cockpit. Some 30 research programs have been run on this airplane. The majority of them were general flying qualities research; ten programs were on specific airplanes. A T-33 variable-stability airplane also had a very productive career, with more than 8,000 flying hours to date. A new application of variable-stability airplanes has been reported from the DLR, in which the ATTAS in-flight simulator investigated manual flight control laws for a future 110-seat Airbus transport airplane.

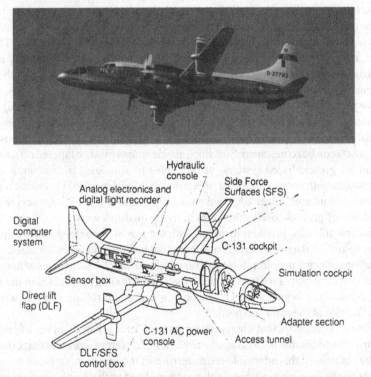

Figure 3.12   The TIFS (Total In-Flight Simulator) variable-stability airplane, built up by Calspan from a Convair C-131B airplane for the Air Force. The TIFS can generate direct lift and side force. (From Phillips, *Jour. of Guidance, Control, and Dynamics*, July–Aug. 1989)

In spite of this impressive record, there are reasons to look for limitations in the future use of variable-stability airplanes in the engineering development of new aircraft. A significant obstacle is the practical difficulty in updating and maintaining the vast computer data bases needed to represent the mathematical models of complex digital flight control and display systems and nonlinear, multivariable aerodynamic data bases. Maintaining current data bases should be inherently easier for locally controlled ground-based simulators, as compared with variable-stability airplanes operated by another agency at a remote site.

Another limitation to the future use of variable-stability airplanes in the engineering development of specific airplanes has to do with the cockpit environment. Correctly detailed controls, displays, and window arrangements, important for a faithful stability and control simulation, may be difficult to provide on a general-purpose variable-stability airplane. Correct matching of accelerations felt by the pilot is also desirable. Although variable-stability airplanes do provide the pilot with both acceleration and visual cues, both cannot be represented exactly, along with airplane motions, unless the variable-stability machine flies at the same velocity as the airplane being simulated and unless the pilot is at the same distance from the airplane's center of gravity in both cases.

Those conditions are rarely satisfied, except in some landing approach simulations. For example, the Princeton University VRA, flying at 105 knots, has been used to simulate the Space Shuttle Orbiter flying at a Mach number of 1.5. Pilot acceleration cues can be retained under a velocity mismatch of this kind by a transformation of variable-stability airplane outputs that amounts to using a much higher yaw rate (Stengel, 1979). Likewise, pilot location mismatch is conveniently corrected for by a transformation on the sideslip angle. If these transformations are applied to correct pilot acceleration cues, visual cues will be made incorrect. An alternative scheme to provide correct pilot acceleration cues relies on the direct side and normal force capabilities of advanced machines such as the TIFS.

In general, the cockpit environment of a new airplane can be represented fairly readily in a ground-based simulator. Correct visual cues can be provided as well, although there are often troubling lags in projection systems. The major loss in fidelity for ground simulators, as compared with variable-stability airplanes, comes from the compromises or actual losses in pilot motion cues. When these are provided by servo-driven cabs, accelerations must be washed out. That is, to avoid unreasonably large simulator cockpit cab motions, only acceleration onsets can be represented. Sustained accelerations must be tapered off smoothly and quickly in the ground-based systems, or they must be simulated by pressures applied to the pilot's bodies with servo-controlled pressure suits. Belsley (1963) provided an early summary paper in this area. Later on, Ashkenas (1985) and Barnes (1988) reviewed the utility and fidelity of ground-based simulators in flying qualities work.

There is a debatable size problem involved with the use of variable-stability airplanes. W. H. Phillips points out that in Robert Gilruth's original handling qualities studies, contrary to the expectations of many people, pilots were satisfied with much lower values of maximum rolling velocity on large airplanes than on small ones. This finding is reflected in the pb/2V criterion of acceptability, which allows half as much maximum rolling velocity when the wing span is doubled at the same airspeed.

Again, pilots of small airplanes choose lower control forces than do pilots of large airplanes. Phillips concludes that pilots adapt to airplanes of different sizes and that erroneous results may be obtained if this adaptable characteristic of the human pilot is not accounted for. This might be the case when a large airplane is simulated with a much smaller variable-stability airplane, or vice versa.

A counterargument is that two fundamental airplane dynamics properties affecting airplane feel vary systematically with airplane size, giving the pilot a cue to the size of the airplane, even if all that the pilot sees of the airplane is the cockpit and the forward view out of the windshield. Short-period pitch natural frequency shows a systematic trend downward with increasing airplane weight and size. The roll time constant, the time required for an airplane to attain final rolling velocity after step aileron inputs, shows a systematic trend upward with increasing airplane size.

Thus, a small variable-stability airplane whose dynamics match those of a large airplane may well feel like the large one to the pilot. W. O. Breuhaus (1991) reports that this seems to be the case:

> the pilot must be able to convince himself that he is flying the assigned mission in the airplane being simulated... one of the variable-stability B-26's was used to simulate the roll characteristics of the much larger C-5A before the latter airplane was built. The results of those tests showed a less stringent roll requirement for the C-5A than was being specified for the airplane, and these results were verified when the C-5A flew.

The relative merits of variable-stability airplanes as compared with ground-based simulators for representing airplane flying qualities are still being debated; each has its proponents. However, it is a fact that sophisticated ground-based simulators are now absolutely integral to the development of new aircraft types, such as the Northrop B-2 and the Boeing 777. Typically, ground-based simulators handy to the engineering staff are in constant use during airplane design development. At the same time, variable-stability airplanes remain important tools for design validation and for the development of generalized flying qualities requirements.

The question of when variable-stability airplane simulation is really necessary is taken up by Gawron and Reynolds (1995). They provide a table of ten flight conditions that seem to require in-flight simulation, together with evidence for each condition. An example condition is a high gain task. Evidence for this is the space shuttle approach and landing and other instances such as YF-16 and YF-17 landings.

The Air Force operates the new VISTA/F-16D variable-stability airplane (Figure 3.13) and the Europeans are running impressive programs of their own. However, in-flight simulation was not considered for the Jaguar fly-by-wire, the EAP (Experimental Aircraft Programme), or for the Eurofighter. Shafer (1993) provides a history of variable-stability airplane operations at the NASA Dryden Flight Research Center, with an extensive bibliography.

3.15   The V/STOL Case

Vertical or short takeoff and landing (V/STOL) airplane flying qualities requirements present special problems because V/STOL airplane technology covers a large range of possibilities. So far, we have seen tilt rotor, lift fan, vectored thrust, blown flaps, and convertible rotor wing versions. Although the military services have taken up the challenge and in 1970 issued a V/STOL flying qualities specification, MIL-F-83300, there is a danger that the requirements are specific to individual designs, those available for testing at the time.

MIL-F-83300 recognizes three airspeed regimes, from hover to 35 knots, from 35 knots to an airspeed Vcon where conventional flying qualities requirements apply, and airspeeds above Vcon. Requirements are either for small perturbations about some fixed operating point or for accelerated or transitional flight. The V/STOL small-perturbation longitudinal

Figure 3.13   The Air Force Wright Laboratory's VISTA, or multiaxis thrust-vectoring airplane, a variable-stability machine based on the General Dynamics F-16D. Thrust is vectored up to 17 degrees in pitch and yaw, primarily for high-angle-of-attack research. (From *Aerospace America*, Dec. 1993)

dynamics requirements take the familiar MIL-8785 form of acceptable and unsatisfactory boundaries in terms of real and imaginary parts of the system roots. So do the lateral-directional requirements resemble those for conventional airplanes, as requirements on the shape of the bank angle versus time curve for rolls and on permitted adverse yaw.

A complication when applying the familiar period and damping requirements to the roots of V/STOL motions is convergence of the ordinary modes of motion at very low airspeeds. For a powered-lift STOL configuration the longitudinal short-period and phugoid modes merge at an equilibrium weight coefficient, equivalent to the lift coefficient, of 3.5 (Figure 3.14). A similar trend shows up in the lateral case, where an (unstable) spiral mode approaches in time constant the usually much shorter rolling mode at a large value of the equilibrium weight coefficient.

The problem of establishing V/STOL flying qualities requirements that are not tied to specific configurations was taken up again after MIL-F-83300, for the most part with the help of ground simulations and variable-stability airplanes. In 1973, Samuel J. Craig and Robert K. Heffley used analysis and ground simulation to explore the role of thrust vector inclination during STOL landing approaches (Craig and Heffley, 1973).

Still later, in papers delivered in 1982 and 1983, Roger H. Hoh, David G. Mitchell, and M. B. Tischler looked for flying qualities generalizations in VTOL transitions and STOL path control for landings. Precise pitch attitude control at high bandwidth appears to be critical in transitions because of the sensitivity of vertical rate to pitch attitude. However,

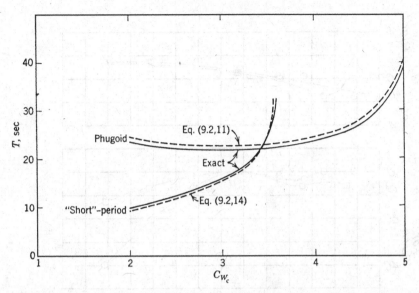

Figure 3.14 A complicating factor in specifying STOL flying qualities. The long- or phugoid and short-period longitudinal modes converge at low airspeeds, where part of the airplane's weight is supported by thrust. (From Etkin, *Dynamics of Atmospheric Flight*, 1972)

the writers found a number of possible requirements for the landing flare control maneuver of STOL airplanes. That series closed out with a major effort to extend the *MIL Prime Standard and Handbook* (Anon., 1987) concept to STOL Landings (Hoh, 1987).

An area that seems to require more attention is the lift loss in the vortex ring state (Glauert, 1934; Coyle, 1996) during increases in descent rate. Vortex rings recirculate the rotor downflow back into the rotor, instead allowing it to descend and produce lift. This is a performance problem for single-rotor helicopters. However, for the tilt-rotor V-22 Osprey, a vortex ring on one of two laterally located rotors is believed to have produced an unrecoverable roll.

The considerable experience gained by DERA and BAE systems in V/STOL projects, leading to the Harrier and the VAAC (Vectored thrust Aircraft Advanced flight Control) Harrier, is summarized by Shanks (1996) and Fielding (2000). One key finding was that eliminating conscious mode changing provides a large reduction in pilot work load. The V/STOL becomes a "conventional aircraft that can hover." Another finding was the need to use closed-loop analysis to specify propulsion system characteristics in terms of bandwidth and response linearity.

Pilot-in-the-loop technology (Chapter 21) has made significant contributions to understanding the special flying qualities requirements of STOL and VTOL airplanes. This approach is especially valuable because it is not closely tied to the design details of specific machines.

3.16   Two Famous Airplanes

NACA measured the flying qualities of the Supermarine Spitfire VA fighter in 1942 and the Douglas DC-3 transport in 1953, both at the Langley Laboratory. These airplanes had been built in large numbers, had served magnificently in World War II, and had inspired great affection among their pilots. Yet neither of these famous airplanes had the specified

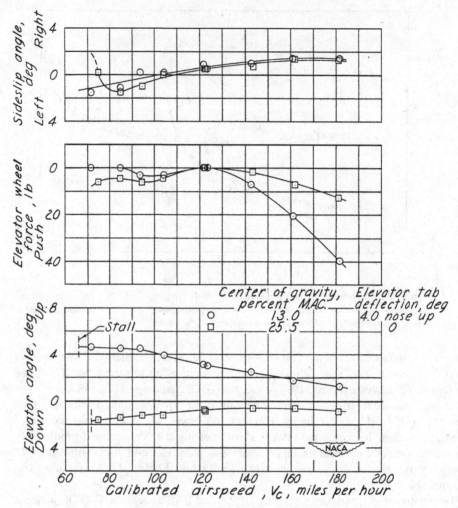

Figure 3.15   The venerable Douglas DC-3 exhibits static longitudinal instability in the normal rated power, clean configuration, at an aft center of gravity position of 25.5 percent MAC. (From Assadourian and Harper, NACA TN 3088, 1953)

level of the most basic stability of them all, static longitudinal stability, as measured by the elevator angles required for steady flight at various airspeeds. This form of stability is often called stick-fixed stability.

The Spitfire shows neutral stick-fixed stability under all flight conditions. The DC-3 is stable only in power-off glides or with cruise power. With normal rated power or in a power approach condition at aft loadings, increasing amounts of down elevator are needed as the airspeed is reduced, along with push column forces (Figure 3.15). For both airplanes there are other less striking deviations from NACA and military stability and control specifications. What should be made from all of this?

The Spitfire and DC-3 cases should not furnish an excuse to dismiss flying qualities requirements. It is reasonable to assume that if the Spitfire and DC-3 were longitudinally stable under all flight conditions, both of these fine airplanes would have been even better. In fact, the Spitfire Mark 22, developed at the end of the war, had a 27 percent increase in tail areas and flew "magnificently," according to one account. The bottom line is that nobody has ever found it feasible to run definitive, statistically valid experiments on the value of good flying qualities in terms of reducing losses in accidents or success in military missions. Instead, we rely on common sense. That is, it is highly plausible that good handling qualities in landing approach conditions will reduce training and operational accidents and that precise, light, effective controls will improve air-to-air combat effectiveness. That plausibility is essentially what energizes the drive for good flying qualities, in spite of apparent inconsistencies, such as for the Spitfire and DC-3.

3.17   Changing Military Missions and Flying Qualities Requirements

Flying qualities requirements for general aviation and civil transport airplanes are predictable in that these airplanes are almost always used as envisioned by their designers. This is not so for military airplanes. The record is full of cases in which unanticipated uses or missions changed flying qualities requirements. Four examples follow.

A4D-1 Skyhawk. The A4D-1, later the A-4, was designed around one large atomic bomb, which was to be carried on the centerline. A really small airplane, the A4D-1 sits high on its landing gear to make room for its A-bomb. The airplane was designed to be carrier-based. However, the A4D-1 was used instead mainly as a U.S. Marine close-support airplane, carrying conventional weapons and operating from single-runway airstrips, often in crosswinds. The vestigial high landing gear meant that crosswinds created large rolling moments about the point of contact of the downwind main tire and the ground. In simpler terms, side winds tried to roll the airplane over while it was landing or taking off. Originally, pilots reported that it was impossible to hold the upwind wing down in crosswinds, even with full ailerons. Upper surface wing spoilers had to be added to the airplane to augment aileron control on the ground.

B-47 Stratojet. This airplane started life as a high-altitude horizontal bomber. Its very flexible wings were adequate for that mission, but not for its later low-altitude penetration and loft bombing missions. Loft bombing requires pullups and rolls at high speed and low altitude. In aileron reversal ailerons act as tabs, applying torsional moments to twist a wing in the direction to produce rolling moments that overpower the rolling moments of the aileron itself. This phenomenon limited the B-47's allowable airspeed at low altitudes.

F-4 Phantom. The F-4 was developed originally for the U.S. Navy as a long-range attack airplane, then as a missile-carrying interceptor. A second crew member was added for the latter role, to serve as a radar operator. Good high angle of attack stability and control were not required for these missions, but then the U.S. Air Force pressed the F-4 into service in Vietnam as an air superiority fighter. Belatedly, leading-edge slats were added for better high angle of attack stability and control.

NC-130B Hercules. This was a prototype C-130 STOL version, fitted with boundary layer control. The airplane's external wing tanks were replaced by Allison YJ56-A-6 turbojets to supply bleed air for the boundary layer control system. At the reduced operating airspeeds made possible by boundary layer control the C-130's unaugmented lateral-directional dynamics, or Dutch roll oscillations, were degraded to unacceptable levels.

"Systems engineering" as a discipline was a popular catchphrase in the 1950s. Airplanes and all their accessories and logistics were to be developed to work together as integrated systems, for very specific missions. The well-known designer of naval airplanes Edward H. Heinemann was not impressed. Heinemann's rebuttal to systems engineering was, "If I build a good airplane, the Navy will find a use for it." Heinemann's reaction to systems engineering seems justified by the four cases cited above, in which flying qualities requirements for the airplanes changed well after the designs had been fixed.

3.18   Long-Lived Stability and Control Myths

The achievements of S. B. Gates, R. R. Gilruth, and others in putting airplane stability and control on a scientific basis have not eliminated a number of early myths attached to the subject. Dr. John C. Gibson (1995) lists no fewer than 15 of these myths and counters them with what we know to be correct. A few of the Gibson's list of 15 myths and corrections follow:

Wing center of pressure (cp) movement affects longitudinal stability. Correction: Wing cp movement with angle of attack is controlled by the wing's zero-lift pitching moment coefficient about its aerodynamic center (1/4 chord), or C_{m0}. This parameter affects only trim for rigid airplanes. Wing cp has been discarded in modern stability and control calculations and replaced by wing aerodynamic center and C_{m0}.

A down tail load is required for stability. Correction: Stability is provided by the change in tail load with change in airplane angle of attack. The change is independent of the direction of the initial load.

Gibson comments that this myth survives in FAA private pilot examinations and in an exhibit at the National Air and Space Museum in Washington. This subject is distinct from the instability caused by tail down load in the presence of propeller slipstream, an effect discussed in Chapter 4, Section 6.

A stable airplane is less maneuverable than an unstable one. Correction: Unstable airplanes are notoriously difficult to control precisely. Given light control forces, a stable airplane can be pitched rapidly to a precise load factor or aiming point. Gibson says, "...the [stable] Hurricane, Typhoon, and Tempest were highly manoeuvrable and were greatly superior as gun platforms to the skittish Spitfire."

The reader is referred to Gibson's 1995 paper for the rest of these interesting myths and their corrections.

CHAPTER 4

Power Effects on Stability and Control

The World War II period 1939–1945 coincided almost exactly with the appearance of power effects as a major stability and control problem. Grumman Navy fighters of that period illustrate the situation. World War II opened with the F4F Wildcat as the Navy's first-line fighter and ended with the debut of the F8F Bearcat. The external dimensions of the two aircraft were almost identical, but the F8F's engine was rated at 2,400 horsepower, compared to 1,350 horsepower for the F4F.

In unpublished correspondence, W. Hewitt Phillips remarks that the appearance of power effects as a major stability and control problem was not entirely the result of growth in engine power:

> these effects have been with us since World War I, but weren't serious then because of the light control forces required to offset these effects, resulting from the low speeds and smaller size of these airplanes. The power effects in terms of thrust and moment coefficients were probably of the same order as in the case of the World War II fighters. These effects would have been somewhat reduced because of the short nose moment arm of these planes, and because of the lower lift coefficients due to the lack of high lift flaps.

The further growth in power and stability effects on military propeller-driven aircraft was of course interrupted by the advent of jets, with a different set of power effects on stability and control, generally of a minor nature. This chapter reviews the history of both propeller and jet power effects on stability and control. Although the days of high-powered propeller-driven military aircraft may be ended, their civil counterparts still exist, with a new set of stability and control problems.

4.1 Propeller Effects on Stability and Control

A remarkable paper by a Cal Tech professor was a source for years afterwards of propeller power effects on static longitudinal and directional stability (Millikan, 1940). Millikan's paper was based primarily on test results from Cal Tech's Guggenheim Aeronautical Laboratories' (GALCIT) low-speed wind tunnel. The tunnel staff and the companies who tested new aircraft models there had the foresight to develop both the hardware and techniques needed to run tests that simulated power-on flight conditions. The primary hardware need was for compact, powerful electric motors that could be mounted inside model fuselages (Figure 4.1). Electric motors of this type with larger diameters and lower power had been designed and installed in British N.P.L. wind tunnel models much earlier (Relf, 1922).

Testing techniques required that model thrust be adjusted to predicted full-scale conditions. This ensured that the slipstream velocities behind the model propellers were in the correct proportions to free-stream velocities. Propeller effects on stability and control arise from two other factors aside from the slipstreams washing over wings, tail surfaces, and fuselages. These are the simple direct-thrust moments in pitch and yaw and the forces and moments acting on the propellers themselves. The following sections trace the development of our understanding of these effects.

Figure 4.1 A typical compact electric induction motor developed in the late 1930s for powered wind-tunnel models. These motors developed 5 to 12 horsepower at 18,000 RPM. (From Millikan, *Jour. of the Aeronautical Sciences*, Jan. 1940)

4.2 Direct-Thrust Moments in Pitch

Static longitudinal stability for a powered airplane is defined as the tendency to return to a trimmed angle of attack and airspeed following a disturbance, *with throttle fixed*. Having the airplane's center of gravity below the line of action of the thrust vector, or the thrust line, provides a stabilizing pitching moment increment under these conditions, whether the thrust comes from propellers or jets.

With fixed throttle, thrust remains more or less constant as airspeed falls below the trim value, actually increasing somewhat in the case of propellers. On the other hand, the aerodynamic forces on the rest of the airframe necessarily decrease at lower airspeeds. The net result is that the diving, or nose-down pitching moment, caused by the thrust line passing above the center of gravity increases relative to the other forces, producing the desired restoring nose-down effect. For single-engine airplanes, tilting the entire engine and propeller assembly nose down by a few degrees raises the thrust line relative to the center of gravity.

The powerful (1,900 horsepower) Curtiss SB2C Helldiver came into production in 1942 and soon became a valuable addition to the U.S. Navy's offensive capability. Along with the Helldiver's speed and bomb-carrying capacity came an unenviable reputation for poor longitudinal stability and handling difficulties at the low speeds used in carrier approaches. Propeller down tilt seemed to be a natural fix for the SB2C.

The propeller down tilt idea was tested on two generic wind tunnel models at the NACA Ames Aeronautical Laboratory; one model was quite similar to the SB2C (Goett and Delany, 1944). The SB2C-like model had a forward, or unstable, shift in the neutral point (center of gravity for neutral stability) of about 10 percent of the mean aerodynamic chord in a 2,100-horsepower climb. This longitudinal stability loss was cut in half with 5 degrees of propeller downward tilt.

The exigencies of wartime prevented the application of down tilt to the SB2C – poor stability was just another hazard that Navy pilots had to contend with in those days. Many of the youngsters who flew SB2Cs from carriers had graduated into the type direct from training aircraft of the North American SNJ Texan class. The SB2C was the first really high-powered airplane they encountered.

These relatively inexperienced pilots apparently thought poor longitudinal stability, manifested by difficulty in establishing a fixed trim airspeed during carrier approaches, was what one should expect on a big, fast airplane. In discussing possible SB2C propeller tilt with Navy fleet pilots, Bureau of Aeronautics engineers were met with "Leave it alone! It flies just fine!" Although the SB2C missed out on propeller down tilt, this feature was used later to increase stability for three single-engine high-powered propeller airplanes. The Douglas AD Skyraider and the Grumman F6F Hellcat and F8F Bearcat all had nose-down tilted engines.

4.3   Direct-Thrust Moments in Yaw

For a multiengine airplane whose engines are mounted on the wings, when all engines are running and developing about the same power, there is no unbalanced yawing moment due to power. Failure of a wing-mounted engine of course sets up a thrust-caused yawing unbalance that must be counteracted by an equal and opposite aerodynamic yawing moment. The more engines on a multiengine airplane, the less effect will the failure of a single one have on yawing moments. Flight crews of Boeing B-29 and B-50 four-engine airplanes had the strange experience of losing engines during normal cruise flight and being unaware of it for many seconds. RPM for the dead engine would drop very little at first because of propeller windmilling. The directional stability of both airplanes was high enough to keep the ships close to course, initially.

Current Boeing Company design practice requires that a twin-engine jet transport be able to continue a climbout after takeoff, rudder-free, with one engine failed. This accounts for the generously sized vertical tails on the 737, 757, 767, and 777 models.

4.4   World War II Twin-Engine Bombers

The situation was quite different for the high-powered twin-engine bombers of World War II, such as the Martin B-26 Marauder, the Douglas A-20 Havoc and A-26 Invader, and the North American B-25 Mitchell. Loss of one engine on these airplanes, especially at low airspeeds, produced rapid and dangerous changes in yaw and sideslip, unless promptly corrected by rudder control. Remember that these airplanes were heavy, large, and fast, and that hydraulic power-assisted controls had not yet been introduced. The limiting factor in keeping these airplanes under control when an engine failed was not insufficient rudder control power but high rudder pedal force.

Rudder pedal forces to counteract engine failure on airplanes of the B-25 and B-26 class were generally aggravated by the poor design of the rudder aerodynamic balance. Suppose for example that there was a loss of power in the right engine. The airplane's nose would quickly swing to the right, in a right yaw. Momentum would carry the airplane's flight path along its previous direction, causing the relative wind to come from the left side. This is a condition of left sideslip. This direction of the relative wind would cause the rudder (or rudders) to "float" or trail with its trailing edge to the right, giving right rudder.

But to regain control, left rudder would be needed, to give a left yawing moment in opposition to the thrust of the working left engine. The pilot would have to apply a large amount of left pedal force just to center the rudder from its "floated" position and then an additional amount of pedal force to get the required left rudder. The amount of left rudder required could be minimized by holding the wing with the working engine low, in a slight bank, but the net pedal force was generally the critical factor, determining the minimum airspeed at which these airplanes could be flown with one engine dead.

Reduction or elimination of rudder float in sideslip was available to the designers of these airplanes through tailoring of the rudder's aerodynamic balance. Specifically, rudder horn balances would have that effect (Figure 4.2). Rudder horn balances, used as far back

Figure 4.2 Experimental rudder horn balance-fitted to the Martin B-26 Marauder. This design reduced the rudder forces required for flight after failure of one engine. It was never put into production. (From U.S. Army Air Corps photo 108769, 1942)

in aviation history as the Blériot monoplane, were probably considered somewhat archaic to the Martin, Douglas, and North American designers. There was also a practical objection in that the projecting horns could conceivably snag parachute lines if the crew had to bail out. In any case, the A-20, A-26, B-25, and B-26 high-powered twins got through World War II without rudder horn balances.

4.5   Modern Light Twin Airplanes

The situation is different again in the case of the modern light twin airplanes. The first of these planes was the five-to-seven place Aero Commander 520, introduced by the Aero Design and Engineering Corporation of Culver City, California, in 1950. A year or so later Beech introduced its Model 50 Twin Bonanza, Piper its Model PA-23 Twin Stinson (later called Apache), and Cessna its Model 310 twin. These aircraft and their successors have a great deal of appeal to aviators who regularly fly on instrument flight plans into bad weather and those who want the extra safety of a second engine.

Yet by the early 1980s the safety records compiled by the modern light twins did not bear out this expectation. Writing in the *AOPA Pilot* of January 1983, Barry Schiff pointed out that the fatality rate following engine failure in light twins was four times that for engine failures in single-engine airplanes. It seems that relatively low-time private pilots were being trapped by the yaw and roll caused by the failure of one engine at low speeds and altitudes.

The Beech Model 95 Travelair and its higher power military derivative, the U.S. Army's T-42A, are good examples of what could happen. After several fatal stall-spin accidents following power loss on one engine a courageous Army pilot made a series of T-42A stall tests, with symmetric and antisymmetric power. His report told of moderate wing drop in symmetric stalls, but of vicious behavior in stalls with one engine idling. The airplane would roll nearly inverted, clearly headed for a spin.

The response of the Federal Aviation Administration (FAA) to this generic light twin hazard was not to require design changes, but to warn pilots and to stress recognition and compensation for single-engine failure during training and flight tests for multiengine pilot ratings. Pilots are drilled to instantly recognize the failed engine by the mantra "Dead foot, dead engine." Since accidents occur during the incessant single-engine drills in training, there is a special minimum airspeed for "intentionally rendering one engine inoperative in flight for pilot training."

This is Vsse, the fourth of the special airspeeds the poor pilot has to memorize in order to legally operate multiengine airplanes. The others are Vmc or Vmca, the minimum airspeed for control with the critical engine's propeller windmilling or feathered, the other delivering takeoff power; and Vxse and Vyse, the best angle of climb and rate of climb airspeeds with one engine inoperative. Vyse has its own marking on the airspeed dial, a blue line usually used as the landing approach airspeed under normal conditions. Evidently, if an engine fails on landing approach, one wants the airplane to be already at its best airspeed to climb away or to lose as little altitude as possible. The four special airspeeds for multiengine airplane operation are added to *nine* other special airspeeds (six if the airplane has no wing flaps or retractable landing gear) to be remembered.

In spite of the FAA's apparent disinterest in obliging light twin builders to design safe single-engine behavior into their airplanes, there have been some attempts made in this direction. There is an FAA-approved design retrofit of vortex generators for the upper wing surfaces of some light twins. The installation reduces Vmca, the minimum airspeed for control with an engine out (Figure 4.3). Vortex generators are tiny (about 2 inches square)

Figure 4.3 Vortex generators fitted to the upper wing surface of a Piper PA-31-350 Chieftan light twin-engine airplane, to reduce minimum single-engine control speed Vmca. This installation of 43 generators on each wing was designed by Boundary Layer Research, Inc., of Everett, WA.

low-aspect ratio wings that stick out of a surface. The tip vortices from a spanwise row of generators set at angles of attack energize the surface's boundary layer by mixing in with it high-energy air from the surrounding flow. The energized boundary layer tends to remain attached, avoiding separation or stall.

According to John G. Lee (1984), vortex generators were invented by "an introspective and rather unapproachable loner" named Hendrik Bruynes, who used eight vortex generators to correct separation from the walls of the diffuser in a new 18-foot United Aircraft Research Department wind tunnel. While Bruynes was named in the vortex generator patent, Lee credits Henry H. Hoadley with the key idea of reversing the angles of alternate generators. The *Forty-Second Annual Report* of the NACA, dated 1956, flatly credits H. D. Taylor of United Aircraft as the developer of vortex generators; no mention is made of either Bruynes or Hoadley.

4.6   Propeller Slipstream Effects

Of all the propeller or jet power effects on stability and control, those due to the propeller slipstream or wake are the most difficult to deal with, analytically or in test. The British engineer William Froude (1810–1879) laid some initial groundwork in his application of momentum theory to establish propeller (or ship screw) slipstream velocity. But the nominally cylindrical slipstream is distorted badly as it passes over wings and fuselages. As a result, slipstream effects on wings, tails, and fuselages are difficult to predict. The Smelt and Davies (1937) and Millikan (1940) papers are fair starts on prediction, but most designers still rely on educated guesses, based on test data from earlier projects.

A few generalities exist for propeller slipstream effects on static longitudinal stability. The relative increment in slipstream velocity over that of the free stream increases as airspeed is reduced, at fixed throttle setting. If the horizontal tail carries a net down load and is exposed to the slipstream, relative higher velocities as airspeed is reduced increases that down load, pitching the airplane nose up, a destabilizing effect.

Therefore, net down tail loads on propeller-driven airplanes, such as are required to trim out landing flaps, are destabilizing under power-on conditions. This effect is avoided only if the horizontal tail is mounted high enough on the vertical tail so as to be out of the propeller slipstreams at all angles of attack. Although down-load destabilization must have been known to designers before 1948, the first discussion of this effect published in the United States dates from that year (Phillips, 1948).

In addition to down tail load destabilization, increased downwash at the horizontal tail due to power is also a destabilizing factor. This factor was also first noted by Phillips. A semiempirical correlation of the increased downwash due to running propellers was made shortly afterwards by Weil and Sleeman, based on the power-off downwash and the propeller's thrust coefficient (Figure 4.4).

Rotation of the slipstream behind propellers of single-engine airplanes creates side loads on vertical tails at low airspeeds, requiring that pilots apply counteracting rudder deflections to hold a straight course. This can be unnerving at high power levels, such as in takeoff runs and in waveoffs from carrier landings. Some manufacturers offset fin leading edges to the left for U.S. right-hand propeller rotation when viewed from the rear, to minimize required rudder angles at high power levels and low airspeeds.

However, offset fins result in large and rapidly changing rudder pedal forces in high-speed dives, requiring high pedal forces to stay on a dive-bombing target. On the other hand,

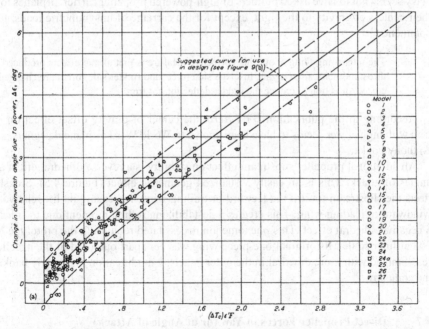

Figure 4.4 Correlation of increment in downwash $\Delta\epsilon$ with propeller power for single-engine airplanes. F is a taper ratio factor and ϵ' is the power-off downwash. (From Weil and Sleeman, NACA Rept. 941, 1949)

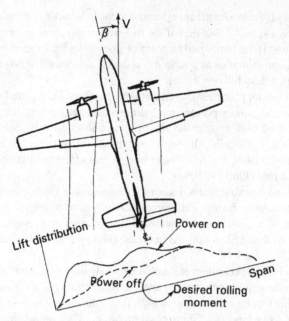

Figure 4.5 Unstable dihedral effect (forward-wing-down) due to slipstream loads.

lateral offset of an airplane's center of gravity is effective in reducing the rudder deflections required for trim under high power conditions while causing only a constant, rather than a rapidly, increasing pedal force to trim in dives (Phillips, Crane, and Hunter, 1944). The U.S. Navy seems not to have asked builders of high-powered propeller carrier airplanes to offset their centers of gravity to the right, except for the carriage of unsymmetric loads. Phillips comments:

> The interesting thing about this corrective measure is that all the effects produced (direct thrust moment, aileron yaw, fuselage side force, and rudder side force) are changed in the correct direction to reduce the rudder deflection for trim.

According to the publication *WWI Aero*, the center of gravity offset effect on rudder trim with high power seems to have been known to World War I airplane designers, such as Anthony Fokker.

An additional propeller slipstream effect on stability is a reverse dihedral effect at high angles of attack. This is an easily visualized geometric effect (Figure 4.5). Sideslip deflects the slipstream toward the leeward or trailing wing, increasing its lift relative to the windward or leading wing. This creates a destabilizing rolling moment due to sideslip, or a negative dihedral effect. This phenomenon necessitated rebuilding the prototype Martin 202 airliner so that its original one-piece wing was cut apart outboard of the nacelles and reassembled with more dihedral in the outer panels, a modification that nearly bankrupted the company.

4.7   Direct Propeller Forces in Yaw (or at Angle of Attack)

In contrast to the somewhat unsatisfactory state of the theory for propeller slip-stream effects, the theories for direct propeller forces in yaw are well established, and those theories were around as early as needed. According to Dr. Herbert S. Ribner:

It was realized as early as 1909 that a propeller in yaw develops a side force like that of a fin. In 1917, R. G. Harris expressed this force in terms of the torque coefficient for the unyawed propeller.

A 1914 British R & M by Relf, Bramwell, Fage, and Bryant presented experimental results on propeller side forces. Two 1945 NACA reports by Ribner are usually taken as the definitive modern work on the subject. These reports provide a blade-element analysis applicable to any single- or dual-rotation system, sample calculations for two representative propellers with an interpolation scheme for other propellers, experimental verification of the blade-element method, and, finally, a remarkably simple rule-of-thumb side force estimate for preliminary design. This is to take the yawed propeller as a fin of area equal to the projected side area of the propeller. This fin's effective aspect ratio is taken as 8, and the effective dynamic pressure at the fin is that for the propeller disk augmented by inflow. The side force for a propeller in yaw or sideslip is clearly the same as the propeller normal force at angle of attack.

For tractor airplanes, direct propeller forces in yaw act as a fin ahead of the airplane's center of gravity. This is a major destabilizing contribution to static directional stability, especially at large propeller blade angles. The destabilizing effect depends on the propeller plane distance to the center of gravity, which is relatively greater for single-engine airplanes than for multiengine airplanes with wing-mounted engines. The same can be said for the effects of propeller normal force at angle of attack, in relation to static longitudinal stability. The classical NACA design method for satisfactory longitudinal stability (Gilruth, 1941) accounts for idling power effects using a propeller normal force calculation.

4.8   Jet and Rocket Effects on Stability and Control

Except for the V/STOL case, jet and rocket effects on airplane stability and control tend to be small compared with those for high-powered propeller airplanes. This is because of the absence of slipstream and direct propeller effects. Yet, they are not negligible. By the time the first jet aircraft were being tested, the necessary theory was in place. Two new factors needed to be accounted for – jet intake normal force and airstream deviation due to inflow into the jet or rocket exhaust.

4.8.1   *Jet Intake Normal Force*

The NACA engineer Dr. Herbert Ribner, who was also an important contributor to the body of knowledge on propeller forces in yaw, or propeller normal forces, provided the analogous theory for jet air intakes, or an algorithm for jet normal force (Ribner, 1946). Ribner's jet intake normal force formulation is based on the mass flow of air into each jet intake and the angle turned by this air to get into the duct. This method neatly avoids having to estimate or measure pressure distributions inside and outside of the intake ducts, although the resultant normal forces must be generated by those pressures.

One interesting refinement is to take into account the upflow before each jet intake caused by the wing's lifting system. In sideslip, the corresponding correction is for any sidewash ahead of the intake. Fortunately, wing upflow angles are readily available in chart form for any wing planform.

Herbert Ribner was active until 2001, spending winters at the Langley Research Center as a NASA Distinguished Research Associate and the rest of the year at the University of Toronto.

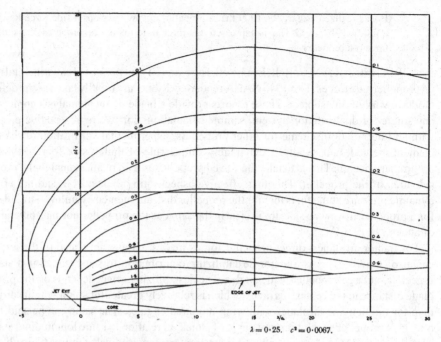

$\lambda = 0 \cdot 25. \quad c^2 = 0 \cdot 0067.$

Figure 4.6 Calculated isoclines of the surrounding flow into a jet wake. Destabilizing downwash for tails mounted above the jet wake is indicated by the positive deviation angles. (From Squire and Trouncer, British R & M 1974, 1944)

4.8.2   *Airstream Deviation Due to Inflow*

A jet or rocket stream issuing from a nozzle acts like a hydrodynamic sink on the surrounding free-stream cold air flow. H. B. Squire and J. Trouncer (1944) produced beautiful isocline maps of the free-stream deviation angles around a jet (Figure 4.6). The sense of the deviation angles is for the surrounding free-stream flow to feed into the jet. Squire and Trouncer's calculated deviation angles are parameterized in terms of the ratio of jet to free-stream velocities. The larger the velocity ratio, the larger is the deviation angle.

If airspeed is reduced from a trim value at a fixed throttle setting, the ratio of jet to free-stream velocity increases. This increases the free-stream deviation angles into the jet at any given location. In the common case in which the jet passes under the horizontal tail, this increases the effective downwash angle as the speed is reduced. This in turn provides a nose-up pitching moment at speeds below trim, a destabilizing effect. Forward neutral point shifts of as much as 10 percent of the wing mean chord are found for airplanes whose jet exhausts are forward of the horizontal tail. Conversely, only minor stability effects are measured for jet exhausts behind the horizontal tail.

Squire and Trouncer's calculated stream deviation angles into a jet are for a jet stream at the same temperature as a free stream. A correction is needed to apply their data to the heated jets that come from actual jet or rocket engines. The equivalent cold jet velocity ratio is related to the actual jet velocity ratio by a function involving the ratio of the jet temperature to free-stream temperature.

4.9   **Special VTOL Jet Inflow Effects**

Deflected jet VTOL airplanes such as the Hawker-Siddeley Harrier and the McDonnell Douglas AV-8B Harrier II can have troublesome jet inflow effects on static

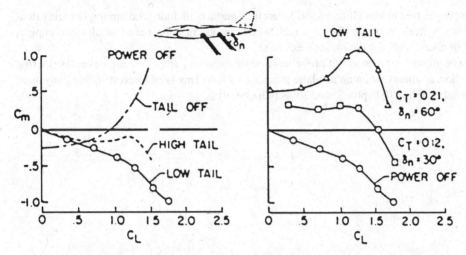

Figure 4.7 The low tail position that produces good stability at high lift coefficients for swept-wing airplanes can destabilize the airplane as a deflected jet VTOL. The left-hand diagram shows stability with jet power off; the right-hand diagram shows instability for a jet deflection angle of 60 degrees. (Reprinted with permission from SAE Paper No. 864A, © 1964, Society of Automotive Engineers, Inc.)

longitudinal stability. Problems can arise at jet deflection angles that are intermediate between hovering and normal flight.

As shown in Figure 4.7, jet deflection angles of about 60 degrees can put the horizontal tail in an effective high position relative to the jet flow field. This causes large downwash angles over the tail (McKinney, Kuhn, and Reeder, 1964). This occurs in spite of an actual low tail position relative to the wing chord plane, which is necessary on swept wings to avoid transonic pitchup. In other words, finding a single horizontal tail vertical position to give good static longitudinal stability under all flight conditions may be difficult for swept-wing deflected-jet VTOL airplanes of the Harrier type.

An additional inflow problem occurs with tilt-rotor VTOL airplanes at high descent rates at low airspeeds. This problem is shared with rotary-wing aircraft. High descent rates can lead to asymmetric loss in lift and uncontrollable roll, because of upflow through propeller disks due to the descent rate. The upflow interferes with the downflow required for lift.

4.9.1   Jet Damping and Inertial Effects

While he was at the Douglas Company plant in El Segundo, California, Hans C. Vetter described a damping effect to be expected from jet air intakes and exhausts. He argued that the air in a jet duct travels in a radial path with respect to the center of gravity when the airplane performs rotational oscillations. Pressure forces on the structure result which are in the direction to resist angular velocities, adding "Coriolis" damping to the aerodynamic damping moments provided by the wing and tail surfaces. Jet damping moments depend on the distances from the center of gravity to the jet intake and exits and on other dimensions (Vetter, 1953).

Artificial yaw and pitch damping, used on almost all modern jet aircraft, tends to swamp out jet damping effects. Furthermore, jet damping is most significant at low airspeeds and high thrust levels, normally encountered only at low altitudes. But the airframe's natural

damping is best at low altitudes. Still, careful designers include jet damping in their calculations. Vetter's theory implies that rocket-powered aircraft also have Coriolis jet damping,
but of course only for the rocket's exhaust.

The angular momentum of propellers and the rotating parts of jet engines create inertial
reaction moments when an airplane pitches or yaws. This is of interest in the analysis of
inertial coupling (Chapter 8) and spins (Chapter 9).

CHAPTER 5

Managing Control Forces

As airplanes evolved from stick and wire contraptions to awesome supersonic machines, the pilot at the center of it all has not changed. Desirable maximum and minimum levels of pilot stick, yoke, and rudder pedal control forces required to steer and maneuver are much the same – but the engineering solutions that bring these forces about have changed with the times.

5.1   Desirable Control Force Levels

In 1936 and 1937, NACA research pilots and engineers Melvin N. Gough, A. P. Beard, and William H. McAvoy used an instrumented cockpit to establish maximum force levels for control sticks and wheels. In lateral control the maximums for one hand are 30 pounds applied at a stick grip and 80 pounds applied at the rim of a control wheel. In longitudinal control the maximums are 35 pounds for a stick and 50 pounds for a wheel. Lower forces are desirable and easily attainable with modern artificial feel systems.

The Federal Aviation Administration (FAA) allows higher forces for transport-category airplanes under FAR Part 25. Seventy-five pounds is allowed for temporary application. However, the data compilation for the handbook accompanying MIL-STD-1797, a current military document, shows that a little over 50 percent of male pilots and fewer than 5 percent of female pilots are capable of this force level.

Gough-Beard-McAvoy force levels are generally used as maximum limits for conventional stick, yoke, and rudder pedal controllers, but much lower control force levels are specified for artificial-feel systems and for side-stick controls operated by wrist and forearm motions.

5.2   Background to Aerodynamically Balanced Control Surfaces

When airplanes and their control surfaces became large and airplane speeds rose to several hundred miles per hour, control forces grew to the point where even the Gough-Beard-McAvoy force limits were exceeded. Pilots needed assistance to move control surfaces to their full travels against the pressure of the air moving past the surfaces. An obvious expedient was to use those same pressures on extensions of the control surface forward of the hinges, to balance the pressure forces that tried to keep the control surfaces faired with the wing.

The actual developmental history of aerodynamically balanced control surfaces did not proceed in a logical manner. But a logical first step would have been to establish a background for design of the balances by developing design charts for the forces and hinge moments for unbalanced control surfaces. That step took place first in Great Britain (Glauert, 1927). Glauert's calculations were based on thin airfoil theory. W. G. Perrin followed in the next year with the theoretical basis for control tab design (Perrin, 1928).

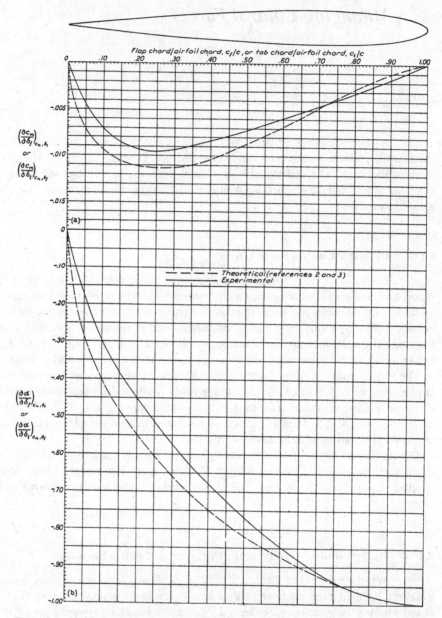

Figure 5.1 Pitching moment and control effectiveness parameters for plain flaps on the NACA 0009 airfoil, derived from pressure distributions. The dashed lines are Glauert's thin airfoil theory. (From Ames and Sears, NACA Rept. 721, 1941)

The next significant step in the background for forces and hinge moments for unbalanced control surfaces was NACA pressure distribution tests on a NACA 0009 airfoil, an airfoil particularly suited to tail surfaces (Ames, Street, and Sears, 1941). Figures 5.1 and 5.2 compare those results with Glauert's theory. The trends with control surface hinge position along the airfoil chord match Glauert's thin airfoil theory exactly, but with lower flap effectiveness and hinge moment than the theoretical values. Ames and his associates developed

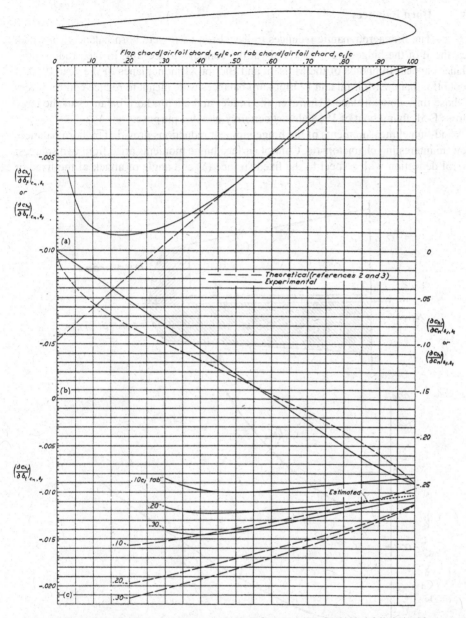

Figure 5.2 Hinge moment parameters for plain flaps on the NACA 0009 airfoil, derived from pressure distributions. The dashed lines are Glauert's thin airfoil theory. (From Ames and Sears, NACA Rept. 721, 1941)

a fairly complex scheme to derive three-dimensional wing and tail surface data from the two-dimensional design charts. That NACA work was complemented for horizontal tails by a collection of actual horizontal tail data for 17 tail surfaces, 8 Russian and 3 each Polish, British, and U.S. (Silverstein and Katzoff, 1940). Full control surface design charts came later, with the publication of stability and control handbooks in several countries (see Chapter 6, Sec. 2.6).

5.3   Horn Balances

The first aerodynamic balances to have been used were horn balances, in which area ahead of the hinge line is used only at the control surface tips. In fact, rudder horn balances appear in photos of the Moisant and Blériot XI monoplanes of the year 1910. It is doubtful that the Moisant and Blériot horn balances were meant to reduce control forces on those tiny, slow airplanes. However, the rudder and aileron horn balances of the large Curtiss F-5L flying boat of 1918 almost certainly had that purpose.

Wind-tunnel measurements of the hinge moment reductions provided by horn balances show an interesting characteristic. Control surface hinge moments arise from two sources: control deflection with respect to the fixed surface (δ) and angle of attack of the fixed or

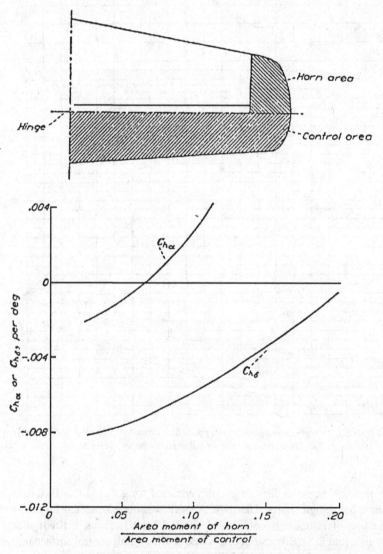

Figure 5.3   Typical hinge moment parameter variations with size for unshielded horn aerodynamic balances. The hinge moment due to angle of attack Ch_α is affected more strongly by the horn balance than by the hinge moment due to surface deflection Ch_δ. (From Phillips, NACA Rept. 927, 1948)

main surface (α). The relationship is given in linearized dimensionless form by the equation $C_h = C_{h_\delta}\delta + C_{h_\alpha}\alpha$, where the hinge moment coefficient C_h is the hinge moment divided by the surface area and mean chord aft of the hinge line and by the dynamic pressure. C_{h_δ} and C_{h_α} are derivatives of C_h with respect to δ and α, respectively. Both derivatives are normally negative in sign. Negative C_{h_δ} means that when deflected the control tends to return to the faired position. Negative C_{h_α} means that when the fixed surface takes a positive angle of attack the control floats upward, or trailing edge high.

Upfloating control surfaces reduce the stabilizing effect of the tail surfaces. It was discovered that horn balances produce positive changes in C_{h_α}, reducing the upfloating tendency and increasing stability with the pilot's controls free and the control surfaces free to float (Figure 5.3). This horn balance advantage has to be weighed against two disadvantages. The aerodynamic balancing moments applied at control surface tips twist the control surface. Likewise, flutter balance weights placed at the tips of the horn, where they have a good moment arm with respect to the hinge line, lose effectiveness with control surface twist.

A horn balance variation is the shielded horn balance, in which the horn leading edge is set behind the fixed structure of a wing or tail surface. Shielded horn balances are thought to be less susceptible to accumulating leading-edge ice. Shielded horn balances are also thought to be less susceptible to snagging a pilot's parachute lines during bailout.

5.4 Overhang or Leading-Edge Balances

When control surface area ahead of the hinge line is distributed along the span of the control surface, instead of in a horn at the tip, the balance is called an overhang or a leading-edge balance. Overhang design parameters are the percentage of area ahead of the hinge line relative to the total control surface area and the cross-sectional shape of the overhang (Figure 5.4).

Experimental data on the effects of overhang balances on hinge moments and control effectiveness started to be collected as far back as the late 1920s. Some of these early data are given by Abe Silverstein and S. Katzoff (1940). Airplane manufacturers made their own correlations of the effects of overhang balances, notably at the Douglas Aircraft Company (Root, 1939). As in many other disciplines, the pressure of World War II accelerated these developments. Root and his group at Douglas found optimized overhang balance proportions for the SBD-1 Dauntless dive bomber by providing for adjustments on hinge line location and overhang nose shape on the SBD-1 prototype, known as the XBT-2.

Root wrote a NACA Advance Confidential Report in May 1942 to document a long series of control surface and other modifications leading to flying qualities that satisfied Navy test pilots. For example, in 1 of 12 horizontal tail modifications that were flight tested, the elevator overhang was changed from an elliptical to a "radial," or more blunt, cross-section, to provide more aerodynamic balancing for small elevator movements. This was to reduce control forces at high airspeeds.

Overhang aerodynamic balance, in combination with spring tabs, continue in use in Douglas transport airplanes, from the DC-6 and DC-7 series right up to the elevators and ailerons of the jet-powered DC-8. The DC-8's elevator is balanced by a 35-percent elliptical nose overhang balance. Remarkably constant hinge moment coefficient variations with elevator deflection are obtained up to a Mach number of 0.96.

George S. Schairer came to the Boeing Company with an extensive control surface development background at Convair and in the Cal Tech GALCIT 10-foot wind tunnel. Although early B-17s had used spring tabs, Schairer decided to switch to leading-edge

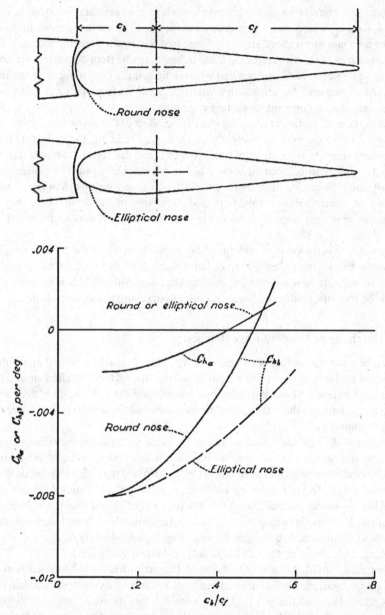

Figure 5.4   Typical hinge moment parameter variation with size for leading-edge or overhang aerodynamic balances. The round nose is more effective in reducing hinge moment due to surface deflection Ch_δ than the elliptical nose, which does not protrude into the airstream as much when the surface is deflected. (From Phillips, NACA Rept. 927, 1948)

balances for the B-17E and the B-29 bombers. The rounded nose overhang balances on the B-29s worked generally well, except for an elevator overbalance tendency at large deflection angles. Large elevator angles were used in push-overs into dives for evasive action. William Cook remarks, "A World War II B-29 pilot friend of mine was quite familiar with this characteristic, so the fact that he got back meant this must have been tolerable." However, overhang balance was not effective for the B-29 ailerons. Forces were excessive.

The wartime and other work on overhang aerodynamic balance was summarized by the NACA Langley Research Department (Toll, 1947). The Toll report remains a useful reference for modern stability and control designers working with overhang aerodynamic balances and other aerodynamic balance types as well.

5.5   Frise Ailerons

The hinge line of the Frise aileron, invented by Leslie George Frise, is always at or below the wing's lower surface. If one sees aileron hinge brackets below the wing, chances are that one is looking at a Frise aileron (Figure 5.5). Frise ailerons were used on many historic airplanes after the first World War, including the Boeing XB-15 and B-17, the Bell P-39, the Grumman F6F-3 and TBF, and the famous World War II opponents – the Spitfire, Hurricane, and Focke-Wulf 190 fighters. Frise ailerons were applied to both the Curtiss-Wright C-46 Commando and the Douglas C-54 Skymaster during World War II, to replace the hydraulic boost systems used in their respective prototypes.

With the hinge point below the wing surface, an arc drawn from the hinge point to be tangent to the wing upper surface penetrates the wing lower surface some distance ahead of the hinge line, thus establishing an overhang balance. The gap between the aileron and wing can be made as narrow as desired by describing another arc slightly larger than the first. This in fact is typical of the Frise aileron design. The narrow wing-to-aileron gap reduces air flow from the high-pressure wing under surface to the lower pressure wing upper surface, reducing drag. The Frise aileron is less prone to accumulate ice for that same reason. It was promoted by the U.S. Army Air Corps *Handbook for Airplane Designers* as an anti-icing aileron.

The relatively sharp Frise aileron nose develops high velocities and low static pressures when projecting below the wing lower surface, when the aileron goes trailing-edge up. This generally overbalances the up-going aileron. On the other hand, the overbalanced up aileron is connected by control cables or pushrods to the down-going aileron on the other side of the wing. The sharp Frise nose on that side is within the wing contour; the down aileron is underbalanced. By connecting the up and down sides through the pilot's controls the combination is made stable, with lowered control forces relative to ailerons without aerodynamic balance.

The sharp nose of the Frise aileron, protruding below the wing's lower surface for trailing-edge-up deflections, has been thought to help reduce adverse yaw when rolling. The trailing-edge-up aileron is on the down-going wing in a roll. In adverse yaw, the down-going wing moves forward, while the airplane yaws in a direction opposite to that corresponding to a

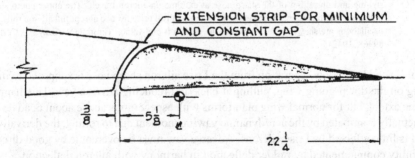

Figure 5.5   A Frise aileron design used on the Douglas SBD-1 Dauntless. This design was the seventh and final configuration tested in 1939 and 1940. Nose shape, wing-to-aileron gap, hinge line position, and gap seal parameters were all varied. (From Root, NACA W-81, 1942)

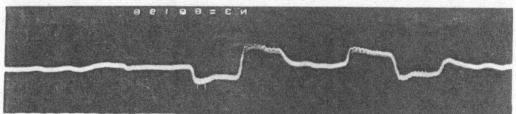

Figure 5.6 Flight test evidence of Frise aileron oscillations on a Waco XCG-3 glider due to alternate stalling and unstalling of the sharp nose at extreme up-aileron travels. The upper photo shows the bulky roll rate recorder. The lower photo is a rate of roll trace for two abrupt full aileron rolls. Aileron oscillations are shown by the ripples at the peak roll rate values. (From U.S. Army Air Corps photo 89368, 1942)

coordinated turn. Flow separation from the Frise aileron sharp nose is supposed to increase drag on the down-going wing, pulling it back and reducing adverse yaw. This happens to some extent, but for normal wing plan forms with aspect ratios above about 6, adverse yaw is actually dominated by the aerodynamic yawing moment due to rolling, the derivative C_{n_p}, and is little affected by Frise ailerons. Adverse yaw must be overcome by good directional stability complemented by rudder deflection in harmony with aileron deflection.

Frise ailerons turned out to have problems on large airplanes, where there is a long cable run from the control yoke to the ailerons. In the development of the Waco XCG-3 glider in

1942, the sharp nose of its Frise ailerons alternately stalled and unstalled when the ailerons were held in a deflected position. This created severe buffeting. The aileron nose stalled at the largest angle, reducing the balancing hinge moment. Control cable stretch allowed the aileron to start back toward neutral. But as the aileron angle reduced the nose unstalled, the aerodynamic balance returned, and the aileron started back toward full deflection, completing the cycle (Figure 5.6).

The fix for the XCG-3 was to limit up-aileron angles from 30 to 20 degrees and to round off the sharp nose to delay stalling of the nose. Modified Frise ailerons, with noses raised to delay stalling, had been tested in Britain by A. S. Hartshorn and F. B. Bradfield as early as 1934. The advantages of raised-nose Frise ailerons were verified in NACA tests on a Curtiss P-40 (Goranson, 1945). Beveled trailing edges were added to the raised-nose Frise ailerons on the P-40, to make up for loss in aerodynamic balance at small deflections. Lateral stick force remained fairly linear and very low up to a total (sum of up and down) aileron deflection of 48 degrees, giving a remarkably high dimensionless roll rate pb/2V of 0.138 at 200 miles per hour.

5.6   Aileron Differential

The larger travel of one aileron relative to the other is called aileron differential (Figure 5.7). Aileron differential is a method of reducing control forces by taking advantage of hinge moment bias in one direction (Jones and Nerkin, 1936; Gates, 1940). At positive wing angles of attack, the hinge moment acting on both ailerons is normally trailing-edge-up, and we say the ailerons want to float up. Assume that the up-going aileron is given a larger travel than the down-going aileron for a given control stick or wheel throw. Then, the work done by the trailing-edge-up hinge moment acting on the up-going aileron can be nearly as great as the work the pilot does in moving the down-going aileron against its up-acting hinge moment, and little pilot force is needed to move the combination. The differential appropriate for up-float is more trailing-edge-up angle than down. Typical values are 30 degrees up and 15 degrees down. The floating hinge moment can be augmented, or even reversed, by fixed tabs.

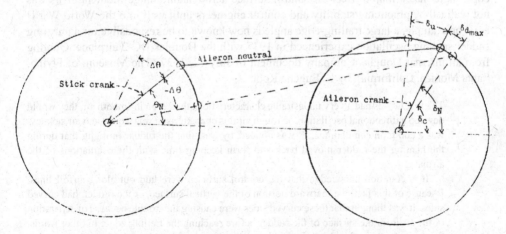

Figure 5.7   The principle of aileron differential, or unsymmetrical up and down travels. Stick crank motions $\Delta\Theta$ of the same amount on each wing cause larger up-aileron deflections δ_u than down-aileron deflections δd_{max}. (From Jones and Nerkin, NACA TN 586, 1936)

Aileron up-float, associated with negative values of the hinge moment derivative $C_{h\alpha}$, is greatest at high wing angles of attack. Neglecting accelerated flight, high wing angles of attack occur at low airspeeds. Thus, aileron differential has the unfortunate effect of reducing aileron control forces at low airspeeds more than at high airspeeds, where reductions are really needed. In addition to the force-lightening characteristic of aileron differential, increased up relative to down aileron tends to minimize adverse yaw in aileron rolls, which is the tendency of the nose to swing initially in the opposite direction to the commanded roll.

Adverse yaw in aileron rolls remains a problem for modern airplanes, especially those with low directional stability, such as tailless airplanes. Where stability augmentation (Chapter 20) is available, it is a more powerful means of overcoming adverse yaw than aileron differential.

5.7   Balancing or Geared Tabs

Control surface tabs affect the pressure distribution at the rear of control surfaces, where there is a large moment arm about the hinge line. A trailing-edge-up tab creates relative positive pressure on the control's upper surface and a relative negative pressure peak over the tab-surface hinge line. Both pressure changes drive the control surface in the opposite direction to the tab, or trailing-edge-down.

When a tab is linked to the main wing so as to drive the tab in opposition to control surface motion, it is called a balancing or geared tab. Balancing tabs are used widely to reduce control forces due to control surface deflection. They have no effect on the hinge moments due to wing or tail surface angle of attack. Airplanes with balancing tabs include the Lockheed Jetstar rudder, the Bell P-39 ailerons (augmenting Frise ailerons), and the Convair 880M.

5.8   Trailing-Edge Angle and Beveled Controls

The included angle of upper and lower surfaces at the trailing edge, or trailing-edge angle, has a major effect on control surface aerodynamic hinge moment. This was not realized by practicing stability and control engineers until well into the World War II era. For example, a large trailing-edge angle is now known to be responsible for a puzzling rudder snaking oscillation experienced in 1937 with the Douglas DC-2 airplane. Quoting from an internal Douglas Company document of July 12, 1937 (The Museum of Flying, Santa Monica, California), by L. Eugene Root:

> The first DC-2s had a very undesirable characteristic in that, even in smooth air, they would develop a directional oscillation. In rough air this characteristic was worse, and air sickness was a common complaint.... It was noticed, by watching the rudder in flight, that during the hunting the rudder moved back and forth keeping time with the oscillations of the airplane.
>
> It is common knowledge that the control surfaces were laid out along airfoil lines. Because of this fact, the rearward portion of the vertical surface, or the rudder, had curved sides. It was thought that these curved sides were causing the trouble because of separation of the air from the surface of the rudder before reaching the trailing edge. In other words, there was a region in which the rudder could move and not hit "solid" air, thus causing the movement from side to side. The curvature was increased towards the trailing edge of the rudder in such a way as to reduce the supposedly "dead" area.... The change that

we made to the rudder was definitely in the wrong direction, for the airplane oscillated severely. . . . After trying several combinations on both elevators and rudder, we finally tried a rudder with straight sides instead of those which would normally result from the use of airfoil sections for the vertical surfaces. We were relieved when the oscillations disappeared entirely upon the use of this type of rudder.

The Douglas group had stumbled on the solution to the oscillation or snaking problem, reduction of the rudder floating tendency through reduction of the trailing-edge angle. Flat-sided control surfaces have reduced trailing-edge angles compared with control surfaces that fill out the airfoil contour. We now understand the role of the control surface trailing-edge angle on hinge moments. The wing's boundary layer is thinned on the control surface's windward side, or the wing surface from which the control protrudes. Conversely, the wing's boundary layer thickens on the control surface's leeward side, where the control surface has moved away from the flow. Otherwise stated, for small downward control surface angles or positive wing angles of attack the wing's boundary layer is thinned on the control surface bottom and thickened on the control's upper surface.

The effect of this differential boundary layer action for down-control angles or positive wing angles of attack is to cause the flow to adhere more closely to the lower control surface side than to the upper side. In following the lower surface contour the flow curves toward the trailing edge. This curve creates local suction, just as an upward-deflected tab would do. On the other hand, the relatively thickened upper surface boundary layer causes the flow to ignore the upper surface curvature. The absence of a flow curve around the upper surface completes the analogy to the effect of an upward-deflected tab. The technical jargon for this effect is that large control surface trailing-edge angles create positive values of the derivatives C_{h_α} and C_{h_δ}, the floating and restoring derivatives, respectively.

The dynamic mechanism for unstable lateral-directional oscillations with a free rudder became known on both sides of the Atlantic a little after the Douglas DC-2 experience. Unstable yaw oscillations were calculated in Britain for a rudder that floated into the wind (Bryant and Gandy, 1939). This was confirmed in two NACA studies (Jones and Cohen, 1941; Greenberg and Sternfield, 1943). The aerodynamic connection between trailing-edge angle and control surface hinge moment, including the floating tendency, completed the story (Jones and Ames, 1942).

Following the success of the flat-sided rudder in correcting yaw snaking oscillations on the Douglas DC-2, flat-sided control surfaces became standard design practice on Douglas airplanes. William H. Cook credits George S. Schairer with introducing flat-sided control surfaces at Boeing, where they were used first on the B-17E and B-29 airplanes. Trailing-edge angles of fabric-covered control surfaces vary in flight with the pressure differential across the fabric (Mathews, 1944). A Douglas C-74 transport was lost in 1946 when elevator fabric bulging between ribs increased the trailing-edge angle, causing pitch oscillations that broke off the wing tips. C-74 elevators were metal-covered after that.

Understanding of the role of the trailing-edge angle in aerodynamic hinge moments opened the way for its use as another method of control force management. Beveled control surfaces, in which the trailing-edge angle is made arbitrarily large, is such an application (Figure 5.8). Beveled control surfaces, a British invention of World War II vintage, work like balancing tabs for small control surface angles.

The beveled-edge control works quite well for moderate bevel angles. As applied to the North American P-51 Mustang, beveled ailerons almost doubled the available rate of roll at high airspeeds, where high control forces limit the available amount of aileron deflection.

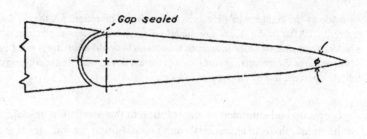

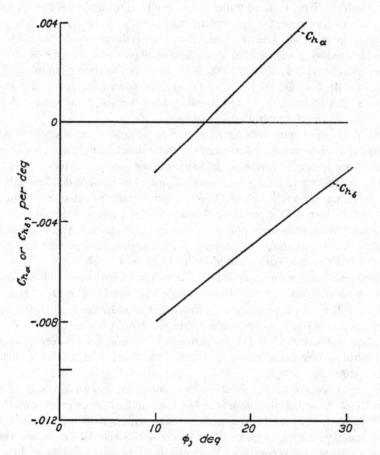

Figure 5.8 Typical hinge moment parameter variation with bevel angle ϕ for a beveled control surface. (From Phillips, NACA Rept. 927, 1948)

But large bevel angles, around 30 degrees, acted too well at high Mach numbers, causing overbalance and unacceptable limit cycle oscillations (Figure 5.9). Beveled controls have survived into recent times, used for example on the ailerons of the Grumman/Gulfstream AA-5 Tiger and on some Mooney airplanes.

5.9   Corded Controls

Corded controls, apparently invented in Britain, are thin cylinders, such as actual cord, fastened to control surfaces just ahead of the trailing edge. They are used on one

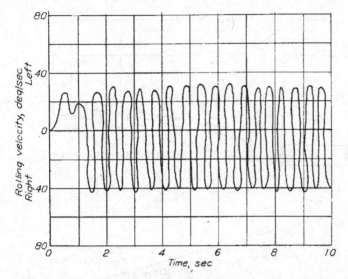

Figure 5.9 Overbalance at small deflections, the downside of control surface bevel. Beveled ailerons with an excessive angle of 32 degrees go through a limit cycle oscillation on an XP-51 airplane. The oscillation is poorly traced from an original. (From Toll, NACA Rept. 868, 1947)

or both sides of a control surface. Corded controls are the inverse of beveled controls. Bevels on the control surface side that projects into the wind produce relative negative pressures near the bevel that balance the control aerodynamically, reducing operating force. On the other hand, cords on the control surface side that projects into the wind create local positive pressures on the surface just ahead of the cord. This increases control operating force.

Cords on both sides of a control surface are used to eliminate aerodynamic overbalance. On one side they act as a fixed trim tab. Very light control forces have been achieved by cut and try by starting with aerodynamically overbalanced surfaces, caused by deliberately oversized overhang balances. Quite long cords correct the overbalance, providing stable control forces. In the cut and try process the cords are trimmed back in increments until the forces have been lightened to the pilot's or designer's satisfaction. Adjustable projections normal to the trailing edge, called Gurney flaps, act as one-sided cord trim tabs.

5.10 Spoiler Ailerons

Spoiler ailerons project upward from the upper surface of one wing, reducing lift on that wing and thus producing a rolling moment (Figure 5.10). Spoiler ailerons are often the same surfaces used symmetrically to reduce lift and increase drag on large jet airplanes for rapid descents and to assist braking on runways. Spoiler ailerons are generally used either to free wing trailing edges for full-span landing flaps or to minimize wing twist due to aileron action on very flexible wings.

The aerodynamic details of spoiler operation are still not completely understood, even after years of experiment and theoretical studies. The aerodynamics of a rapidly opened spoiler has two phases, the opening and steady-state phases.

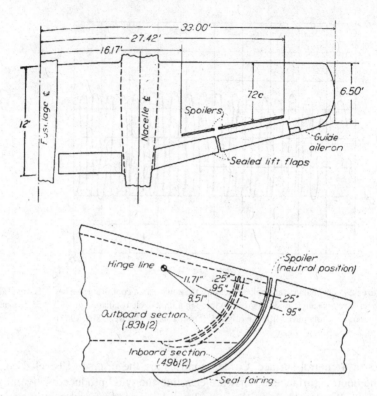

Figure 5.10   Plug-slot spoiler ailerons used on the Northrop P-61 Black Widow airplane. An attempt to devote most of the wing's trailing edge to landing flaps, rather than ailerons. The small guide aileron provides stick force feel. (From Toll, NACA Rept. 868, 1947)

5.10.1   Spoiler Opening Aerodynamics

Experimental or wind-tunnel studies of rapidly opening upper-wing surface spoilers show a momentary increase in lift, followed by a rapid decrease to a steady-state value that is lower than the initial value. At a wind speed of 39 feet per second, the initial increase is over in less than a half-second, and steady-state conditions appear in about 3 seconds (Yeung, Xu, and Gu, 1997). Results from the computational fluid dynamics method known as the discrete vortex method also predict the momentary increase in lift and associate it with a vortex shed from the spoiler upper edge in a direction that increases net airfoil circulation in the lifting direction. A subsequent shed vortex from the wing trailing edge in the opposite direction reduces circulation to the steady-state value. While suggestive, experimental flow visualization results do not exist that confirm this vortex model.

The Yeung, Xu, and Gu experiments show that providing small clearances between the spoiler lower edge and the wing upper surface reduces the momentary increase in lift following spoiler extension. This is consistent with a small shed vortex from the spoiler lower edge of opposite rotation to the vortex shed at the upper edge. A clearance between spoiler and wing surface of this type has also been used to reduce buffet.

5.10.2   Spoiler Steady-State Aerodynamics

Separation behind an opened spoiler on a wing upper surface causes distortion of the external or potential flow that is similar to the effect of a flap-type surface with trailing-edge-up deflection. In the latter case, streamlines above the wing are raised toward the wing trailing edge. The effective wing camber is negative in the trailing-edge region, causing a net loss in circulation and lift. The difference in the two cases is that the effective wing trailing edge in the spoiler case is somewhere in the middle of the separated region, instead of at the actual trailing edge, as in the flap-type surface case.

5.10.3 Spoiler Operating Forces

The hinge moments of ordinary hinged-flap and slot lip spoiler ailerons are high; brute hydraulic force is used to open them against the airstream. Retractable arc and plug spoiler ailerons are designed for very low hinge moments and operating forces. Although aerodynamic pressures on the curved surfaces of these ailerons are high, the lines of action of these pressures are directed through the hinge line and do not show up as hinge moments. Hinge moments arise only from pressure forces on the ends of the arcs and from small skin friction forces on the curved surfaces.

5.10.4 Spoiler Aileron Applications

A very early application of plug ailerons was to the Northrop P-61 Black Widow, which went into production in 1943. The P-61 application illustrates the compromises that are needed at times when adapting a device tested in a wind tunnel to an actual airplane. The plug aileron is obviously intended to work only in the up position. However, it turned out not to be possible to have the P-61 plug ailerons come to a dead stop within the wing when retracting them from the up position. The only practical way to gear the P-61 plug ailerons to the cable control system attached to the wheel was by extreme differential. Full up-plug aileron extension on one side results in a slight amount of down-plug aileron angle on the other side. The down-plug aileron actually projects slightly from the bottom surface of the wing. Down-plug aileron angles are shielded from the airstream by a fairing that looks like a bump running spanwise.

Plug-type spoiler ailerons are subject to nonlinearities in the first part of their travel out of the wing. Negative pressures on the wing's upper surface tend to suck the plugs out, causing control overbalance. Centering springs may be needed. There can be a small range of reversed aileron effectiveness if the flow remains attached to the wing's upper surface behind the spoiler for small spoiler projections. Nonlinearities at small deflections in the P-61 plug ailerons were swamped out (as an afterthought) by small flap-type ailerons, called guide ailerons, at the wing tips.

Early flight and wind-tunnel tests of spoilers for lateral control disclosed an important design consideration, related to their chordwise location on the wing. Spoilers located about midchord are quite effective in a static sense but have noticeable lags. That is, for a forward-located spoiler, there is no lift or rolling moment change immediately after an abrupt up-spoiler deflection. Since airfoil circulation and lift are fixed by the Kutta trailing-edge condition, the lag is probably related to the time required for the flow perturbation at the forward-located spoiler to reach the wing trailing edge. Spoilers at aft locations, where flap-type ailerons are found, have no lag problems (Choi, Chang, and Ok, 2001).

Another spoiler characteristic was found in early tests that would have great significance when aileron reversal became a problem. Spoiler deflections produce far less wing section pitching moment for a given lift change than ordinary flap-type ailerons. The local section pitching moment produced by ailerons twists the wing in a direction to oppose the lift due

Figure 5.11 Open slot-lip spoilers on the Boeing 707. Note the exposed upper surface of the first element of the flaps. The open spoilers destroy the slot that ordinarily directs the flow over the flap upper surface, reducing flap effectiveness. The reduced lift improves lateral control power when the spoilers are used asymmetrically or the airplane's braking power when deployed symmetrically on when the ground. (From Cook, *The Road to the 707*, 1991)

to the aileron. This is why spoilers are so common as lateral controls on high-aspect ratio wing airplanes, as discussed in Chapter 19, "The Elastic Airplane."

Slot-lip spoiler ailerons are made by hinging the wing structure that forms the upper rear part of the slot on slotted landing flaps. Since a rear wing spar normally is found just ahead of the landing flaps, hinging slot-lip spoilers and installing hydraulic servos to operate them is straightforward. There is a gratifying amplification of slot-lip spoiler effectiveness when landing flaps are lowered. The landing flap slot is opened up when the slot-lip spoiler is deflected up, reducing the flap's effectiveness on that side only and increasing rolling moment (Figure 5.11).

5.11   Internally Balanced Controls

Another control surface balance type that appeared about the same time as beveled controls was the internally balanced control. This control is called the Westland-Irving internal balance in Great Britain. Internally balanced controls are intended to replace the external aerodynamic balance, a source of wing drag because of the break in the wing contour. In the internally balanced control the surface area ahead of the hinge line is a shelf contained completely within the wing contour (Figure 5.12). Unless the wing is quite thick and has its maximum thickness far aft, mechanical clearance requires either that the shelf be made small, restricting the available amount of aerodynamic balance, or control surface throws be made small, restricting effectiveness.

By coincidence, internally balanced controls appeared about the same time as the NACA 65-, 66-, and 67-series airfoil sections. These are the laminar flow airfoils of the 1940s and 1950s. Internally balanced ailerons are natural partners of laminar flow airfoil sections, since aerodynamic balance is obtained without large drag-producing surface cutouts for the overhang. Not only that, but the 66 and 67 series have far aft locations of wing maximum thickness. This helps with the clearance problem of the shelf inside of the wing contour.

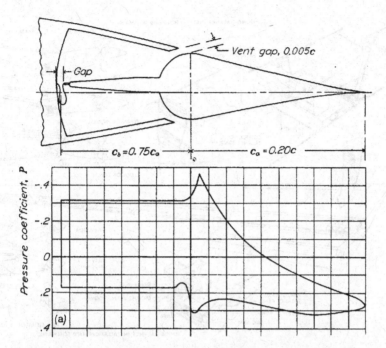

Figure 5.12 The internally balanced control surface, used to reduce drag by eliminating the wing cutouts needed with overhang aerodynamic balances. Pressures at the upper and lower vent gaps are delivered to the sealed chamber, balancing the surface about its hinge. Pressure coefficients shown are for a 5-degree down-surface deflection. (From Toll, NACA Rept. 868, 1947)

An internal balance modification that gets around the mechanical clearance problem on thin airfoils is the compound internal balance. The compound shelf is made in two, or even three, hinged sections. The forward edge of the forward shelf section is hinged to fixed airplane structure, such as the tail or wing rear spar (Figure 5.13). The first application of the compound internal balance appears to have been made by William H. Cook, on the Boeing B-47 Stratojet. Internally balanced elevators and the rudder of the Boeing B-52 have compound shelves on the inner sections of the control surfaces and simple shelves on the outer sections.

Compound internal balances continue to be used on Boeing jets, including the 707, 727, and 737 series. The 707 elevator is completely dependent on its internal aerodynamic balance; there is no hydraulic boost. According to Cook, in an early Pan American 707, an inexperienced co-pilot became disoriented over Gander, Newfoundland, and put the airplane into a steep dive. The pilot, Waldo Lynch, had been aft chatting with passengers. He made it back to the cockpit and recovered the airplane, putting permanent set into the wings. In effect, this near-supersonic pullout proved out the 707's manual elevator control. The 707's internally balanced ailerons are supplemented by spoilers, as described in Chapter 19, "The Elastic Airplane."

The later Boeing 727 used dual hydraulic control on all control surfaces, but internal aerodynamic balance lightens control forces in a manual reversion mode. An electrically driven adjustable stabilizer helps in manual reversion. At least one 727 lost all hydraulic power and made it back using manual reversion.

Internally balanced controls were used on a number of airplanes of the 1940s and 1950s. The famous North American P-51 Mustang had internally balanced ailerons, but they were unsealed, relying on small clearances at the front of the shelf to maintain a pressure

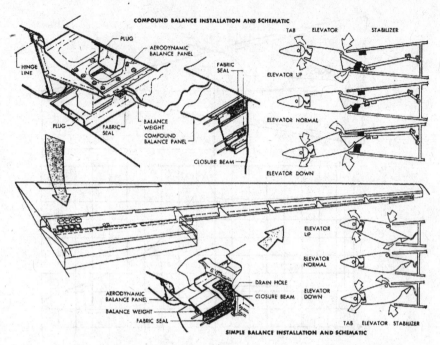

Figure 5.13   Simple and two-element compound internal aerodynamic balances on the Boeing B-52 elevator. The compound balance segment is inboard. (From B-52 *Training Manual*, 1956)

differential across the shelf. The Curtiss XP-60 and Republic XF-12 both used internally balanced controls, not without operational problems on the part of the XP-60. Water collected on the seal, sometimes turning to ice.

5.12   Flying or Servo and Linked Tabs

Orville R. Dunn (1949) gave 30,000 pounds as a rule-of-thumb upper limit for the weight of transport airplanes using leading-edge aerodynamic balance. Dunn considered that airplanes larger than that would require some form of tab control, or else hydraulically boosted controls. The first really large airplane to rely on tab controls was the Douglas B-19 bomber, which flew first in 1941. The B-19 used pure flying or servo tab control on the rudder and elevator and a plain-linked tab on the ailerons. In a flying tab the pilot's controls are connected only to the tab itself. The main control surfaces float freely; no portion of the pilot's efforts go into moving them. A plain-linked tab on the other hand divides the pilot's efforts in some proportion between the tab and the main surface. The rudder of the Douglas C-54 Skymaster transport uses a linked tab.

Roger D. Schaufele recalls some anxious moments at the time of the B-19's first flight out of Clover Field, California. The pilot was Air Corps pilot Stanley Olmstead, an experienced hand with large airplanes. This experience almost led to disaster, as Olmstead "grabbed the yoke and rotated hard" at liftoff, as he had been accustomed to doing on other large airplanes. With the flying tab providing really light elevator forces, the B-19 rotated nose up to an estimated 15 to 18 degrees, in danger of stalling, before Olmstead reacted with forward control motion.

Flying tabs are quite effective in allowing large airplanes to be flown by pilot effort alone, although the B-19 actually carried along a backup hydraulic system. A strong disadvantage is the lack of control over the main control surfaces at very low airspeeds, such as in taxi, the early part of takeoffs, and the rollout after landing. The linked tab is not much better in that the pilot gets control over the main surface only after the tab has gone to its stop. Still, by providing control for the B-19, the world's largest bomber in its time, flying and linked tabs, and the Douglas Aircraft Company engineers who applied them, deserve notice in this history.

An apocryphal story about the B-19 flying tab system illustrates the need for a skeptical view of flying tales. MIT's Otto Koppen was said to have told of a B-19 vertical tail fitted to a B-23 bomber, an airplane the size of a DC-3, to check on the flying tab scheme. The point of the story is that the B-23 flew well with its huge vertical tail. Koppen said this proved that a vertical tail could not be made too large. Unfortunately, this never occurred. Orville Dunn pointed out that (1) the B-23 came years after the B-19, and (2) it didn't happen.

5.13   Spring Tabs

Spring tabs overcome the main problem of flying tabs, which do not provide the pilot with control of the main surface at low speeds, as when taxiing. In spring tabs, the pilot's linkage to the tab is also connected to the main surface through a spring. If the spring is quite stiff, good low-speed surface control results. At the same time, a portion of the pilot's efforts goes into moving the main surface, increasing controller forces.

Spring tabs have the useful feature of decreasing control forces at high airspeeds, where control forces usually are too heavy, more than at low airspeeds. At low airspeeds, the spring that puts pilot effort into moving the main surface is stiff relative to the aerodynamic forces on the surface; the tab hardly deflects. The reverse happens at high airspeeds. At high airspeeds the spring that puts pilot effort into moving the main surface is relatively weak compared with aerodynamic forces. The spring gives under pilot load; the main surface moves little, but the spring gives, deflecting the tab, which moves the main surface without requiring pilot effort.

The earliest published references to spring tabs appeared as Royal Aircraft Establishment publications (Brown, 1941; Gates, 1941). NACA publications followed (Greenberg, 1944; Phillips, 1944). But the credit for devising a generalized control tab model that covers all possible variations (Figure 5.14) belongs to Orville R. Dunn (1949). The Dunn model uses three basic parameters to characterize spring tab variations, which include the geared tab, the flying tab, the linked tab, and the geared spring tab.

Although the derivation of pilot controller force equations for the different tab systems involve only statics and the virtual work principle, the manipulations required are surprisingly complex. As is typical for engineering papers prepared for publication, Dunn provides only bare outlines of equation derivations. Readers of the 1949 Dunn paper who want to derive his final equations should be prepared for some hard labor.

Dunn concluded that spring tabs can produce satisfactory pilot forces on subsonic transport-type airplanes weighing up to several million pounds. At the time of Dunn's paper, spring tabs had indeed been used successfully on the Hawker Tempest, the Vultee Vengeance rudder, all axes of the Canberra, the rudder and elevator of the Curtiss C-46 Commando, the Republic XF-12, and the very large Convair B-36 bomber. They also would be used later on the Boeing B-52 Stratofortress. Dunn's account of the DC-6 development tells of rapid, almost overnight, linkage adjustments during flight testing. The major concerns in spring tab applications are careful design and maintenance to minimize control system static friction and looseness in the linkages.

The B-19 experience encouraged Douglas engineers to use spring tabs for many years afterwards. Both the large C-124 and C-133 military transports were so equipped. The

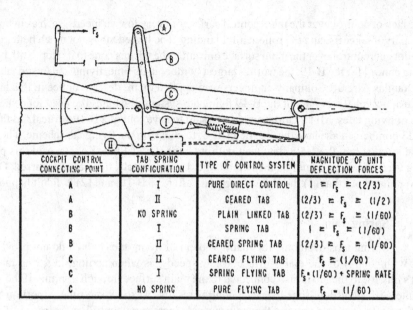

COCKPIT CONTROL CONNECTING POINT	TAB SPRING CONFIGURATION	TYPE OF CONTROL SYSTEM	MAGNITUDE OF UNIT DEFLECTION FORCES
A	I	PURE DIRECT CONTROL	$1 \gtrsim F_s \gtrsim (2/3)$
A	II	GEARED TAB	$(2/3) \gtrsim F_s \gtrsim (1/2)$
B	NO SPRING	PLAIN LINKED TAB	$(2/3) \gtrsim F_s \gtrsim (1/60)$
B	I	SPRING TAB	$1 \gtrsim F_s \gtrsim (1/60)$
B	II	GEARED SPRING TAB	$(2/3) \gtrsim F_s \gtrsim (1/60)$
C	II	GEARED FLYING TAB	$F_s \gtrsim (1/60)$
C	I	SPRING FLYING TAB	$F_s \cdot (1/60) + \text{SPRING RATE}$
C	NO SPRING	PURE FLYING TAB	$F_s \sim (1/60)$

Figure 5.14 Orville R. Dunn's generalized tab control system schematic. Varying the cockpit connecting point and tab spring configurations gives rise to seven different tab control schemes, ranging from the pure flying tab to the geared spring tab. (From Dunn, I.A.S.- R.Ae.S. Proc., 1949)

DC-6, -7, -8, and -9 commercial transports all have some form of spring tab controls, the DC-8 on the elevator and the DC-9 on all main surfaces, right up to the latest MD-90 version. In that case, the switch was made to a powered elevator to avoid increasing horizontal tail size to accommodate the airplane's stretch. A powered elevator avoids tab losses and effective tail area reductions because tabs move in opposition to elevator travel.

The Douglas DC-8 and -9 elevator control tabs are actually linked tabs, in which pilot effort is shared between the tab and the elevator. This gives the pilot control over the elevator when on the ground. The DC-8 and -9 elevator linked tabs are inboard and rather small. The inboard linked tabs are augmented by outboard geared tabs, which increase the flutter margin over single large linked tabs. The DC-9 elevator controls are hybrid in that hydraulic power comes in when the link tab's deflection exceeds 10 degrees. Spring tabs serve a backup purpose on the fully powered DC-8 ailerons and rudder and on the DC-9 rudder. The tabs are unlocked automatically and used for control when hydraulic system pressure fails. The same tab backup system is used for the Boeing 727 elevator.

The spring tab design for the elevators of the Curtiss C-46 Commando was interesting for an ingenious linkage designed by Harold Otto Wendt. Elevator surfaces must be statically balanced about their hinge lines to avoid control surface flutter. Spring tabs should also be statically balanced about their own hinge lines. Spring tab balance weights and the spring mechanisms add to the elevator's weight unbalance about its hinge line. Wendt's C-46 spring tab linkage was designed to be largely ahead of the elevator hinge line, minimizing the amount of lead balance required to statically balance the elevator.

Spring tabs appear to be almost a lost art in today's design rooms. Most large airplanes have hydraulic systems for landing gear retraction and other uses, so that hydraulically operated flight controls do not require the introduction of hydraulic subsystems. Furthermore, modern hydraulic control surface actuators are quite reliable. Although spring tab design requires manipulation of only three basic parameters, designing spring tabs for a new airplane entails much more work for the stability and control engineer than specifying parameters for hydraulic controls. Computer-aided design may provide spring tabs with a new future on airplanes that do not really need hydraulically powered controls.

5.14   Springy Tabs and Downsprings

Sometimes called "Vee" tabs, springy tabs first appeared on the Curtiss C-46 Commando twin-engine transport airplane. Their inventor, Roland J. White, used the springy tab to increase the C-46's allowable aft center of gravity travel. White was a Cal Tech classmate of another noted stability and control figure, the late L. Eugene Root. Springy tabs increase in a stable direction the variation of stick force with airspeed. A springy tab moves in one direction, with the trailing edge upward. It is freely hinged and is pushed from neutral in the trailing-edge-upward direction by a compression spring (Figure 5.15). An NACA application mounted the springy tab on flexure pivots.

The springy tab principle of operation is that large upward tab angles are obtained at low airspeeds, where the aerodynamic moment of the tab about its own hinge line is low compared with the force of the compression spring. Upward tab angle creates trailing-edge-down elevator hinge moment, which must be resisted by the pilot with a pull force. Pull force at low airspeed is required for stick-free stability.

The C-46 springy tabs were called Vee tabs because the no-load-up deflection was balanced aerodynamically by the same down rig angle on a trim tab on the opposite elevator (Figure 5.15). The C-46 springy tabs were also geared in the conventional sense. The compression spring that operated the C-46's springy tab was a low-rate or long-travel spring with a considerable preload of 52 pounds. Tab deflection occurred only after the preload was exceeded, making the system somewhat nonlinear.

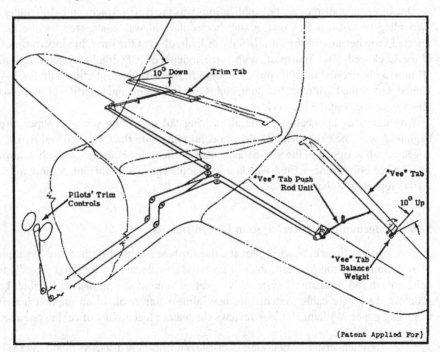

Figure 5.15 Schematic diagram of the elevator trim and vee-tab installations on the Curtiss C-46 Commando. The vee tab augments static longitudinal stick-free stability. (From Rumph and White, Curtiss Rept. 20-Y48, 1945)

Springy tabs were also used successfully on the Lockheed Electra turboprop. Although White is considered the springy tab's inventor and was the applicant for a patent on the device, it may have been invented independently by the late C. Desmond Pengelly. Springy tabs are not in common use currently because of potential flutter. Irreversible tab drives are preferred to freely hinged tabs from a flutter standpoint.

A flutter-conservative means of accomplishing the same effect as a springy tab is the downspring. This is a long-travel spring connected between the elevator linkage and airplane fixed structure. The stick or yoke is pulled forward by the long-travel spring with an essentially constant force. Elevator aerodynamic hinge moment, which would normally fair the elevator to the stabilizer, is low compared with the spring force, and the pilot is obliged to use pull force to hold the elevator at the angle required for trim. As with the springy tab, this provides artificial stick-free stability. Downsprings are often found in light airplanes. If the yoke rests against its forward stop with the airplane parked, and a pull force is needed to neutralize yoke travel, either a downspring is installed or, less likely, the elevator has mass unbalance.

5.15   All-Movable Controls

All-movable tail surfaces became interesting to stability and control designers when high Mach number theory and transonic wind-tunnel tests disclosed poor performance of ordinary flap-type controls. Effectiveness was down, and hinge moments were up. More consistent longitudinal and directional control over the entire speed range seemed possible with all-moving surfaces. However, application of all-moving or slab tail surfaces had to await reliable power controls.

One of the first all-moving tail applications was the North American F-100 Super Sabre. According to William E. Cook, a slab horizontal tail was considered for the B-52 and rejected only because of the unreliability of hydraulics at the time. In modern times, there is the Lockheed 1011 transport, with three independent hydraulic systems actuating its all-moving horizontal tail. Of course, modern fighter airplanes, starting with the F-4 in the United States; the Lightning, Scimitar, and Hawk in Britain; and the MiG-21 in Russia, have all-moving horizontal tails.

An interesting application is the all-moving tail on a long series of Piper airplanes, beginning with the Comanche PA-24 and continuing with the Cherokee and Arrow series. A geared tab is rigged in the anti-balance sense. The geared tab adds to both control force and surface effectiveness. Fred Weick credits John Thorp with this innovation, inspired by a 1943 report by Robert T. Jones.

5.16   Mechanical Control System Design Details

Connections between a pilot and the airplane's control surfaces are in a rapid state of evolution, from mechanical cables or push rods, to electrical wires, and possibly to fiber optics. Push rod mechanical systems have fallen somewhat into disuse; flexible, braided, stainless steel wire cable systems are now almost universal. In an unpublished Boeing Company paper, William H. Cook reviews the mature technology of cable systems:

> The multi-strand 7×19 flexible steel cables usually have diameters from 1/8 to 3/16 inch. They are not easily damaged by being stepped on or deflected out of position. They are usually sized to reduce stretch, and are much over-strength for a 200-pound pilot force. The swaged end connections, using a pin or bolt and cotter pin, are easily checked. The

turnbuckles which set tension are safety-wired, and are easily checked. A Northwest Airlines early Electra crashed due to a turnbuckle in the aileron system that was not secured with safety wire wrap.

Since the cable between the cockpit and the control is tensioned, the simplest inspection is to pull it sideways anywhere along its length to check both the tension and the end connections. In a big airplane with several body sections this is good assurance. To avoid connections at each body section joint, the cable can be made in one piece and strung out after joining the sections. The avoidance of fittings required to join cable lengths also avoids the possibility of fittings jamming at bulkheads. Since the cable is rugged, it can be installed in a fairly open manner.... Deterioration of the cables from fatigue, as can happen in running over pulleys, or from corrosion, can be checked by sliding a hand over its length. If a strand of the 7×19 cable is broken, it will "draw blood."

A recurrent problem in all mechanical flight control systems is possible rigging in reverse. This can happen on a new airplane or upon re-rigging an old airplane after disassembly. Modern high-performance sailplanes are generally stored in covered trailers and are assembled only before flying. Sailplane pilots have a keen appreciation of the dangers of rigging errors, including reversals. Preflight checks require the ground crew to resist pilot effort by holding control surfaces and to call out the sense of surface motions, up or down, right or left.

A few crossed cable control accidents have occurred on first flights. The aileron cables were crossed for the first flight of Boeing XB-29 No. 2, but the pilot aborted the takeoff in time. Crossed electrical connections or gyros installed in incorrect orientations are a more subtle type of error, but careful preflight procedures can catch them, too.

5.17   Hydraulic Control Boost

Control boost by hydraulic power refers to the arrangement that divides aerodynamic hinge moment in some proportion between the pilot and a hydraulic cylinder. A schematic for an NACA experimental boosted elevator for the Boeing B-29 airplane shows the simple manner in which control force is divided between the pilot and the hydraulic boost mechanism (Figure 5.16). Boosted controls were historically the first hydraulic power assistance application.

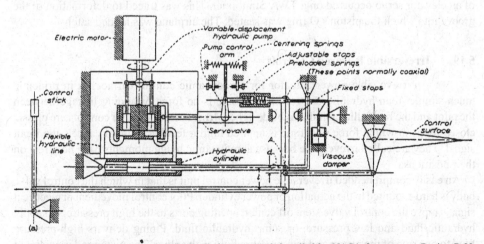

Figure 5.16   A very early hydraulic-boost control, installed by NACA for test on a Boeing B-29 elevator. Boost ratio l/d is varied by adjusting the location of point A. (From Mathews, Talmage, and Whitten, NACA Rept. 1076, 1952)

By retaining some aerodynamic hinge moments for the pilot to work against two things are accomplished. First, the control feel of an unaugmented airplane is still there. The pilot can feel in the normal way the effects of high airspeeds and any buffet forces. Second, no artificial feel systems are needed, avoiding the weight and complexity of another flight subsystem. Hydraulic power boost came into the picture only at the very end of World War II, on the late version Lockheed P-38J Lightning, and only on that airplane's ailerons. After that, hydraulic power boost was the favored control system arrangement for large and fast airplanes, such as the 70-ton Martin XPB2M-1 Mars flying boat, the Boeing 307 Stratoliner, and the Lockheed Constellation series transports, until irreversible power controls took their place.

5.18   Early Hydraulic Boost Problems

Early hydraulic boosted controls were notoriously unreliable, prone to leakage and outright failures. Among other innovative systems at the time, the Douglas DC-4E prototype airplane had hydraulic power boost. Experience with that system was bad enough to encourage Douglas engineers to face up to pure aerodynamic balance and linked tabs for the production versions of the airplane, the DC-4 or C-54 Skymaster.

A similar sequence took place at the Curtiss-Wright plant in St. Louis, where the Curtiss C-46 Commando was designed. At a gross weight of 45,000 pounds, the C-46 exceeded O. R. Dunn's rule of thumb of 30,000 pounds for the maximum weight of a transport with leading-edge aerodynamic balance only. Thus, the CW-20, a C-46 prototype, was fitted initially with hydraulic boost having a 3:1 ratio, like those on the Douglas DC-4E Skymaster prototype and the Lockheed Constellation. However, maintenance and outright failure problems on the C-46's hydraulic boost were so severe that the Air Materiel Command decreed that the airplane be redesigned to have aerodynamically balanced control surfaces. The previous successful use of aerodynamic balance on the 62,000-pound gross weight Douglas C-54 motivated the Air Corps decree. This was the start of the "C-46 Boost Elimination Program," which kept one of this book's authors (Larrabee) busy during World War II.

Another airplane with early hydraulically boosted controls was the Boeing 307 Stratoliner. Hydraulic servos were installed on both elevator and rudder controls. Partial jamming of an elevator servo occurred on a TWA Stratoliner. This was traced to deformation of the groove into which the piston's O ring was seated. The airplane was landed safely.

5.19   Irreversible Powered Controls

An irreversible power actuator for aerodynamic control surfaces is in principle much simpler than hydraulic control boost. There is no force balancing linkage between the pilot and the hydraulic cylinder to be designed. Irreversible powered controls are classic closed loops in which force or torque is applied until a feedback signal cancels the input signal. They are called irreversible because aerodynamic hinge moments have no effect on their positions.

An easily comprehended irreversible power control unit is that in which the control valve body is hard-mounted to the actuation or power cylinder. Pilot control movement or electrical signals move the control valve stem off center, opening ports to the high pressure, or supply hydraulic fluid and low pressure, or sump hydraulic fluid. Piping delivers high-pressure fluid to one side of the piston and low-pressure fluid to the other. The piston rod is anchored to structure and the power cylinder to the control surface. When the power cylinder moves with respect to structure in response to the unbalanced pressure it carries the control valve body along with it. This centers the control valve around the displaced stem, stopping the motion. The airplane's control surface has been carried to a new position, following up the input to the control valve in a closed-loop manner.

The first irreversible power controls are believed to have been used on the Northrop XB-35 and YB-49 flying wing airplanes. Irreversibility was essential for these airplanes because of the large up-floating elevon hinge moment at high angles of attack, as the stall was approached. This was unstable in the sense that pilot aft-yoke motion to increase the angle of attack would suddenly be augmented by the elevon's own up-deflection. One of the N9M flying scale models of the Northrop flying wings was lost due to elevon up-float (Sears, 1987). The YB-49's irreversible actuators held the elevons in the precise position called for by pilot yoke position, eliminating up-float. Other early applications of irreversible power controls were to the de Havilland Comet; the English Electric Lightning P1.A, which first flew in 1954; and the AVRO Canada CF-105 Arrow, which first flew in 1958.

Howard (2000) believes that the Comet application of irreversible powered controls was the first to a passenger jet. The U.K. Air Registration Board "made the key decision to accept that a hydraulic piston could not jam in its cylinder, a vital factor necessary to ensure the failure-survivability of parallel multiple-power control connections to single surfaces."

While irreversible power controls are simple in principle, it was several years before they could be used routinely on airplanes. The high powers and bandwidths associated with irreversible power controls, as compared with earlier boosted controls, led to system limit cycling and instabilities involving support structures and oil compressibility. These problems were encountered and solved in an ad hoc manner by mechanical controls engineer T. A. Feeney for the Northrop flying wings on a ground mockup of the airframe and its control system, called an iron bird. An adequate theory was needed for power control limit cycle instability, to explain the roots of the problem. This was presented by D. T. McRuer at a symposium in 1949 and subsequently published (Bureau of Aeronautics, 1953).

The post–World War II history of gradual improvements in the design of irreversible power controls is traced by Robert H. Maskrey and W. J. Thayer (1978). They found that Tinsley in England patented the first two-stage electromechanical valve in 1946. Shortly afterwards, R. E. Bayer, B. A. Johnson, and L. Schmid improved on the Tinsley design with direct mechanical feedback from the second-stage valve output back to the first stage.

Engineers at the MIT Dynamic Analysis and Controls Laboratory added two improvements to the two-stage valve. The first was the use in the first stage of a true torque motor instead of a solenoid. The second improvement was electrical feedback of the second-stage valve position. In 1950, W. C. Moog, Jr., developed the first two-stage servovalve using a frictionless first-stage actuator, a flapper or vane. Valve bandwidths of up to 100 cycles per second could be attained. The next significant advance was mechanical force feedback in a two-stage servovalve, pioneered by T. H. Carson, in 1953. The main trends after that were toward redundancy and integration with electrical commands from both the pilot and stability augmentation computers.

In general, satisfactory irreversible power control designs require attention to many details, as described by Glenn (1963). In addition to the limit cycling referred to previously, these include minimum increment of control, position and time lags, surface positioning accuracy, flexibility, springback, hysteresis, and irreversibility in the face of external forces.

5.20 Artificial Feel Systems

Since irreversible power controls isolate the pilot from aerodynamic hinge moments, artificial restoration of the hinge moments, or "artificial feel," is required.

Longitudinal artificial feel systems range in complexity from simple springs, weights, and stick dampers to computer-generated reactive forces applied to the control column by servos.

A particularly simple artificial feel system element is the bobweight. The bobweight introduces mass unbalance into the control circuit, in addition to the unbalances inherent in the basic design. That is, even mass-balanced mechanical control circuits have inertia that tends to keep the control sticks, cables, and brackets fixed while the airplane accelerates around them. Bobweights are designed to add the unbalance, creating artificial pilot forces proportional to airplane linear and angular accelerations. They also have been used on airplanes without irreversible power controls, such as the Spitfire and P-51D.

The most common bobweight form is a simple weight attached to a bracket in front of the control stick. Positive normal acceleration, as in a pullup, requires pilot pull force to overcome the moment about the stick pivot of increased downward force acting on the bobweight. There is an additional pilot pull force required during pullup initiation, while the airplane experiences pitching acceleration. The additional pull force arises from pitching acceleration times the arm from the center of gravity to the bobweight. Without the pitching acceleration component, the pilot could get excessive back-stick motions before the normal acceleration builds up and tends to pull the stick forward.

In the case of the McDonnell Douglas A-4 airplane's bobweight installation, an increased pitching acceleration component is needed to overcome overcontrol tendencies at high airspeeds and low altitudes. A second, reversed bobweight is installed at the rear of the airplane. The reversed bobweight reduces the normal acceleration component of stick force but increases the pitching acceleration component.

Another interesting artificial feel system element is the q-spring. As applied to the Boeing XB-47 rudder (White, 1950) the q-spring provides pedal forces proportional to both pedal deflection and airplane dynamic pressure, or q. Total pressure (dynamic plus static) is put into a sealed container having a bellows at one end. The bellows is equilibrated by static pressure external to the sealed container and by tension in a cable, producing a cable force proportional to the pressure difference, or q. Pilot control motion moves an attachment point of that cable laterally, providing a restoring moment proportional to control motion and to dynamic pressure.

It appears that a q-spring artificial feel system was first used on the Northop XB-35 and B-49 flying wing elevons, combined with a bobweight. Q-spring artificial feel system versions have survived to be used on modern aircraft, such as the elevators of the Boeing 727, 747, and 767; the English Electric Lightning; and the McDonnell Douglas DC-10. Hydraulic rather than pneumatic springs are used, with hydraulic pressure made proportional to dynamic pressure by a regulator valve. In many transport airplanes the force gradient is further modulated by trim stabilizer angle. Stabilizer angle modulation, acting through a cam, provides a rough correction for the center of gravity position, reducing the spring force gradient at forward center of gravity positions. Other modulations can be introduced.

Advanced artificial feel systems are able to modify stick spring and damper characteristics in accordance with a computer program, or even to apply forces to the stick with computer-controlled servos.

5.21   Fly-by-Wire

In fly-by-wire systems control surface servos are driven by electrical inputs from the pilot's controls. Single-channel fly-by-wire has been in use for many years, generally through airplane automatic pilots. For example, both the Sperry A-12 and the Honeywell

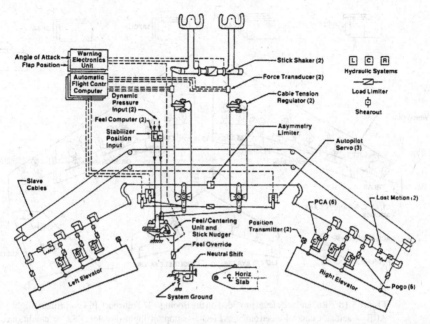

Figure 5.17 Schematic of the Boeing 767 elevator control system, possibly the last fly-by-cable or mechanical flight control system to be designed for a Boeing transport. Each elevator half is powered by three parallel hydromechanical servo actuators. Cam overrides (Pogos) and shear units allow separation of jammed system components. (Reprinted with permission from SAE Paper No. 831488, © 1983, Society of Automotive Engineers, Inc.)

C-1 autopilots of the 1940s provided pilot flight control inputs through cockpit console controls. However, in modern usage, fly-by-wire is defined by multiple redundant channel electrical input systems and multiple control surface servos, usually with no or very limited mechanical (cable) backup.

According to Professor Bernard Etkin, a very early application of fly-by-wire technology was to the Avro Canada CF-105 Arrow, a supersonic delta-winged interceptor that first flew in 1958. A rudimentary fly-by-wire system, with a side-stick controller, was flown in 1954 in a NASA-modified Grumman F9F (Chambers, 2000). The NASA/Dryden digital fly-by-wire F-8 program was another early development. Readers can consult Schmitt (1988) and Tomayko (2000) for the interesting history of airplane fly-by-wire.

The Boeing 767 is probably the last design from that company to retain pilot mechanical inputs to irreversible power control actuators, or fly-by-cable. The 767 elevator control schematic shows a high redundancy level, with three independent actuators on each elevator, each supplied by a different hydraulic system (Figure 5.17). Automatic pilot inputs to the system require separate actuators, since the primary surface servos do not accept electrical signals.

The Boeing 777 is that company's first fly-by-wire (FBW) airplane, in which the primary surface servos accept electrical inputs from the pilot's controls. With the Boeing 777, fly-by-wire can be said to have come of age in having been adopted by the very conservative Boeing Company. Fly-by-wire had previously been operational on the Airbus A320, 330, and A340 airplanes

Figure 5.18 (Osder 1999) shows the redundancy level provided on the Boeing 777 control actuators. In this figure PFC refers to primary flight control computers, the ACE are actuator control electronic units, the AFDC are autopilot flight director Controls, the PSA are power

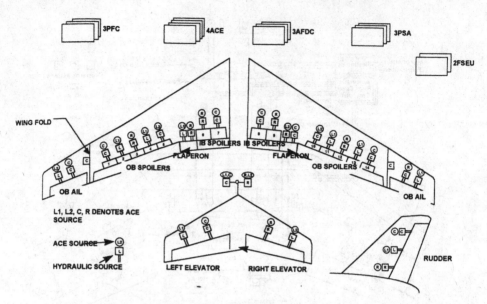

Figure 5.18 Redundancy level provided on the Boeing 777 Transport. PFC = primary flight computer, ACE = actuator control electronics, AFDC = autopilot flight director, PSA = conditioned power, FSEU = flap slatelectronics unit. (From Osder, 1999).

supplies, and the FSEU are secondary control units. Note the cross-linkages of the ACEs to the hydraulic power sources.

McLean (1999) gives interesting details on the 777 and A320 fly-by-wire systems:

> **[Boeing 777]** ... to prevent pilots exceeding bank angle boundaries, the roll force on the column increases as the bank angle nears 35 degrees. FBW enables more complex inter-axis coupling than the traditional rudder crossfeed for roll/yaw coordination which results in negligible sideslip even in extreme maneuvers ... the yaw gust damper (which is independent and separate from the standard yaw damper on the aircraft) ... senses any lateral gust and immediately applies rudder to alleviate loads on the vertical fin. The Boeing 777 has an FBW system which allows the longitudinal static margin to be relaxed – a 6 percent static margin is maintained ... stall protection is provided by increasing column control forces gradually with increases in angle of attack. Pilots cannot trim out these forces as the aircraft nears stall speed or the angle of attack limit.
>
> **[Airbus 320]** ... sidestick controllers are used. The pitch control law on that aircraft is basically a flight path rate command/flight path angle hold system and there is extensive provision of flight envelope protection ... the bank angle is limited to 35 degrees. ... There is pitch coordination in turns. A speed control system maintains either VREF [a reference airspeed] or the speed which is obtained at engagement. There is no mechanical backup. ... Equipment has to be triplicated, or in some cases quadruplicated with automatic "majority voters" and there is some provision for system reconfiguration.

The two cases illustrate an interesting difference in transport fly-by-wire design philosophy. Boeing 777 pilots are not restricted from applying load factors above the limit, except by a large increase in control forces. Wings could be bent in an emergency pullout. Airbus control logic prevents load factors beyond limit.

The McDonnell Douglas F/A-18 Hornet represents a move in the direction of completely integrated flight control actuators. Pilot inputs to the F/A-18's all-moving horizontal tail or stabilator are made through two sets of dual solenoid-controlled valves, a true "fly-by-wire"

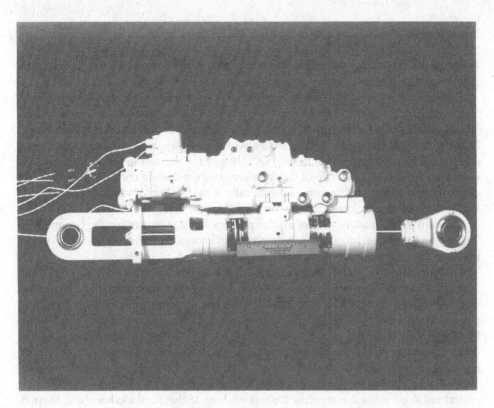

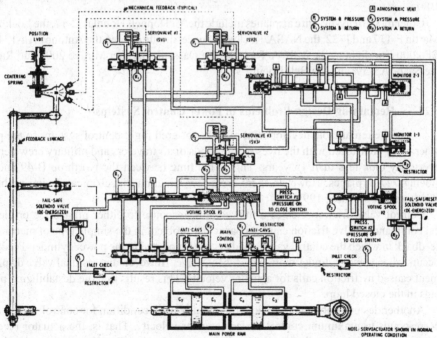

Figure 5.19 Photograph and schematic of the General Dynamics F-16 Integrated Servo Actuator (ISA) made by the National Waterlift Company. This actuator design is typical of an entirely fly-by-wire flight control system. The actuator uses mechanical rate (main valve spool position) and position feedback, although electrical feedback has been tried. Internal hydromechanical failure detection and correction, using three independent servovalves, causes the piping complexity. (Reprinted with permission from SAE Paper No. 831483, © 1983, Society of Automotive Engineers, Inc.)

system. A mechanical input from the pilot is applied only in the event of a series of electrical failures and one hydraulic system failure.

The General Dynamics F-16 is a completely fly-by-wire airplane, incorporating fully integrated servoactuators, known by their initials as ISAs. Each actuator is driven by three electrically controlled servo valves (Figure 5.19). There are no mechanical valve inputs at all from the pilot. Of course, the servo valves also accept signals from a digital flight control computer. The complexity seen in the ISA schematic is due to the failure detection and correction provisions. Only two of the three servo valves operate normally. A first failure of one of these valves shifts control automatically to the third servo valve. A first failure of the third servo valve locks the actuator on the sum of the first two.

The F-16 servoactuators also are used as primary surface actuators on the Grumman X-29A research airplane. Integrated servoactuators of equivalent technology were developed by Moog, Inc., for the Israeli Lavi fighter airplane.

The Northrop/Lear/Moog design for the B-2 Stealth bomber's flight controls represents another interesting fly-by-wire variant. On this quite large airplane part of the servo control electronics that normally resides in centralized flight control computers has been distributed close to the control surfaces. Digital flight control surface commands are sent by data bus to actuator remote terminals, which are located close to the control surfaces. The terminals contain digital processors for redundancy management and analog loop closure and compensation circuits for the actuators. Distributing the flight control servo actuator feedback functions in this manner saves a great deal of weight, as compared with using centralized flight control computers for this function (Schaefer, Inderhees, and Moynes, 1991).

Other modern fly-by-wire airplanes include the McDonnell Douglas C-17, the Lockheed-Martin F-117 and F-22, the NASA/Rockwell Space Shuttle orbiter, the Antonov An-124, the EF 2000 Eurofighter, the MRCA/Tornado, the Dassault Breguet Mirage 2000 and Rafale, the Saab JAS-39, and the Bell Boeing V-22.

5.22   Remaining Design Problems in Power Control Systems

The remarkable development of fully powered flight control systems to the point where they are trusted with the lives of thousands of air travelers and military crew persons every day took less than 15 years. This is the time between the Northrop B-49 and the Boeing 727 airplanes. However, there are a few remaining mechanical design problems (Graham and McRuer, 1991).

Control valve friction creates a null zone in response to either pilot force or electrical commands. Valve friction causes a particular problem in the simple type of mechanical feedback in which the control valve's body is hard-mounted to the power cylinder. Feedback occurs when power cylinder motion closes the valve. However, any residual valve displacement caused by friction calls for actuator velocity. This results in large destabilizing phase lags in the closed loop.

Another design problem has to do with the fully open condition for control valves. This corresponds to maximum control surface angular velocity. That is, the actuator receives the maximum flow rate that the hydraulic system can provide. The resultant maximum available control surface angular velocity must be higher than any demand made by the pilot or an autopilot. If a large upset or maneuver requires control surface angular velocity that exceeds the fully open valve figure, then velocity limiting will occur. Velocity limiting is highly destabilizing. Control surface angles become functions of the velocity limit and the input amplitude and frequency and lag far behind inputs by the human or automatic pilot.

The destabilizing effects of velocity limiting have been experienced during the entire history of fully powered control systems. A North American F86 series jet was lost on landing approach when an air-propeller–driven hydraulic pump took over from a failed engine-driven pump. When airspeed dropped off near the runway, the air-propeller–driven pump slowed, reducing the maximum available hydraulic flow rate. The pilot went into a divergent pitch oscillation, an early pilot-induced oscillation (PIO) event (see Chapter 21). Reported actuator velocity saturation incidents in recent airplanes include the McDonnell Douglas C-17, the SAAB JAS-39, and the Lockheed Martin/Boeing YF-22 (McRuer, 1997).

5.23   Safety Issues in Fly-by-Wire Control Systems

Although fully fly-by-wire flight control systems have become common on very fast or large airplanes, questions remain as to their safety. No matter what level of redundancy is provided, one can always imagine improbable situations in which all hydraulic or electrical systems are wiped out. Because of the very high-power requirements of hydraulic controls, their pumps are driven by the main engines. This makes necessary long high-pressure tubing runs between the engines and the control surfaces. The long high-pressure hydraulic lines are subject to breakage from fatigue; from wing, tail, and fuselage structural deflections; and from corrosion and maintenance operations.

The dangers of high-pressure hydraulic line breakage or leaking, with drainage of the system, could be avoided at some cost in weight and complexity with standby emergency electrically driven hydraulic pumps located at each control surface. An additional safety issue is hydraulic fluid contamination. Precision high-pressure hydraulic pumps, valves, and actuators are sensitive to hydraulic fluid contamination.

In view of rare but possible multiple hydraulic and electrical system failures, not to mention sabotage, midair collisions, and incorrect maintenance, how far should one go in providing some form of last-ditch backup manual control? Should airplanes in passenger service have last-ditch manual control system reversion? If so, how will that be accomplished with side-stick controllers?

In the early days of hydraulically operated controls and relatively small airplanes the answer was easy. For example, the 307 Stratoliner experience and other hydraulic power problems on the XB-47 led Boeing to provide automatic reversion to direct pilot control following loss in hydraulic pressure on the production B-47 airplanes. Follow-up trim tabs geared to the artificial feel system minimized trim change when the hydraulic system was cut out. Also, when hydraulic power was lost, spring tabs were unlocked from neutral.

Manual reversion saved at least one Boeing 727 when all hydraulic power was lost, and a United Airlines Boeing 720 made a safe landing without electrical power. The last-ditch safety issue is less easily addressed for commercial airplanes of the Boeing 747 class and any larger superjumbos that may be built. Both Lockheed L1011 and Boeing 747 jumbos lost three out of their four hydraulic systems in flight. The L1011 had a fan hub failure; the 747 flew into San Francisco approach lights. A rear bulkhead failure in Japan wiped out all four hydraulic systems of another 747, causing the loss of the airplane.

In another such incident the crew, headed by Delta Airlines Captain Jack McMahan, was able to save a Lockheed 1011 in 1977 when the left elevator jammed full up, apparently during flight control check prior to takeoff at San Diego (McMahan, 1983). There is no cockpit indicator for this type of failure on the 1011, and the ground crew did not notice the problem. McMahan controlled the airplane with differential thrust to a landing at Los Angeles. This incident was a focus of a 1982 NASA Langley workshop on restructurable controls.

Workshop attendees discussed the possible roles of real-time parameter identification and rapid control system redesign as a solution for control failures.

Thus, although fully mechanical systems can also fail in many ways, such as cable misrig or breakage, jammed bellcranks, and missing bolts, questions remain as to the safety of modern fly-by-wire control systems. The 1977 Lockheed 1011 incident, a complete loss in hydraulic power in a DC-10 in 1989, and other complete control system losses led to the interesting research in propulsion-controlled aircraft described in Sec. 20.11.

5.24   Managing Redundancy in Fly-by-Wire Control Systems

While redundancy is universally understood to be essential for safe fly-by-wire flight control systems, there are two schools of thought on how to provide and manage redundancy. Stephen Osder (1999) defines the two approaches as physical redundancy, which uses measurements from redundant elements of the system for detecting faults, and analytic redundancy, which is based on signals generated from a mathematical model of the system. Analytic redundancy (Frank, 1990) uses real-time system identification techniques, as in Chapter 14, Sec. 8, or normal optimization techniques.

Physical redundancy is the current technology for fly-by-wire, except for isolated sub-systems. Figure 5.20 is a highly simplified diagram of a generic triplex physically redundant flight control system. The key concept is grouping of all sensors into sets and using the set outputs for each of the three redundant computers. Likewise, each of the computers feeds all three redundant actuator sets. Voting circuitry outputs the midvalue of the three inputs to the voting system. Fail-operability is provided, a necessity for fly-by-wire systems. Figure 5.20 clearly could be extended to quadruple redundant flight controls.

The practical application of physical redundancy requires close attention to communications among the subsystems. Unless signals that are presented to the voting logic are perfectly synchronized in time, incorrect results will occur. In the real world, sensors, computers, and actuators operate at different data rates. Special communication devices are needed to provide synchronization. Additional care is required to avoid fights among the redundant channels resulting from normal error buildup, and not from the result of failures.

The situation with regard to analytic redundancy is still uncertain, since broad applications to production systems have not been made. By replacing some physical or hardware redundant elements with software, some weight savings, better flexibility, and more reliability are promised. However, a major difficulty arises from current limitations of vehicle

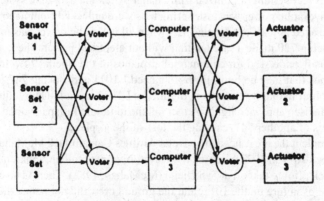

Figure 5.20   Generic triplex-redundant architecture for flight-critical control systems. (From Osder, 1999)

system identification and optimization methods to largely linearized or perturbation models. If an airplane is flown into regions where aerodynamic nonlinearities and hysteresis effects are dominant, misidentification could result. Misidentification with analytic redundancy could also arise from the coupled nature of the sensor, computer, and actuator subsystems. Osder (1999) gives as an example a situation where an actuator position feedback loop opening could be misdiagnosed as a sensor failure, based on system identification.

An analytic redundancy application to reconfiguring a system with multiple actuators is given by Jiang (2000). The proposed system uses (linearized) optimization to reconfigure a prefilter that allocates control among a set of redundant actuators and to recompute feedback proportional and integral gains. A somewhat similar analytic redundancy scheme, using adaptive control techniques, is reported by Hess (2000). Baumgarten (1996) reported on reconfiguration techniques focusing on actuator failures.

The best hope for future practical applications of analytical redundancy rests in heavy investments in improved methods of system identification. This appears to be the goal of several programs at the Institute of Flight Mechanics of the DLR. Several advances at that institute and at other places are noted in Chapter 14, Sec. 8, "Flight Vehicle System Identification from Flight Test."

5.25 Electric and Fly-by-Light Controls

Fully electrical airplane flight control systems are a possibility for the future. Elimination of hydraulic control system elements should increase reliability. Failure detection and correction should become a simple electronic logic function as compared with the complex hydraulic arrangement seen in the F-16's ISA. Fly-by-light control systems, using fiber optic technology to replace electrical wires, are likewise a future possibility. Advanced hardware of this type requires no particular advances in basic stability and control theory.

CHAPTER 6

Stability and Control at the Design Stage

In the preliminary layout of a new airplane, the stability and control engineer is generally guided by some well-known principles related to balance and tail sizing, for example. Once a preliminary design is laid out, its main stability and control characteristics can be predicted entirely from drawings. This includes the neutral point (center of gravity for zero-static longitudinal stability), static directional (weathercock) and lateral (dihedral effect) stability, and assurance that the airplane can be trimmed to zero-pitching moment over its lift coefficient and center of gravity ranges.

In the best of circumstances, the new design has a family resemblance to an earlier design. Then the estimations resemble extrapolations from known, measured characteristics. All airplane manufacturers seem to maintain proprietary aerodynamic handbook collections and correlations of stability and control data from previous designs. This is a great help if the extrapolation route is indicated. Aside from these private collections, there is a large body of theory and correlations from generalized wind-tunnel data that can be called upon for prediction or estimation.

A closely related subject to the prediction of stability and control characteristics entirely from drawings is the problem posed at the next stage in an airplane's development, when wind-tunnel test data have been obtained. In former times, one was often asked to prepare a complete set of predicted flying qualities using the wind-tunnel data and any flight control details that may have been available at the time. Instead, current practice is to plug wind-tunnel test and control system data into a flight simulator, for pilot flying qualities evaluation. Radio-controlled flying scale models are an alternate stability and control source for projects that cannot afford wind-tunnel tests.

The three design-stage topics – layout principles, estimation from drawings, and estimation from wind-tunnel data – are treated in this chapter.

6.1 Layout Principles

6.1.1 Subsonic Airplane Balance

Subsonic tail-last (not canard) airplanes are generally balanced to bring their centers of gravity near the wing-alone aerodynamic center. This is the point at which the wing's pitching moment coefficient is invariant with angle of attack. For reasonably high wing aspect ratios, the wing-alone aerodynamic center is near the 25-percent point behind the leading edge of chord line passing through the wing's center of area. This chord line is called the wing's mean aerodynamic chord or mac. Figure 6.1 shows the simple geometric construction defining the mac for straight-tapered and elliptical wings.

Tailless airplanes must have their centers of gravity ahead of the wing aerodynamic center or 25-percent mac point to be inherently statically stable. If the wing is swept back, it can be trimmed at a reasonably high lift coefficient with trailing-edge-up deflections of its elevons. The degree of static stability desired and the maximum lift coefficient obtained are interrelated. Tailless airplanes can have their centers of gravity behind the wing

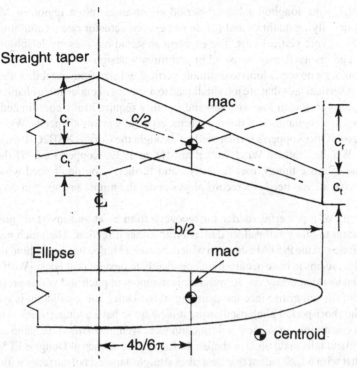

Figure 6.1   Geometrical constructions for the mean aerodynamic chord (mac) on straight-tapered and elliptical wings.

aerodynamic center if static stability is provided by artificial means or stability augmentation (see Chapter 20). Longitudinal trim then requires trailing-edge-down elevon. This increases effective wing camber, with beneficial effects on performance (Ashkenas and Klyde, 1989).

The canard configuration, abandoned after 1910 by its inventors, the Wright brothers, has been revived in recent years, notably by Burt Rutan, in the belief that the arrangement provides natural stall prevention (see Chapter 17, Sec. 2). Also, trimming with an upload is thought to reduce induced drag, although this has been disputed. The neutral point, or center of gravity for neutral stability, of canard airplanes is considerably ahead of the 25-percent point of the wing mac. On the Rutan machines, fuel tanks are fitted in triangular leading-edge extensions to keep the fuel near the airplane's center of gravity.

6.1.2   Tail Location, Size, and Shape

Horizontal and vertical tails are commonly located about a wing semispan behind the center of gravity. While horizontal tail sizes normally range from 15 to 30 percent of the wing area, the actual size is a complex function of desired center of gravity range, ground effect, and other factors. There is a minimum tail size that will trim a neutrally stable airplane at maximum lift in ground effect. Horizontal tails that are larger than this absolute minimum permit a useful operational center of gravity range.

Optimization theory has been proposed to size horizontal tails for particular center of gravity ranges, considering actuator rate and amplitude and flying qualities constraints. A particular application (Kaminer, Howard, and Buttrill, 1997) starts with a particular horizontal tail volume. Then, the most aft center of gravity location and feedback gains are found that (1) put longitudinal short-period eigenvalues into a region of MIL-STD 1797 Level 1 or 2 flying qualities and (2) do not exceed actuator rate or amplitude limits in response to a severe vertical gust. The problem as stated has reasonable solutions. The method, although involved, may be useful in preliminary design.

There seems to be no upper limit to desirable vertical tail size from a stability and control standpoint, but vertical tails that are too small lead to a variety of undesirable characteristics. For example, airplanes with low weathervane stability require heavy coordinated rudder-aileron inputs when beginning and stopping turns, especially at low airspeeds. When Walter Brewer, Professor Otto Koppen's former student, brought the Curtiss XSB2C-1 wind-tunnel model to the Wright Brothers Wind Tunnel at MIT in 1939, Koppen said, "If they build more than one of those things, they're crazy," and further, "You don't need wind-tunnel balances for data, all you need is a record player under the tunnel saying, 'put on a bigger vertical tail!'"

The necessity of a powerful rudder for recovery from erect and inverted spins led to notched elevators to allow full rudder deflection in either direction. The much neater and lower drag solution of the P-51 Mustang, in which the rudder hinge line lies behind the elevator trailing edge, seems to have occurred independently to designers at Focke-Wulf and was rapidly adopted by other designers. Aerodynamic damping in pitch and yaw is proportional to the square of the tail arm. Since the damping of the Dutch roll oscillation is inherently poorer than the short-period pitch oscillation, it is better to have a longer vertical tail arm.

Before the constraints on vertical tail function were well understood, airplane manufacturers built vertical tails in distinctive shapes. L. Eugene Root, then at Douglas El Segundo, changed all that with a U.S. patent that describes straight-tapered tail surfaces with leading edges and hinge lines all at a constant percentage chord.

6.2   Estimation from Drawings

6.2.1   *Early Methods*

The elements of stability and control prediction from drawings started to be available as early as aerodynamic theory itself. That is, aside from elements such as propellers and jet intakes and exhausts, airplane configurations are combinations of lifting surfaces and bodies. However, it took some time before the lift and moment of lifting surfaces and bodies were codified into a form useful for preliminary stability and control design. Simple correlations of lift and moment with geometrical characteristics such as wing aspect and taper ratios and the longitudinal distributions of body volume were needed.

6.2.2   *Wing and Tail Methods*

For stability and control calculations at the design stage, the variations of lift coefficient with angle of attack, or lift curve slope, are needed for airplane wings and tail surfaces. Wing and tail lift curve slopes are to first-order functions of aspect ratio and sweepback angle, and to a lesser extent of Mach number, section trailing-edge angle, and taper ratio. The primary aspect ratio effect is given by Ludwig Prandtl's lifting line theory and can be found as charts of lift curve slope versus aspect ratio in early stability and control research reports. The sweepback effect was added by DeYoung and Harper (1948).

However, classical lifting line theory for wings and tails fails for large sweep angles and low aspect ratios, even at low Mach numbers. A 1925 theory of supersonic airfoils in two-dimensional flow due to Ackeret existed, and also in the 1920s Prandtl and Glauert showed how subsonic airfoil theory could be corrected for subsonic Mach number effects. Both the Ackeret theory and Prandtl-Glauert subsonic Mach number correction theory fail at Mach 1. R. T. Jones (1946) developed a very low aspect-ratio wing theory, valid for all Mach numbers, which applies to highly swept wings, that is, wings whose leading edges are well inside the Mach cone formed at the vertex.

6.2.3   Bodies

A fundamental source for the effects of bodies on longitudinal and directional stability is the momentum or apparent mass analysis of Max M. Munk (1923). This models the flow around nonlifting bodies such as fuselages, nacelles, and external fuel tanks in terms of the growing or diminishing momentum imparted to segments of the air that the body passes through. Pitching and yawing moments as functions of angle of attack and sideslip are found by this method.

6.2.4   Wing–Body Interference

The longitudinal, lateral, and directional stability of wings and bodies in combination are the isolated characteristics plus effects that reflect modification of the flow by interference. In the longitudinal case, upwash ahead of the wing and downwash behind the wing change the body local angles of attack that enter into the Munk momentum theory calculations. Munk's apparent mass theory for bodies was extended by Hans Multhopp (1941) to account for the nonconstant fuselage angle of attack due to the wing's flow field. Gilruth and White (1941) used strip theory for this modification.

Stability and control designers have known for some time that whether an airplane has a high or a low wing influences static directional and lateral stabilities. There was an organized study of this at NACA starting in 1939 as a part of a broader attack on the factors influencing directional and lateral stability. The wing position part of the study was completed in 1941 by House and Wallace.

Distortion of the wing's spanwise lift distribution and trailing vortex system due to sideslip has the following systematic effects:

> Low wing airplanes: Static lateral stability is reduced by about 5 degrees of equivalent wing dihedral as compared with mid-wing airplanes. This rule of thumb has lasted to the present day. Static directional or weathervane stability is increased.

> High wing airplanes: The reverse of the low wing case. Dihedral effect is increased by about 5 degrees, weathervane stability is decreased.

Cross-Flow Concept   The cross-flow concept aids in understanding aerodynamic forces for an airplane in sideslip. The total velocity vector VEL of an airplane in sideslip can be resolved into a component U along the X or longitudinal body axis and a component V along the Y or lateral body axis. The U component gives rise to a symmetric flow, while the V component gives rise to a hypothetical flow at right angles, along the Y body axis. The component flows add together to make up the total streamline pattern of the airplane in sideslip.

The V or cross-flow component is represented in Figure 6.2. This figure provides an explanation for the effects of high and low wing positions on stability. The effects are the result of the distortion of a wing's span load distribution in sideslip. Undistorted wing span load distributions feature sharp gradients of load with spanwise distance at both wing tips.

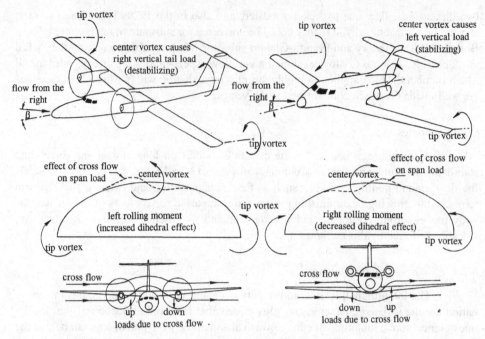

Figure 6.2 Cross-flow explanation of wing vertical position effects on directional stability and dihedral effect. Distortion of the wing span load in right sideslip creates a center vortex that gives destabilizing sidewash for a high wing and stabilizing sidewash for a low wing. The distorted span load gives increased dihedral effect for a high wing and decreased dihedral effect for a low wing.

Local shed vortex strength is proportional to this gradient, resulting in the familiar wing tip vortices. The flow of air from higher to lower pressure determines the sense of vortex rotation. Thus wing tip vortices rotate to create downflow, or downwash, inboard of the wing tip.

The center vortices shown in Figure 6.2 are the result of the local span load distortion due to wing–fuselage interference in sideslip. Center vortex rotations for low and high wing arrangements in sideslip are seen to be consistent with the observed stability changes noted above.

6.2.5   *Downwash and Sidewash*

The flow behind wing–body combinations is deflected from the free-stream values, affecting the stabilizing contributions of the tail surfaces. Downwash is the downward deflection of the free stream behind a lifting surface, a momentum change consistent with the lift itself. Sidewash is a sideward deflection of the free stream, related to the side force on the wing–fuselage combination in side-slipping flow. Sidewash at the vertical tail is dominated by vortices that accompany the downwash when sideslip distorts the pattern.

Wing downwash charts for the symmetric flow (no sidewash) case suitable for preliminary design became available in 1939 from Silverstein and Katzoff. Later investigators broadened the design charts to include the effects of landing flap deflection, ground plane interference, wing sweep, and compressibility.

An interesting sidewash effect is the loss in directional stability experienced by receiver aircraft in close trail to tanker aircraft. Following reports of directional wandering of receiver aircraft, Bloy and Lea (1995) tested tanker–receiver model combinations in a low-speed wind tunnel. These results, together with vortex lattice modeling, confirm the loss in receiver directional stability. Rolled-up tanker wing tip vortices acting on the receiver vertical tail in a low position cause the problem.

6.2.6   Early Design Methods Matured – DATCOM, RAeS, JSASS Data Sheets

New stability and control problems associated with geometries appropriate to transonic and supersonic speeds and their approximate theoretical or empirical consequences led to the creation of handbook data for their solution in a form suitable for the use of airplane designers. Handbooks have been produced by the USAF Wright Air Development Center, the British Royal Aeronautical Society (RAeS), the Japan Society for Aeronautical and Space Sciences (JSASS), and others. The USAF version, called DATCOM, for Data Compilation (Hoak, 1976), is supplemented by a computer version intended to reduce the manual labor in using the rather bulky hard copy version of the material.

The goal of all these compilations is to show the effects of all possible design factors on aircraft forces and moments. Charts and elaborate formulas are used, as in the example of Figure 6.3, from the DATCOM. RAeS data sheets have similar function and appearance, except for a wide use of ingenious carpet plots. Dr. H. H. B. M. Thomas played a key role in the development of the RAeS data sheets.

6.2.7   Computational Fluid Dynamics

Computational fluid dynamic methods apply the power of modern digital computers to the problem of estimating stability and control from drawings at the design stage. Finite-element methods are one form of computational fluid dynamics. The great power of computational fluid dynamics is its ability to deal with arbitrarily shaped airplanes. Even advanced handbook methods such as the DATCOM can fail to represent a truly unusual design.

Computational aerodynamics are not exactly new, in that approximate methods for the calculation of aerodynamic loads on arbitrarily shaped wings in subsonic flow have been available for many years. However, before the advent of the digital computer the number of unknowns in the mathematical solutions for the flow had to be kept low. As pointed out by Sven G. Hedman, one of the inventors of the modern vortex lattice method, the number of unknowns was kept low in the predigital computer era by assumptions for the wing's chordwise and spanwise load distributions. That early work was done by Falkner (1943), who also coined the term *vortex lattice theory*.

Assumed load distributions are not needed for modern finite-element methods using the digital computer. The earliest applications of modern finite-element methods to the calculation of aerodynamic forces and moments appears to have been made at the Boeing Company in about 1960. This was the discrete loading vortex lattice method, developed independently by Sven G. Hedman and P. E. Rubbert. The development of the vortex lattice method is a classic case of research people all over the world contributing steps to a remarkably useful result. For a detailed history, see De Young (1976).

6.2.7.1 Vortex Lattice Methods

When the vortex lattice method is applied to wings, the surface is arbitrarily divided in the chordwise and spanwise directions into panels or boxes. Each panel contains a horseshoe vortex. The vortex-induced flow field for each panel is derived by the Biot-Savart

For A ≥ 1.0:

$$C_{l_\beta} = C_L\left[\left(\frac{C_{l_\beta}}{C_L}\right)_{\Lambda_{c/2}} \cdot K_{M_\Lambda} + \left(\frac{C_{l_\beta}}{C_L}\right)_A\right] + \Gamma\left(\frac{C_{l_\beta}}{\Gamma} K_{M_\Gamma}\right) + \theta\tan\Lambda_{c/4} \frac{\Delta C_{l_\beta}}{\theta\tan\Lambda_{c/4}} \quad \text{(per degree)}$$

For A < 1.0:

$$C_{l_\beta} = C_L\left[-\frac{1}{57.3}\frac{2}{3}\frac{\Gamma}{A}\right] \cdot \Gamma\left(\frac{A}{6}\right) \quad \text{(per degree)}$$

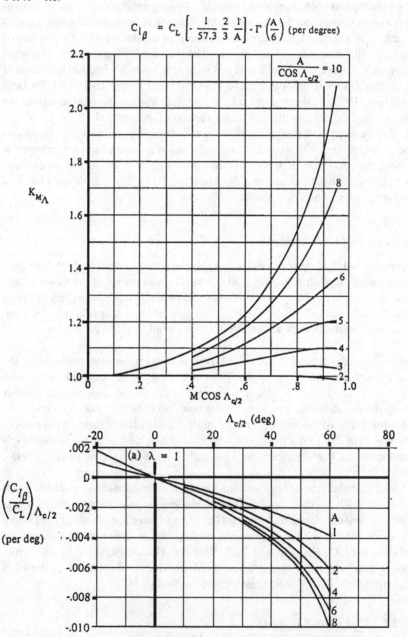

Figure 6.3 Example formula and charts from the USAF DATCOM. This covers only a small part of the material for calculation of the derivative C_{l_β} for straight-tapered wings. RAeS data sheets have similar functions and appearance.

law. While this implies incompressible flow, the Prandtl-Glauert rule can extend the results to subcritical Mach numbers. The boundary condition of no flow across panels is fulfilled at just one control point per panel. Angle of attack and load distributions for the panels are found from a system of simultaneous linear equations that are easily solved on a digital computer. Distortions in data due to Reynolds' number mismatches, jet boundary corrections, and model attachment problems in real wind tunnels are replaced with the necessary approximations of computational fluid dynamics.

When the panels lie in a flat plane and occupy constant percentage chord lines on an idealized straight-tapered wing at more or less arbitrary spanwise locations, and when each panel contains a line vortex across its local quarter chord point and trailing vortices along its side edges, whose collective vorticity provides tangential flow at every panel local three-quarter–chord point, the bound vorticity in each panel can be found by desktop methods, as in the Weissinger method. However, when panels or a mesh cover a complete airplane configuration, automatic machine computation methods become necessary. Depending on the method used, the computer defines the vortex strength for each panel.

6.2.7.2 *Generalized Panel Methods*

Following the pioneering vortex lattice work, computational fluid dynamics programs of increasing complexity have been developed, such as PAN AIR by Boeing, QUADPAN at Lockheed, Analytical Methods, Inc.'s, VSAERO, and MCAERO at McDonnell-Douglas. These approaches have included the Neumann problem in potential flow (Smith, 1962), inviscid Euler methods (Jameson, 1981), and full-blown Navier-Stokes equation solutions (Pulliam, 1989).

Vortex lattice, Euler, and Navier-Stokes methods are now used to generate airplane stability and control data at the preliminary design stage in much the same way that wind-tunnel models were used in earlier times. The computer defines and stores the three-dimensional panel geometry approximating the airframe shape, as in Figure 6.4. Aircraft lift curve slopes, static longitudinal and lateral stability, control effectiveness, and even rotary derivatives are well predicted for small angles of attack, sideslip, and control deflection.

6.3   Estimation from Wind-Tunnel Data

Manufacturers of transport and military airplanes spend a great deal of money and engineering effort on wind-tunnel testing in developing new designs. These costs are rarely questioned anymore; one just budgets wind-tunnel testing at a generous level. Yet, how well can one expect wind-tunnel test results to match stability and control flight test results? This question was dealt with in an early NACA study (Kayten and Koven, 1945). Both engineers later led the stability and control branch in the U.S. Naval Air Systems Command.

Kayten and Koven compared wind-tunnel and flight test measurements for the Douglas A-26 Invader twin-engine attack airplane. The discrepancies were larger than one might have expected. Most of the discrepancies could be explained after the fact, but one is left with the uneasy feeling that wind-tunnel tests can give engineers a distinctly cloudy crystal ball. The factors that led to discrepancies in the case of the A-26 were

1. The geometric wing dihedral was greater in flight than in the wind tunnel due to upward bending under load. This problem could be dealt with by giving tunnel models extra wing dihedral based on calculated bending deflections.

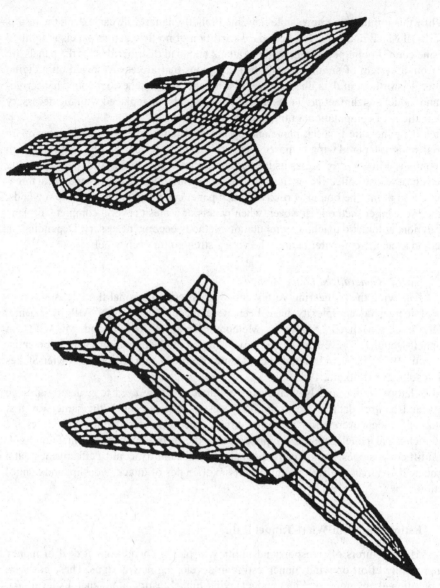

Figure 6.4   PAN AIR panel geometry for a computational fluid dynamic analysis of a complete airplane configuration. (From Tinico, Boeing Commercial Airplane Group, 1992)

2. Control surface contours in flight differed from the wind-tunnel model because of fabric distortion. This problem may have effectively vanished, since fabric-covered control surfaces are now rarely used.

3. There was premature inboard wing stalling in flight that was not present on the smooth, well-faired wind-tunnel model wing. This last problem is of the type that is difficult to deal with in advance. However, the current approach might be to clean up the airplane's premature wing stalling by refairing or vortex generators, incidentally bringing about better agreement between the wind-tunnel and flight data.

In spite of discrepancies such as these, designers ignore unfavorable wind-tunnel results at their peril. For example, before it was flown, power model tests of the Martin 202 showed that its one-piece wing would have negative effective dihedral in the power approach condition. The results were dismissed with the comment, "It's only a wind- tunnel test," but an expensive redesign was needed later.

CHAPTER 7

The Jets at an Awkward Age

Performance of the first jet aircraft outstripped stability and control technology. Their high performance called for two stability and control technologies that were still quite crude – power controls and electronic stability augmentation.

The high transonic Mach numbers reached by the early jets, such as the McDonnell XF-88 and the North American FJ-1, led to large and generally unpredictable control surface hinge moments and the possibility of control surface flutter. Redundant, irreversible hydraulic control actuators on all surfaces were really needed. With irreversible controls, the normal aerodynamically generated control stick forces would be replaced by artificial forces generated by such things as springs, weights, bellows, and closed-loop force generators.

Likewise, the operating altitudes in the 30,000- and 40,000-foot range that jet power had made possible required that the normal source of damping of oscillatory yaw, pitch, and roll motions be augmented. Satisfactory damping of the Dutch roll and short-period longitudinal oscillations comes naturally near sea level from forces generated on an aircraft's wings and tail surfaces with control surfaces fixed. However, at the higher altitudes, control surfaces need to be driven by electronic stability augmentation systems in a series fashion, added to the pilot's inputs and not especially apparent.

7.1   Needed Devices Are Not Installed

In spite of an evident need for redundant, irreversible power controls and electronic series-type stability augmenters, these devices were rarely used in the early jet aircraft. Stability and control designers and their chief engineers were quite justifiably reluctant to do so for reliability reasons, but also to avoid high cost and weight penalties. What was done instead? Some stability and control case histories of jets in that awkward age are given in what follows.

The case histories are of interest not only as history but as cautionary tales for the stability and control designers of future advanced general-aviation aircraft. These case histories tell mostly of failure and shortcomings. But since this poor record was made by some of the brightest stability and control designers of the 1950s, future designers of high-performance aircraft should be wary of trying to avoid irreversible power controls and series-type stability augmentation, in the name of simplicity and cost saving.

7.2   F4D, A4D, and A3D Manual Reversions

In the first two of these early Douglas jets, hydraulic power-assisted controls were indeed used in the original layouts, but only in single channels. That is, in the not-infrequent case of a flight failure of some part of the hydraulic system, the pilot could pull an emergency lever that disconnected the hydraulics. Control would revert to ordinary manual connection of the stick and pedals to the control surfaces.

This was fine if the failure was a jammed hydraulic valve that would interfere with control. However, there were a few regrettable cases in which a pilot incorrectly diagnosed a control difficulty as a hydraulic system failure and made a one-way reversion to manual control. For cases in which this happened at high airspeeds, where manual control was only marginal, manual reversion made a bad situation far worse. Redundant hydraulic power controls, in which several hydraulic actuators in parallel supplied the needed hinge moments, could not come soon enough for the fast jets.

Dual aileron hydraulic boost was used for the A3D Skywarrior (Gunston, 1973), but only to avoid excessively high boost ratios. Two 20:1 boost systems were used instead of one 40:1 boost. One-way manual reversion was used, as in the other two Douglas jets. Because of the A3D's relative large size, even emergency manual control was not possible without a shift in wheel-to-aileron gearing. For manual reversion the pilot shifted gears by 2:1, requiring twice as much wheel throw for a given aileron deflection.

7.3   Partial Power Control

Another control system compromise made during the jet's awkward age was to try to get by with direct manual control for one or more surfaces. The Douglas F4D Skyray's rudder was a good example. The F4D was a small, single-engine jet whose demands for rudder controllability seemed minimal. Of course, there were no asymmetric power conditions to consider. Rudder control in cross-wind landings and takeoffs and to make coordinated turn entries and recoveries was shown to require only modest amounts of rudder deflection and pedal force.

The F4D could be spun, and was required to have good spin recovery characteristics. Ordinarily, this would require full rudder in opposition to the spin, and the corresponding pedal force for a manually controlled rudder would be high. However, the F4D's inertia distribution made the elevons the primary spin recovery control. The rudder, in any case fully shielded from the airflow by the wing at spin attitudes, could be in any position without affecting spin recovery.

All was fine until F4D test pilot Robert O. Rahn inadvertently entered an inverted spin. The rudder was now unshielded. The air flow direction in the spin drove the rudder in the pro-spin direction. Not only that, but the unshielded rudder's effectiveness in the inverted spin was high enough to require that it be deflected in the opposite, or anti-spin, direction for a satisfactory recovery. With no hydraulic power assistance, the best Rahn could do with an estimated 300 pounds of pedal force was to neutralize the rudder (and then use the emergency spin chute for recovery). This unanticipated demand for rudder deflection meant that the original decision to save the cost and complexity of hydraulic power for the rudder was not justified.

7.4   Nonelectronic Stability Augmentation

Really ingenious nonelectronic stability augmentation systems came out of the jet's awkward age, as designers tried to have artificial damping without the heavy, costly, and, above all, unreliable electronics of the period. A mechanical yaw damper, invented by Roland J. White and installed on early Boeing B-52 Stratofortresses, is a good example of the genre.

Imagine a rudder tab that is free to rotate on low friction bearings. Instead of being connected to an electric actuator, or to cables leading to the cockpit, the free tab is driven by

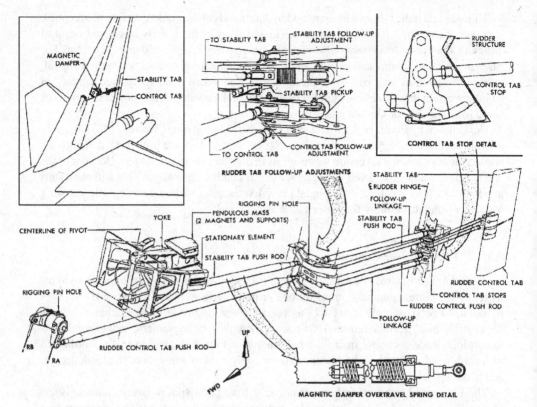

Figure 7.1 Boeing B-52 rudder control linkages. R. J. White's magnetically phased bobweight yaw damper operates the stability tab. (From B-52 *Training Manual*, 1956)

inertia forces acting on a small bobweight located ahead of the hinge line (Figure 7.1). Tab position is further modified dynamically by an eddy current damper, providing damping hinge moments proportional to tab rotational velocity.

As the airplane goes through a typical lateral or Dutch roll oscillation, the vertical tail assembly swings from side to side, accelerating the tab bobweight. Without the eddy current damper it is clear that the tab will take up deflections in phase with the lateral acceleration at the vertical tail. However, ideally, tab positions should be phased with respect to yawing velocity in such a way as to drive the rudder in opposition. This is the classic yaw damping action, right rudder in opposition to left yawing velocity. The function of the eddy current damper is to "tune" tab deflections to create exactly that phasing. In 1952, a similar approach was taken by M. J. Abzug and Hans C. Vetter of the A3D Skywarrior design team at Douglas Aircraft, to provide nonelectronic yaw damping for that airplane. The design method was cut and try on the analog computer, to find the proper combination of bobweight mass and damper size that would phase the tab, creating effective yaw damping.

The obvious practical problem with the B-52 and A3D yaw dampers is one that is faced with any purely mechanical system, as compared with a modern electromechanical control system. In the mechanical system, the result or output depends critically on the condition of each component. If the free tab's bearings deteriorate over time or are invaded by grit, or if the eddy current damper's effectiveness is changed, tab phasing will be thrown off.

In the extreme case, tab action could actually add to the airplane's lateral oscillation, instead of damping it.

In a July 1994 letter Roland White describes such a situation that actually occurred on a B-52, as follows:

> A rudder tail shake on a test airplane caused the magnetic damper to lose its damping. A serious accident would have occurred if the bobweight did not jam due to a mechanical failure. After that I found when going to work the next day your friends will ask if you still work here.

A modern, electromechanical yaw damper drives the rudder in opposition to the measured rate of yaw. It does so by comparing the current rudder position with the desired value and continuing to exert torque on the rudder until that value is reached, overriding mechanical obstacles such as sticky bearings or even losses in performance of the motor that drives the rudder.

The practical shortcomings of purely mechanical yaw damping were not unknown to the Boeing and Douglas design staffs. When a chance appeared to get a yaw damper function electronically, that option was taken instead. In the case of the B-52, the spring-tab–controlled rudders were replaced by powered rudders, allowing Boeing to use the electro-mechanical yaw damper design developed successfully for the B-47.

In the Douglas case, electromechanical yaw damping was installed using components of the airplane's well-proved Sperry A-12 automatic pilot. The Sperry Gyroscope Company's DC-3 "dogship" proved the concept in test flights at the Sperry plant in Long Island, New York. Signals from the outer, or yaw, gimbal of the A-12's free directional gyro were electronically differentiated through a lead network and sent to the rudder servo. Differentiated yaw angle is of course yaw rate.

This worked well when the system was transferred to the A3D and flown routinely at Edwards Air Force Base. Then one day a test pilot bringing an A3D back for landing dove at the runway and pulled up into a chandelle, a natural thing to do for a high-spirited test pilot with an airplane he likes. The A3D, with yaw damper on, responded by applying bottom rudder during the nearly vertical bank, diving the ship back toward the ground.

The pilot regained control and an investigation started at once. The A-12 and yaw damper function were found to be in perfect order. The culprit turned out to be what had been called for years "gimbal error." The A-12 directional gyro is a conventional two-gimbal free gyro, with yaw measured on the outer gimbal. The rotor, spinning in the inner gimbal, is slaved to magnetic north and the inner gimbal itself is erected to gravity by a bubble level system. The angle between the outer gimbal and the instrument's case is true yaw or heading angle as long as the outer and inner gimbals are at right angles to each other. This holds only for zero bank angle. At the sharp bank angles of the chandelle, or in any steep turn, the yaw reading picks up errors that depend on the heading angle (Figure 7.2).

During turns, differentiation with respect to time of the erroneous yaw angle exaggerates the ordinary gimbal errors. The A3D experience proves dramatically that one cannot in general differentiate free gyro signals to produce damping signals for stability augmentation, at least for airplanes that maneuver radically. After the all-mechanical and free-gyro A3D yaw damper designs were proved faulty the airplane was finally fitted with what is now the standard design, a single-degree-of-freedom yaw rate gyro driving the rudder servo.

A rather more successful nonelectronic stability augmentation system was developed at the Naval Weapons Center, China Lake, for the AIM-9 Sidewinder missile. The Sidewinder

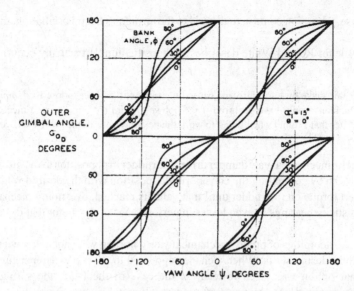

Figure 7.2   Gimbal angles of the outer gimbal of the Sperry A-12 directional gyro, as a function of bank and yaw angles. The outer gimbal rate fluctuates strongly in turns at steep bank angles. Differentiating outer gimbal angle to obtain yaw rate caused a near-crash of a Douglas A3D-1 Sky Warrior. (From Abzug, *Jour. of the Aero. Sciences*, July 1956)

derives roll damping from nonelectronic, air-driven flywheels mounted at the tips of the missile's ailerons, producing gyroscopic torques that drive the ailerons to oppose roll rate. The flywheel torques are evidently high enough to override variations in aileron bearing friction. There seems to have been no application of this all-mechanical damping system to airplanes.

7.5   Grumman XF10F Jaguar

The variable-sweep XF10F Jaguar was Grumman's attempt to avoid fully powered pitch controls during the jet's awkward age. The Jaguar's horizontal tail was on a large streamlined pod that pivoted in pitch on top of the vertical tail (Figure 7.3). This pivoted assembly was in effect a separate canard airplane, trimmed in pitch by a direct connection between the pilot's stick and the canard.

The trim lift coefficient attained by the pivoted canard tail assembly was of course the tail lift coefficient for the XF10F as a whole. This ingenious system amounted to a tab control system, since the canard control surface attached to the pilot's stick was a relatively small surface with low hinge moments. Yet the tail loads for the XF10F's stability and control were provided by an all-moving surface that was not subject to control surface effectiveness losses and unpredictable hinge moments at transonic Mach numbers.

In Navy flight tests, the XF10F ran out of nose-up longitudinal control for landing at some loadings, due to insufficient down deflection of the canard. Another serious problem with the canard airplane horizontal tail was a low natural frequency at low airspeeds. This produced a large time lag between pilot stick motion, tail, then airplane, response. Pilots complained that at low airspeeds they had no idea of the tail's incidence angle.

The Navy canceled production plans for the variable-sweep XF10F in 1953. The airplane never received the engine it was supposed to have, leaving it with poor performance. Also, the advent of slanted deck carriers and steam catapults allowed fixed-sweep airplanes to be

Figure 7.3 The Grumman XF10F-1 Jaguar, an attempt to avoid powered pitch controls. A freely pivoted canard-controlled body and wing is mounted on top of the vertical tail. The pilot's stick moves the canard, which controls the angle of attack of the pivoted body. (From National Air and Space Museum)

operated from carriers. The XF10F's ingenious canard airplane horizontal tail design has not been used on later airplanes.

7.6   Successful B-52 Compromises

The B-52 Stratofortress appeared in 1951, during the jet's awkward age. However, the compromises made by Boeing Company engineers to get around the period's (well-deserved) lack of confidence in flight control hydraulics and electronics were so successful that this airplane continues to have an active service life nearly 50 years later. These compromises are suited to the B-52's mission in that required maneuvers are modest and spin recovery is not required.

7.6.1   *The B-52 Rudder Has Limited Control Authority*

The B-52's rudder has an exceptionally narrow chord, just 10 percent of the fin chord (Figure 7.4). For a given control surface span or length, control hinge moments are proportional to the chord length squared. This gives the B-52 rudder small operating hinge moments for an airplane of its size, making feasible manual operation through a spring tab. But how about rudder power or effectiveness? How can asymmetric thrust effects and cross-wind takeoffs and landings be made with that exceedingly small rudder?

George S. Schairer explained that the original Boeing design called for an all-moving vertical tail. This was abandoned because of doubts as to the reliability of the hydraulic actuators. Instead, the solution was to incorporate a yaw-adjustable cross-wind landing gear. The yaw-adjustable cross-wind landing gear is preset by the flight crew before takeoff or

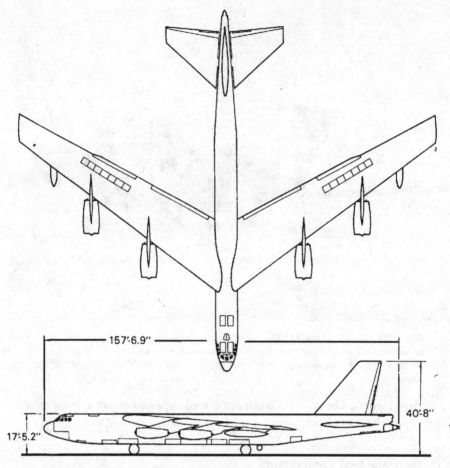

Figure 7.4 B-52G Stratofortress. A seventh wing spoiler segment has been added, the feeler ailerons eliminated. Note the extremely narrow (10-percent chord) rudder and elevators. (From Loftin, NASA SP-468, 1985)

on a landing approach, through a range of 20 degrees to either side of neutral (Figure 7.5). Ground rolls can be made at zero sideslip and bank, the airplane crabbed into the relative wind. The preset landing gear angle is based on the reported wind direction and velocity relative to runway orientation.

Since the airplane has eight engines, asymmetric thrust conditions set up by shutting down an engine are low enough to be handled by the narrow rudder.

7.6.2 *The B-52 Elevator Also Has Limited Control Authority*

The B-52's elevator is as narrow in chord as is the rudder. It depends on help from an adjustable stabilizer for long-term trim and airspeed changes. As in the case of the vertical tail, the original Boeing design called for an all-moving horizontal tail, but this was abandoned because of doubts as to hydraulic actuator reliability.

The B-52's adjustable stabilizer is driven by two independent hydraulic motors through an irreversible screw jack mechanism. One motor drives the jackscrew and the other the live nut on the driven screw thread (Figure 7.6). The control valve for each hydraulic motor is worked

Figure 7.5 B-52 Stratofortress in a crosswind landing attitude. The landing gears are pointed down the runway while the airplane is yawed to the left, presumably into the relative wind. Crosswind landing gear reduces the need for rudder power. (From Loftin, NASA SP-468, 1985)

either by an electric motor or by a backup cable drive from the cockpit. The electric motors are controlled in turn by the usual push-button arrangement on the pilot's control yoke.

With all of this redundancy, stabilizer adjustment failures can still occur, but the B-52 is landable in an emergency with elevator control alone, regardless of stabilizer position. Some center of gravity adjustment by fuel pumping is necessary for this to work.

7.6.3 The B-52 Manually Controlled Ailerons Are Small

The B-52 has only the smallest of ailerons, in the conventional sense. The ailerons are of conventional chord, but their span is only about equal to their chord. They are quite aptly called "feeler ailerons," in that their main function is to supply control forces to the pilot's yoke. Spring tabs are used on the feeler ailerons. Six upper surface spoiler segments on each wing provide the real roll control power for the airplane. The spoiler actuators get their signals to come open from the rotation of the pilot's control wheel, requiring no pilot effort to operate.

The spoiler aileron system adopted for the B-52 was originally tested on a B-47, after that airplane exhibited a marked loss in conventional outboard flap-type aileron power due to wing twist. The spoiler system worked well on the B-47, but the Air Force declined to make the change on that airplane. The B-52 can be landed using the feeler ailerons alone, if all

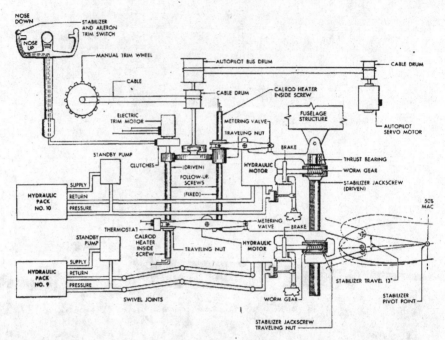

Figure 7.6 Schematic of the Boeing B-52's stabilizer trim controls. Two independent hydraulic motors drive the stabilizer, one through the jackscrew, the other through the jackscrew's traveling nut. The valve of each hydraulic motor is controlled by electric trim motors, with mechanical backup. (From B-52 *Flight Manual*, 1956)

spoilers are inoperative due to hydraulic power failures, for example. Successful landings are possible under benign wind and turbulence conditions.

In the late model B-52G the feeler ailerons have been eliminated and an extra spoiler segment has been added. The B-52G flight manual advises that limited lateral control is available by sideslipping the airplane with the rudder, if all spoilers are inoperative. Landings are "not advised" by the flight manual, meaning that the crew is expected to bail out if all spoilers become inoperative.

CHAPTER 8

The Discovery of Inertial Coupling

Airplanes that fly near the speed of sound are designed with thin, stubby wings. Most of their masses are concentrated in the center, in long, slender fuselages. When these airplanes are rolled rapidly the fuselage masses tend to swing away from the direction of flight and become broadside to the wind. This tendency, essentially a gyroscopic effect, is called inertial coupling.

8.1   W. H. Phillips Finds an Anomaly

The distinction of having discovered inertial or roll coupling in airplanes and then explaining it mathematically in the open literature belongs to W. Hewitt Phillips, then working in the Flight Research Division of the NACA Langley Laboratory. In a 1992 paper Phillips said, "When the [XS-1] model was dropped, it was observed in the optical tracker to be rolling, as shown by flashing of light from the wings.... In examining the records further the oscillation . . . was found to represent a violent pitching in angle of attack from the positive to the negative stall" (Figure 8.1).

Phillips analyzed the problem as a gyroscopic effect, publishing his results in an NACA Technical Note (Phillips, 1948). In those days NACA used the category of Technical Notes for "the results of short research investigations and the results of studies of specific detailed problems which form parts of long investigations." Well, nobody's perfect – the NACA could hardly be blamed for missing the fundamental importance of Phillips' inertial coupling results when so many other people took little notice. In hindsight, the inertial coupling analytical work clearly merited publication in the more exalted category of NACA Technical Reports as the "results of fundamental research in aeronautics."

8.2   The Phillips Inertial Coupling Technical Note

Electronic digital computers were still years away when Phillips did his inertial coupling research. For numerical solutions that would attack the problem Phillips was obliged to simplify the equations with a series of ingenious mathematical steps. His successive transformations led to inertial coupling stability boundaries derived by a simple quadratic equation.

For generality, Phillips nondimensionalized aircraft static stability parameters in terms of rolling frequency, or number of complete roll cycles per second. That is, the levels of both longitudinal and directional static stability or stiffness are characterized by their respective nonrolling natural frequencies, in short-period longitudinal and Dutch roll modes. These frequencies, expressed as cycles per second, are divided by the rolling frequency, as defined above, for the Phillips charts (Figure 8.2).

The remarkable but strong mathematical transformations added to the academic flavor of the charted results obscured the work's significance to the hard-pressed stability and control engineers working in aircraft plants in the late 1940s, who should have paid more attention to Phillips' results. Had a 1980s type digital computer been available to Phillips in 1947, permitting a few time histories of forthcoming fighter aircraft full-aileron rolls to

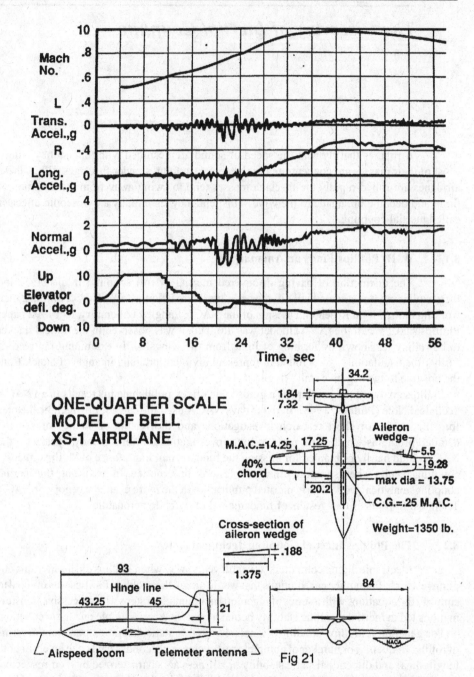

Figure 8.1 The smoking gun – The XS-1 flight record that gave evidence of rapid oscillations in normal and lateral accelerations during steady rolling. The XS-1 drop model had an aileron wedge designed to make it roll steadily. (From Phillips, *Jour. of the Amer. Aviation Historical Soc.*, Summer 1992)

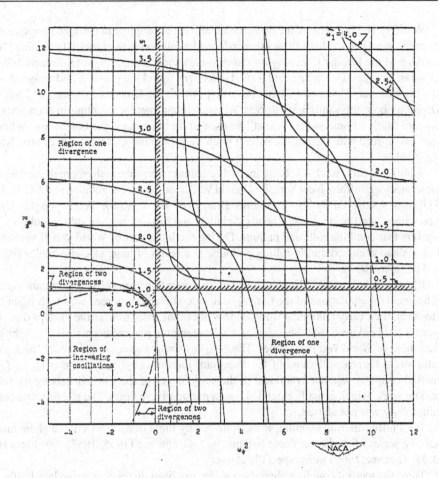

Contour lines of nondimensional oscillation frequencies
of rolling aircraft on a plot of ω_θ^2 against ω_ψ^2 for the case

$I_X = I_Y$, $\zeta_\theta = \zeta_\psi = 0$. Regions of diagram free from

divergence or increasing oscillations indicated by cross-hatching.

Figure 8.2 An example of the W. H. Phillips inertial coupling stability boundaries. (From Phillips, NACA TD 1627, 1948)

be calculated and presented, the airplane stability and control community would have taken notice.

Interesting background on Phillips' inertial coupling work was contained in a 1994 letter from him. An excerpt from the letter reads:

> In thinking about the subject lately, I have concluded that my approach was based on my training at MIT. In the courses that I took, particularly by Prof. Koppen, the derivations did not start with the complete equations of motion. The equations had already been divided into lateral and longitudinal groups and linearized. In Prof. Draper's courses on instrumentation, much emphasis was placed on nondimensionalizing the results in terms of natural frequency. I did not read Bryan's report, which starts from basic principles, for many years after that. If I had known the complete equations of motion, I might have been discouraged from attempting a solution.

While W. H. Phillips gave the first account of inertial coupling in the open literature, there seems to have been at least three other independent discoveries of inertial coupling. While working at the Boeing Company on the then-classified Bomarc missile, Roland J. White, Dunstan Graham, D. Murray, and R. C. Uddenberg found the problem and reported it in a Boeing Company document dated February 1948. At the Douglas Company's El Segundo Division about the same time, Robert W. Bratt found inertial coupling in drop tests of a dummy Mark 7 bomb shape. A small amount of fin twist made the bomb spin. When the spin rate agreed with the bomb's natural pitch frequency the spin went flat, or broadside to the wind.

Additional early work involving inertial coupling took place at the Cornell Aeronautical Laboratory in Buffalo, New York, by Donald W. Rhoads, John M. Schuler, and J. C. O'Hara. This was sponsored by the Structures Branch of the U.S. Air Force Wright Aircraft Laboratory, starting in 1949. Rhoads, Schuler, and O'Hara studied rolling pullouts, maneuvers that combine rolls and pullups. During the latter part of World War II vertical tail failures had occurred during rolling pullouts, as a result of large side-slip angles (Rhoads and Schuler, 1957).

Rhoads, Schuler, and O'Hara included inertial coupling terms in their study, among other refinements. Calculations of the critical peak side-slip angles agreed well with flight tests. However, their early numerical work, done at about the same time as the Phillips discovery, was for the Lockheed P-80 Shooting Star, whose inertial parameters are not much different than those of World War II airplanes. The P-80 has straight wings of moderately high aspect ratio and a fairly small value of the important inertial coupling parameter $(I_x - I_y)/I_z$. Inertial coupling was not prominent in the early stages of the Cornell Laboratory rolling pullout work, which actually extended over a period of five years. The stability and control community was not alerted.

The Phillips inertial coupling work, followed by flight occurrences of the phenomenon, led to a series of studies in Great Britain. W. J. G. Pinsker (1955, 1957, 1958) and H. H. B. M. Thomas (1960) were especially active.

Thus, the inertial coupling phenomenon, having been discovered in the late 1940s, was ignored by airplane designers until it was rediscovered in flight in the early 1950s. By 1956, the U.S. industry was roused enough to turn out for a conference on the subject held at Wright Field.

8.3   The First Flight Occurrences

According to NACA, inertial coupling was first experienced in manned flight by NACA test pilot Joe Walker on the Douglas X-3 research airplane. However, Norman Bergrun and Paul Nickel of the NACA Ames Aeronautical Laboratory had made an earlier flight confirmation of the Phillips theory (1953). They found proof of the theory in the motions of an unmanned rolling body–tail combination tested at NACA's High-Speed Flight Station at Edwards, California. As with the Douglas Mark 7 bomb test, the Bergrun-Nickel model diverged in angle of attack and sideslip when the spin or roll rate agreed with the model's natural frequency in pitch and yaw (Figure 8.3).

While NACA engineers were still studying the X-3 flight records, inertial coupling occurred again, at NACA's Flight Station in California. This was in mid-1954, on a North American F-100A Super Sabre being flown by a NACA research pilot. With two-thirds of full aileron, angles of attack and sideslip diverged and the airplane became uncontrollable (Seckel, 1964).

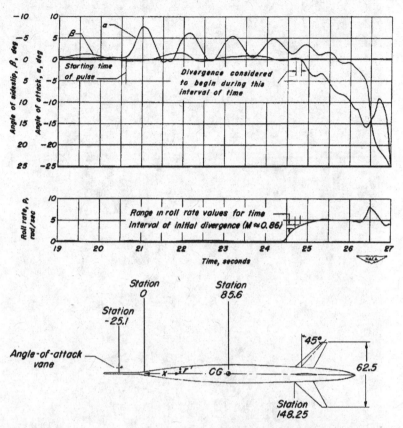

Figure 8.3 First flight confirmation of the Phillips inertial coupling theory, made on a rolling body–tail combination. A divergence begins at a roll rate of 3.5 radians per second. (From Bergrun and Nickel, NACA TN 2985, 1953)

The F-100 was a prime candidate for inertial coupling, with very thin (7-percent thick) wings swept back at 45 degrees. Roll inertia was less than 20,000 slug-feet squared, compared with pitch and yaw inertias of 66,000 and 81,000 slug-feet squared, respectively. There was no artificial yaw or pitch damping. The pilot had plenty of control authority for making rapid rolls, with powered controls all around and simple spring artificial feel systems.

The F-100 was just going into service, and Air Force pilots encountered inertial coupling, as well. Several airplanes were lost and the Air Force grounded the fleet. North American engineers under the direction of the late John Wykes found in analog simulation of the F-100 that, in agreement with the principles laid down by Phillips, an increased level of static directional stability reduced the coupling. Yaw damping alone was found to make little improvement. Larger vertical tails fitted to the F-100A, F-100C, and later versions increased yawing natural frequency, reducing inertial coupling in rolls enough to allow the planes to return to service (Figure 8.4).

By coincidence, larger vertical tails were needed for the F-100 anyway, to prevent static directional instability in supersonic dives. Beyond a Mach number of 1, the static directional stability contribution of the vertical tail was reduced because of the normal reduction in lift curve slope at increasing supersonic speeds. In addition, at all high

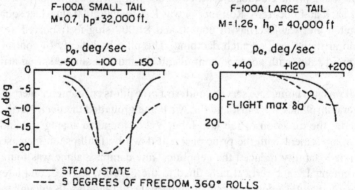

Figure 8.4 The effect of enlarging the F-100A vertical tail. The enlarged tail postpones side-slip angle divergence in rolls from roll rates of about 75 degrees per second to about 180 degrees per second, still not a very good rate. The agreement between the solid and dashed lines shows that the Schmidt-Bergrun-Merrick steady-state analysis method is a reasonably good predictive tool. The photo is of the F-100F, a late model that has the large vertical tail. (From Schmidt, Bergrun, and Merrick, WADC Conf. 56WCLC-1041, 1956)

airspeeds aeroelastic bending and torsion further reduced the fin's contribution to directional stability.

The F-100's inertial coupling problem illustrates a particular finding in Phillips' 1948 analysis. Coupling is most severe when there is a large discrepancy between the levels of static longitudinal and directional stabilities, as shown by their respective natural frequencies. With its original small vertical tail, F-100 yaw natural frequency was particularly low. The reverse was true for the XS-1 drop model, whose flight-recorded oscillations led to the Phillips analysis. The XS-1 model had low pitch and high yaw natural frequencies at transonic speeds.

The Phillips inertial coupling charts show there is a window of stability for the case in which the natural frequencies are equal and there is some damping on each axis. Richard Heppe checked the variations of pitch and yaw natural frequencies for the Lockheed F-104 with airspeed, altitude, and loading, with discouraging results, as follows:

> Maintenance of equal natural frequencies in pitch and yaw over wide speed and altitude ranges by aerodynamic means only will not be possible generally for practical airplane configurations. This implies that satisfactory roll response will become more difficult to obtain the wider the speed–altitude spectrum of the aircraft.

8.4   The 1956 Wright Field Conference

When inertial coupling first appeared on the scene with the Douglas X-3 research airplane and then the F-100A Super Sabre, interest in the subject grew quickly among those responsible for fighter airplane stability and control. Although first-line fighter flight test results were classified either confidential or secret, the grapevine was hard at work, and information started to circulate on this new potential for uncontrolled maneuvers and structural failures.

U.S. Air Force and Naval engineers saw the need for groups grappling with the unexpected inertial coupling problem to convene and exchange information for the common good. A closed-door, classified conference was therefore called at Wright Field for February 1956. Papers were invited from industry, NACA, and MIT. Because of the urgency and national importance of the subject, authors and attendees from industry were expected to give open accounts of their results, putting aside competitive considerations.

The list of speakers at the since-declassified Wright Field Conference, formally called "Wright Air Development Center Conference on Inertia Coupling of Aircraft," included many of the important stability and control researchers and designers of fighter aircraft of that period:

> They were: Robert Bratt and Charles DaRos of Douglas; Frederick Curtis, Mamoree Masaki, and Dewey Mancuso of Convair; John Gautraud, James Flanders, Thomas Parsons, and Lloyd Wilkie of MIT; Richard Heppe of Lockheed; Wayne Huff and Cecil Carter of Chance Vought; Henry Kelley, Hans Hinz, and Robert Kress of Grumman; Darrel Parke of McDonnell; Jerry Pavelka of Republic; Stanley Schmidt, Norman Bergrun, Robert Merrick, Leonard Sternfield, Joseph Weil, and Richard Day of NACA; and John Wykes of North American.

Charles Westbrook chaired the conference and edited the proceedings (Westbrook, 1956). The lively interest in inertial coupling brought no fewer than 184 conference attendees. This conference on a serious stability problem held in the halls of their chief customer brought out

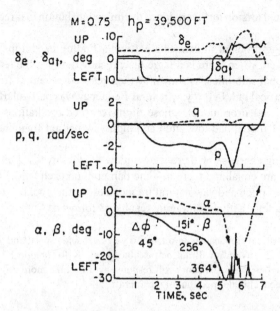

Figure 8.5   Time history of a classic inertial coupling example. The YF-102 diverges to a negative angle of attack and left (negative) side-slip angle in a rapid roll. (From Weil, WADC Conf. 56WCLC-1041, 1956)

a certain defensiveness in the speakers from industry. No blanket criticism is now intended, since it is understandable that airplane designers should want to put their products in the best light. Still, the transcript shows statements such as these:

> In all of these [roll] tests the airplane response has been normal to the pilot and safe from every flight standpoint.

> ... serious difficulty due to inertial coupling is not to be anticipated for the - - - - .

> .... the generally satisfactory roll behavior of the airplane was most welcome....

These benign, reassuring words were accompanied by hair-raising simulation and flight records, in several cases, such as the Convair YF-102 (Figure 8.5).

8.5   Simplifications and Explications

It is not surprising that the mathematical complexity of the inertial coupling problem and the relatively crude state of early analog and digital computers led researchers and designers to try to capture the essence of the problem with simplified mathematical models. With current computer technology, most engineering groups faced with such a problem would probably run the problem without simplification, but automate it so as to use a minimum amount of engineering time.

A strategy made possible by modern digital computers would be to step through automatically all combinations of rolling velocity, Mach number, altitude, center of gravity, initial load factor, stability augmentation feedback, and control usage. This would generate vast numbers of 5- or 6-degree-of-freedom transients. While still in computer memory, the transients could be screened by algorithms that look for excessive air loads or divergences. Only the interesting cases would be retrieved for engineering examination.

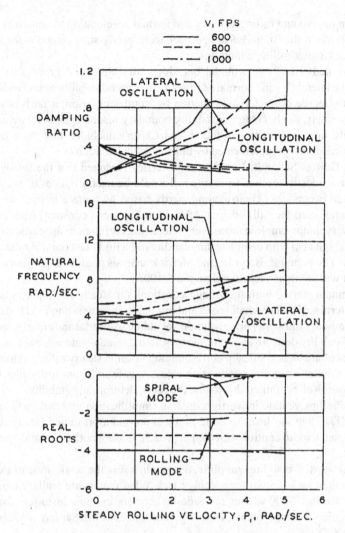

Figure 8.6 Effect of steady rolling on small-perturbation modes of motion of the Douglas F4D-1. Damping of the longitudinal short-period oscillation is halved at high roll rates. (From Abzug, *Jour. of the Aero. Sciences*, Nov. 1954)

However, back in the late 1940s and 1950s, ingenious simplifications were the order of the day rather than the modern head-on approach hypothesized above. Leading the simplification parade in the early days was W. H. Phillips' original work, in which he squeezed the inertial coupling problem down to the solution of a simple quadratic equation. Six of the engineers who came shortly afterward tried to improve on the simplified Phillips model.

An early approach linearized the full 6-degree-of-freedom equations of motion under conditions of steady rolling (Abzug, 1954). Calculations for the F4D Skyray showed a reduction in damping and increase in frequency of the longitudinal oscillation as roll rate increased beyond about 150 degrees per second (Figure 8.6). The succeeding simplified approaches are documented in the proceedings of the *Wright Field Inertial Coupling Conference* (Westbrook, 1956).

Robert Bratt and Charles DaRos reduced the inertial coupling problem to the solution of a quartic, whose roots they determined as functions of rolling velocity. This required setting lateral acceleration to zero and rolling velocity and normal acceleration to constants. Very interesting by-products of the Bratt-DaRos approach were steady-state solutions for angles of sideslip and attack under rolling conditions.

Cecil Carter returned to the Phillips model, but added two additional degrees of freedom, translation along the lateral (Y) and normal (Z) axes. Carter found stability boundaries from the resultant fourth-degree characteristic equation by Routh's criterion, a fairly standard procedure. Unfortunately, in checking the stability boundary results against 20 actual roll time histories, only fair agreement was found, and Carter concluded that the modified Phillips method needed to be supplemented by more exact calculations.

Like Bratt and DaRos, Schmidt, Bergrun, and Merrick reasoned that the steady-state angles of attack and sideslip reached in steady roll would be useful indices of the severity of any unwanted excursions. Their method, partly based on a more limited study by Uddenberg at Boeing, used the full 5-degree-of-freedom (airspeed constant) equations of motion with only very minor simplifications. Very interesting correlations appeared with not only Phillips' work, but also with complete simulations and a flight test point for the North American F-100A. The Schmidt, Bergrun, and Merrick analysis predicted quite accurately the gains obtained with a larger vertical tail on the F-100A.

Kelley at Grumman started with the Abzug equations for steady rolling, eliminating the cyclic gravity terms and other small terms and implicitly incorporating F-11F stability derivatives. A preliminary linear algebraic solution, including some trial and error because of the dependence of stability derivatives on the answers, found steady-state values of pitching velocity and angles of attack and sideslip corresponding to some steady rolling velocity. A final perturbation solution from the steady-state values was found from fifth-order linear differential equations. Kelley limited this to just the roots, determining stability.

The sixth post-Phillips venture in inertial coupling simplification was that by Gautraud and Flanders at MIT. They too linearized the problem around a presumed steady rolling condition, but adopted a more controls-oriented approach, finding both poles and zeros for the Convair F-102.

Augmenting the inertial coupling simplification studies was the work done to explain the phenomenon and to pick it apart for specific cases. What factors are really important? John Wykes dissected the F-100 yawing and sideslip acceleration time histories, showing the relative importance of aerodynamic, engine gyroscopic, and inertial terms as the roll proceeded (Figure 8.7)

Engine angular momentum enters into the inertial coupling problem as a bias in one direction of roll. Divergence occurs at a slightly lower roll rate when rolling in a sense opposite to engine rotation (Pinsker, 1957). Propeller and engine gyroscopic effects occur as well in other flight regimes, especially when aerodynamic forces and moments are relatively low, such as in spins and in takeoff.

8.6   The F4D Skyray Experience

At the Douglas Aircraft Company's El Segundo plant the F4D (later F-6) Skyray was being delivered in service test quantities at the height of the inertial coupling excitement. While it was becoming clear that the coupling problem could be eased, if not avoided, by artificial pitch and yaw damping, these devices were still heavy and unreliable. In any case, properly applied, they are placed in series into a control circuit. That is, artificial damping devices are not a parallel installation, as are many automatic pilots, but are an integral part of the control run from stick to control surface. A series-type artificial damping device cannot be easily retrofitted to a production airplane.

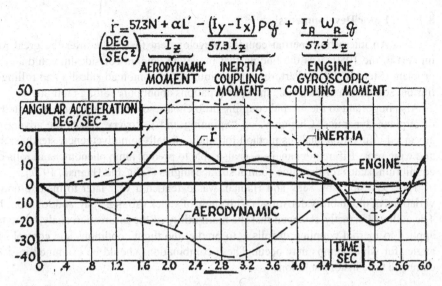

Figure 8.7 The relative importance of aerodynamic, inertial coupling, and engine gyroscopic terms in yawing acceleration during a rapid roll in the North American F-100. In this example, aerodynamic moment is the major factor initially; then inertial coupling overpowers it. (From Wykes, WADC Conf. 56WCLC-1041, 1956)

High levels of static longitudinal and directional stability were likewise known to be helpful, increasing the roll rates that would cause coupling to occur. In the case of the tailless F4D, high levels of static longitudinal and directional stability could hardly have been designed into the basic airframe, even if the need had been known when the ship was laid out.

The only remaining option for the F4D at that stage was to determine safe maneuver limits, the points at which inertial coupling effects did not cause problems. The airplane could then be restricted by placard, with its pilots depended upon to operate within these limits. The problem of determining F4D safe maneuver operating limits fell into the hands of Robert W. Bratt, who was able to collect the large body of numerical data needed on Douglas Accounting Department digital computers.

The analog computers belonging to the Douglas Flight Controls Department could have in principle run the problem, except for two factors. First, the inertial coupling equations of motion include a large number of mathematical nonlinearities, such as trigonometric functions, needed to resolve gravity along airplane axes, and multiplications, needed to solve the Euler, or moment, equations. But representing nonlinearities on the analog computers of those days was a severe problem. The best analog trigonometric function generators and multipliers were small electromechanical servomechanism devices that were costly and somewhat slow. A few systems boasted "quarter-square" multipliers, which had no moving parts but were inaccurate.

The second drawback to having used analog computers for the F4D inertial coupling problem was that accuracy and repeatability were poor. Bratt's problem was to cover systematically in small increments all combinations of lateral and longitudinal stick throws, including phasing or sequence between the two, trimmer settings, and airspeeds and altitudes, in order to establish the points of coupling onset. Vacuum-tube analog computers were not only subject to frequent breakdowns, but also to drifts and gain changes that could make a hash out of a systematic study. These considerations led to using the digital computer, an early use for these machines in flight dynamics.

8.7   Later Developments

An interesting inertial coupling development that came after the great rush of interest in the 1950s was the finding that moderate amounts of sideslip could add to the problem (Stengel, 1975). Perturbation motions about combined sideslip and rolling equilibrium solutions are less stable than perturbations about pure rolling motions.

Also important to the inertial coupling problem are some developments in related fields of airplane dynamics. Chapter 9, on "Spinning and Recovery," notes the advent of the advanced bifurcation analysis method for study of stall–spin divergence, steady spinning, and wing rock. Bifurcation analysis is also able to predict jump phenomena in rolls or two equilibrium states for the same control surface angles (Schy and Hannah, 1977).

The 1977 study by Schy and Hannah was extended a year later to include nonlinear variations of the stability derivatives with angle of attack (Young, Schy, and Johnson, 1978). The authors correctly observed that the main utility of the bifurcation analysis method as applied to inertial coupling in rolls is to predict the flight conditions and control surface angles for which jumps may occur. These combinations should be examined in detail in complete time history solutions.

8.8   Inertial Coupling and Future General-Aviation Aircraft

Inertial coupling has been generally tamed as a potential problem in modern fighter aircraft. Even the most austere of these are equipped with stability augmentation systems that can provide the required feedbacks to minimize excursions in rapid rolls. The McDonnell Douglas F/A-18A is typical in having feedbacks that minimize kinematic coupling in rolls. This means that when the pilot applies roll control, pitch and yaw control are fed in to make the airplane roll about the velocity vector rather than about the longitudinal axis. Thus, angle of attack is not converted into sideslip angle, reducing sideslip in rolls at high angles of attack.

But what about future general-aviation aircraft? The answer is that the problem could conceivably be rediscovered by general-aviation designers the hard way a few years from now, as it was stumbled upon by fighter designers in the early 1950s, some years after the basic theory had already been developed by W. H. Phillips.

There have been a few fighter-type general-aviation designs already, such as the Bede Jet Corporation's BD-10 and the Chichester-Miles Leopard four-seat jet. The BD-10 is a two-seat kit airplane that weighs 4,400 pounds and uses an engine with a thrust of nearly 3,000 pounds. The flight control system is entirely manual, with no provisions for stability augmentation.

The BD-10 has the classic inertial coupling-prone design: small, thin wings and a long, heavily loaded fuselage. We have only to imagine the advent in a few years of inexpensive, reliable, jet engines in the BD-10's thrust class, or even smaller. If this happens, designers will certainly produce fast, agile, personal jet aircraft that would be ripe for inertial coupling problems.

CHAPTER 9

Spinning and Recovery

Spins are uncontrolled rotations of a fully stalled airplane. In aviation's early years, when spins were first encountered, spinning airplanes descended more or less straight down. The motion was mainly yawing and quite stable. Stability and control engineers were concerned only with recovery from spins into unstalled flight.

The coming of jet airplanes saw mass distribution changes that caused spins to be oscillatory. Emphasis shifted somewhat to the entry phase of spins and design features that made spin entry less likely during flight operations. This chapter traces the changing nature of airplane spinning from the early days and the corresponding engineering responses.

9.1   Spinning Before 1916

The spinning experience in the early days of aviation is described by B. Melvill Jones (1943):

> In the early days of flying – before 1916 – the spin generally ended fatally, because what later proved to be the most effective means of checking it was in some respects contrary to the natural reaction of the pilot to the realization that he was diving towards the earth. About 1916 it was discovered that an effective way of checking the type of spin which was common in those days was to thrust the control stick forward and apply rudder in the sense opposed to the rotation. For some time after this knowledge had become general, relatively few fatalities due to spinning occurred, provided that there was enough air-room for the spin to be checked and the resulting steep dive converted into horizontal flight; the spin then became an ordinary manoeuvre.

Jones goes on to tell of the first flat spins, which occurred around 1919. Previously, spins had been steep in pitch attitude, with corresponding low stalled angles of attack of 25 to 35 degrees. On the other hand, the new flat spins had low pitch attitudes and high angles of attack, 45 degrees or higher, and high rotation rates. The flat spins were more dangerous than the early variety. An interesting speculation is that the invention of the parachute increased the number of survivors who could give reports of spins that had become uncontrollable, thus accounting for a seeming increase in the number of flat spins.

9.2   Advent of the Free-Spinning Wind Tunnels

NACA research on airplane spinning began about the year 1926 (Zimmerman, 1936). At first, dynamically scaled airplane models were launched from the top of a balloon shed and observed as they fell. Hartley A. Soulé and N. A. Scudder are associated with these early tests. Similar work went on in Britain at that time. A. V. Stephens writes (1966):

> The technique of using balsa wood [drop] models for spinning research had originated in America and had promptly been abandoned. It was taken up at the Royal Aircraft Establishment [RAE] by K. V. Wright and the author, who launched the models from the catwalk

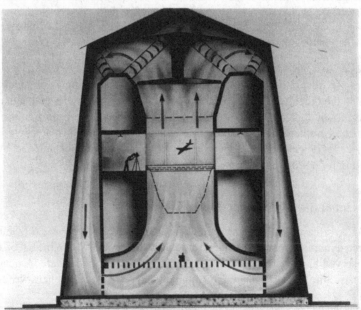

Figure 9.1 Exterior and cross-sectional views of the NASA Langley 20-Foot Free-Spinning Wind Tunnel, built in 1941. (From Neihouse, Klinar, and Scher, NASA Rept. R-57, 1960)

in the roof of the balloon shed and observed their fall. Just as the technique was beginning to yield useful results it was decided to demolish the balloon shed. At this point R. McKinnon Wood put forward a characteristically original suggestion that such experiments could be done in a large vertical wind tunnel. Pessimists argued that the models would immediately run into the tunnel walls, but it was soon demonstrated in a model tunnel that in fact they had a slight tendency to keep away.

Drop models were to make a comeback when helicopters became available, but in the early days, aside from the destruction of the RAE balloon shed, the limited test times in free drops from buildings were a severe handicap. McKinnon Wood's vertical wind tunnel got around the limited test time problem. The RAE model tunnel worked so well that Stephens had a 12-foot-diameter vertical tunnel built, which saw years of service.

The first NACA spin tunnel, built in 1935, had a 5-foot diameter. It was used by Millard J. Bamber and R. O. House, in addition to Charles Zimmerman. Force and moment measurements on a rotary balance could be taken in this little tunnel, in addition to free-spinning tests. However, when a new 15-foot free-spinning wind tunnel was opened in 1936, there was no rotary balance.

The current 20-foot NASA Spin Tunnel dates from 1941 (Figure 9.1). There were comparable vertical wind tunnels built at Wright Field, at IMFL in France, at TRDI in Japan, at the National Research Laboratories of Canada, and at the RAE in Britain. There is even a privately owned spin tunnel in the woods near Neuburg, Germany, operated by Bihrle Applied Research. The balsa and spruce spin models covered with silk tissue paper used in the early days have given way to sturdier vacuum-formed plastic and fiberglass scale models. However, a visitor to any of the modern spin tunnels is aware of a difference from the heavy industry feeling of an ordinary large wind tunnel. Models are launched from a balcony at the top of the tunnel structure and are recovered into nets strung around the bottom. Damaged models are frequently patched and reused.

The free-spinning wind tunnels are essentially analog computers. Their main use is for the study of developed spins and the recovery from developed spins. Models are launched into spin tunnels by hand, with a sort of Frisbee skimming motion. When the initial launch transient has died away, the model is expected to settle into the fully developed erect or inverted, steady or oscillatory, spin for which it is trimmed (Figure 9.2). A clockwork or radio-controlled mechanism applies preprogrammed spin recovery controls or deploys a spin chute. A chief result is the number of turns required before recovery, if there is a recovery, but other parameters are measured as well (Figure 9.3).

An interesting concept of "satisfactory recovery" emerged from the NACA spin tunnel experience. This accounts for the human factor by requiring that recovery into unstalled, straight flight takes no more than 2 1/4 turns after recovery controls are applied. The reasoning is that pilots cannot be expected to stay with what their handbook tells them is the correct recovery procedure after that many turns, but will try something else or leave the airplane if they have parachutes or ejection seats. An additional bow to the human factor in defining satisfactory recovery in the spin tunnel is to use no more than two-thirds of full control travel in the recovery sense.

Spin tunnels are valuable in that aerodynamic forces and moments are correctly represented at large values of angles of attack and sideslip and airplane angular velocities, except of course for scale and compressibility effects. This is no small consideration. Those researchers who try to avoid the use of spin tunnels because of scale or compressibility effects, or to supplement spin tunnels by calculating spinning motions on digital computers, face formidable data base problems, of which we will discuss more later. Where

Figure 9.2   Model spinning in the test section of the NASA 20-Foot Free-Spinning Wind Tunnel. (From Neihouse, Klinar, and Scher, NASA Rept. R-57, 1960)

spin entry characteristics are an important consideration or fully developed spins are not expected, dropped models are favored over spin tunnels.

9.3   Systematic Configuration Variations

Spin researchers recognized quite early the problems in forming engineering generalities for spinning airplanes. During all spin phases – the entry, the developed spin, and the recovery – airplanes operate in nonlinear ranges of angle of attack, control surface angle, and angular velocities, not to mention inertial moments. Nonlinear behavior means that generalizations require a large body of data obtained by systematic variations in design parameters.

Not long after the NACA 20-foot spin tunnel was put into operation, the veteran spin investigators Oscar Seidman and Anshal I. Neihouse began that process of systematic spin data collection. In a series of NACA Technical Notes and a technical report dating from 1937 to 1948, they reported on the effects of systematic wing, tail, relative density, and mass distribution changes on spin characteristics and recoveries.

With the coming of jet- and rocket-powered airplanes having long, slender, heavily loaded fuselages this group, now augmented by Walter J. Klinar and Stanley H. Scher, again picked up the problem of generalizing on spin characteristics by making systematic variations in design parameters. In the new series, the effects of mass distribution were again reviewed, but also the complex aerodynamics of long noses, strakes, and canards (Neihouse, Klinar, and Scher, 1960).

9.4   Design for Spin Recovery

Simple preliminary design rules that would increase the chances for an airplane to have satisfactory recovery characteristics from spins were an important product of the NACA spin tunnel group. Figure 9.4 reproduces the best-known set of preliminary design rules (Neihouse, Lichtenstein, and Pepoon, 1946). Two separate parameters are used. One

[Recovery attempted by full rudder reversal unless otherwise noted (recovery attempted from, and developed-spin data presented for, rudder-full-with spins)]

Airplane, Observation	Attitude, Erect	Direction, Right	Loading (see table II) No. 1		Normal loading
Slots, Closed	Flaps, Up		Center-of-gravity position, 25 percent c̅	Altitude, 18,000 ft	Clean condition

Model values converted to full scale   U-inner wing up   D-inner wing down

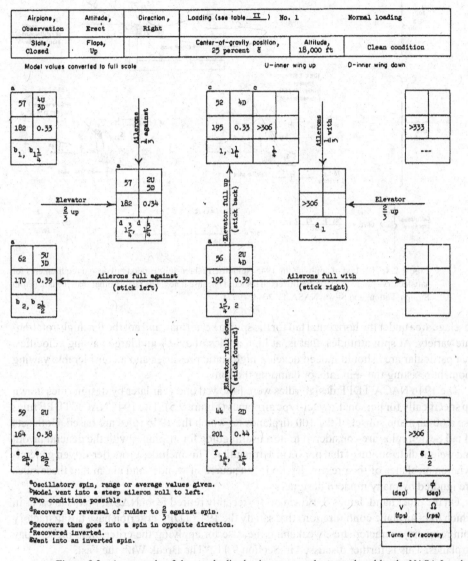

Figure 9.3 An example of the standardized spin recovery charts produced by the NASA Langley spin tunnel. The box locations correspond to control positions for the developed spin that precedes the recovery attempt. Blanks generally correspond to control positions for which the model would not spin. This particular chart is for the Grumman OV-1 Mohawk Army observation airplane. (From Lee, NASA TN D-1516, 1963)

is called TDR, the tail damping ratio, affecting whether the steady spin is steep or flat. The other is called URVC, the unshielded rudder volume coefficient, based on the rudder areas nominally out of the horizontal tail wake and their moment arms. The product of the two parameters is the tail damping power factor, or TDPF.

The 1946 TDPF design rules are a modification of a Royal Aircraft Establishment (RAE) criterion by E. Finn. Both the RAE and NACA criteria are based on empirical results, grounded in the flight mechanics of spins. For example, the TDR rule specifies a minimum

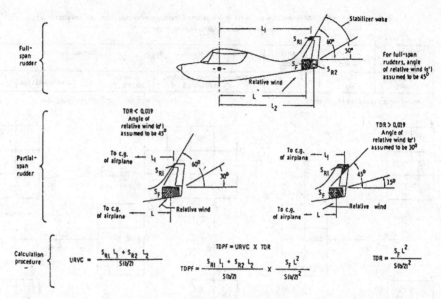

Figure 9.4 Method of applying the 1945 Neihouse/Lichtenstein/Pepoon tail design requirements for satisfactory spin recovery, a controversial standard because it neglects factors other than the tail. (From Stough, Patten, and Sliwa, NASA TP 2644, 1987)

fuselage area under the horizontal tail for the spin to be normal, and not the flat, high-rotation-rate variety. At spin attitudes, that is, at high angles of attack and large yawing velocities, that particular area should indeed develop high static pressures and a considerable yawing moment resisting the spin rate, or damping the spin.

The 1946 NACA TDPF design rules were followed one year later by design rules drawn up specifically for personal-owner-type airplanes (Figure 9.5). The 1947 NACA TDPF rules use a 60-airplane subset of the 100 airplanes on which the 1946 rules are based. Both sets of tail design rules are considered to be a useful guide for airplanes with the general layout and weight distribution of that period in aeronautics. This includes propeller-driven general-aviation airplanes of the present day, so it is a source of wonder and alarm that these rules are ignored by many modern designers.

On the other hand, James S. Bowman, Jr., recently retired from NASA, points to cases in which light airplane configurations that satisfy the 1947 TDPF criterion have unsatisfactory spin recovery characteristics, weakening the case for applying the criterion to present-day airplanes. This is further discussed in Section 9.11, "The Break With the Past."

9.5   Changing Spin Recovery Piloting Techniques

The 1916-era prescription for spin recovery, "thrust the control stick forward and apply rudder in the sense opposed to the rotation," held good for many years afterwards. It was not until airplane mass distributions changed appreciably from those of early airplanes that not only design criteria, but also pilot manipulations, changed for spin recovery.

One interesting change has to do with airplanes with heavy weights along their wings. While early four-engine bombers and transports are in this category, these airplanes are not expected to be spun, even unintentionally, and indeed could fail structurally in spins. Modern light twin airplanes fall somewhat into the wing-heavy weight distribution category, and appropriate spin recovery procedures are less academic. The key result is that down-elevator

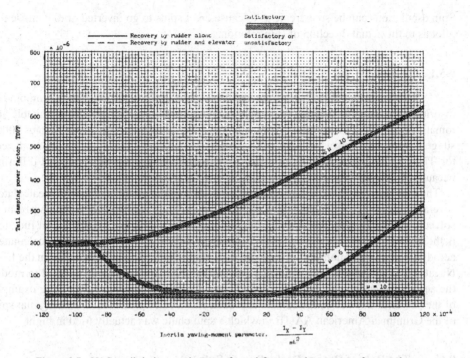

Figure 9.5 NACA tail design requirements for satisfactory spin recovery of personal-owner–type airplanes. The symbol μ is the airplane relative density factor, with typical values of 4.0 for the Cessna 150 and 8.6 for the Grumman/American Yankee. (From Neihouse, NACA TN 1329, 1947)

becomes of primary importance in spin recovery, reducing greatly the need for opposite rudder.

In at least one case, this finding was ignored in the design of a popular light twin airplane (Abzug, 1977). Nominal down-elevator travel of the Rockwell Aero Commander is small, at only 10 degrees. In addition, the elevator stops are located in the cockpit area, rather than at the control surface. This allows control cable stretch at spin attitudes and airspeeds to reduce available down-deflection to an estimated 3 degrees, almost certainly insufficient for spin recovery.

An equally dramatic and significant change in pilot spin recovery techniques is required by the opposite extreme in loadings, or heavy weights along the fuselage. For this loading, representative of all modern thin-wing jet fighters, aileron with the spin becomes the primary spin recovery control, supplemented by the usual rudder against the spin. One of this book's authors (Abzug) briefed the spin test pilot for the Douglas XF4D-1 Skyray at a time when the aileron-with recovery technique had just become known to engineers. Spin tunnel graphs and tables seemed to be getting nowhere in the pilot's ready room at Edwards Air Force Base. Then the pilot perked up. "I see," he said, "Ailerons with the spin will give anti-spin yawing moments because of adverse yaw. Adverse yaw must be large because of the low-aspect-ratio wing" [aspect ratio was just 2.0]. The matter was allowed to rest there, since as long as the controls were to be applied correctly, the misunderstanding would not matter.

Although ailerons with the spin is an accepted spin recovery technique for heavy fuselage loadings and modern power controls ensure that the controls can be applied, spin recovery problems with these airplanes are far from overcome. Heavy fuselage loadings also bring about oscillatory spins and the rather wild departure motions discussed in a later section.

Spin oscillations can be so extreme as to cause erect spins to go inverted or to confuse the pilot as to the actual direction of spin rotation.

9.5.1   *Automatic Spin Recovery*

The problem of recovery from oscillatory spins, where the pilot can be completely disoriented and unable to apply the proper recovery technique, suggests using suitable automatic controls once a spin is recognized. A candidate technique (Lee and Nagati, 2000) suggests applying controls in a direction to cancel the vehicle's total angular momentum vector. The angular momentum vector is usually close to the angular velocity vector, differing because of unequal moments and products of inertia about body axes.

The Lee-Nagati approach is in two parts. First, the angular momentum vector is calculated several times per second during the spin. Then, at each interval, a minimization problem is solved, finding the control surface angles that minimize a cost function. The cost function is the difference between the vehicle's momentum vector and the negative of a calculated aerodynamic control moment vector. Although not fundamental to the concept, in the Lee-Nagati paper a sophisticated regression parameter identification scheme is used to model the aerodynamic control moments used to form the control vector. One striking example of the power of this automatic control approach was a calculated recovery from a flat spin of the Grumman/American AA-1B in which a spin chute was actually used in flight.

9.6   The Role of Rotary Derivatives in Spins

The rotary derivatives are the force and moment coefficient derivatives with respect to dimensionless angular velocity. The rotary derivatives appear in the airplane equations of motion for normal unstalled flight, as well as for spinning flight. However, at the relatively low airspeeds and high angular velocities for spinning flight, the rotary derivatives are much more important than they are for unstalled flight. Physically, under spinning conditions there will be large differences in local flow angles of attack at different parts of the airplane, and possibly local separated flows.

Stated otherwise, the rotary derivatives are generally of secondary importance to flight simulation and flight control design for normal unstalled flight. If the airplane has stability augmentation systems that drive the control surfaces to provide artificial damping, this is even more true; artificial damping swamps out the rotary derivatives that supply natural aerodynamic damping. Thus it is that, at least in modern times, the drive to refine analytic and measurement techniques for the rotary derivatives has come from spinning studies.

The early 1950s saw a rush of 5- and 6-degree-of-freedom inertial coupling computer simulations, as told in Chapter 8, "The Discovery of Inertial Coupling." It is interesting that some of the same investigators, such as Cecil V. Carter, John H. Wykes, and Leo Celniker, who helped crack inertial coupling with their simulations moved on to spin simulation using analog or digital computers. The motivation was there, because the same airplane loading characteristics that lead to inertial coupling also lead to post-stall gyrations and departures, motions not easily studied in free-spinning wind tunnels.

The problem was that this period coincided with a shutdown of rotary balance testing at NACA. The NACA rotary balance was updated in the late 1950s, but it was not used for analytical studies until several years had passed. Thus, the spin computer analysis results reported at the 1957 Wright Air Development Center Airplane Spin Symposium (Westbrook and Doetsch, 1957), made without the benefit of current rotary balance data, came under criticism for using inadequate rotary derivatives by knowledgeable people such as Dr. Irving C. Statler and Ronald F. Sohn.

9.7   Rotary Balances and the Steady Spin

Rotary balances are designed to extract rotary derivatives from wind-tunnel tests. The model is typically held at some fixed angles of attack and sideslip to the relative wind and rotated by an electric motor at a fixed rate (Figure 9.6). Combined aerodynamic, gravity, and inertial forces and moments are measured by a six-component balance internal to the model. The desired aerodynamic forces and moments are obtained by subtracting the other components, as tares. Rotary balance tests in which angles of attack and sideslip remain constant are called "coning tests." The spinning axis is aligned with the tunnel flow direction for coning tests.

Rotary balance testing actually predates the free-spinning tunnel, with E. F. Relf and T. Lavender's 1922 and 1925 measurements in Britain. An earlier paper (Relf and Lavender,

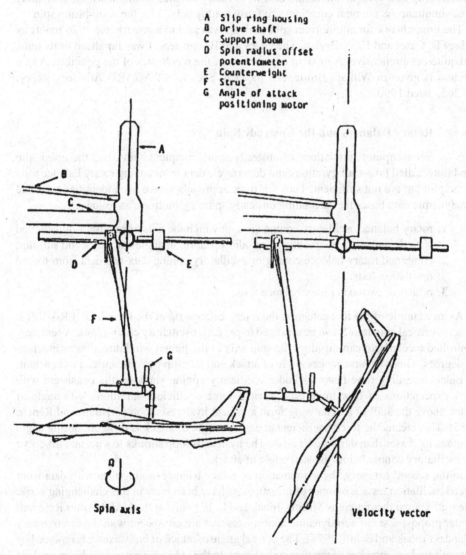

A Slip ring housing
B Drive shaft
C Support boom
D Spin radius offset
   potentiometer
E Counterweight
F Strut
G Angle of attack
   positioning motor

Spin axis

Velocity vector

Figure 9.6   The NASA Langley Research Center's current spin tunnel rotary balance rig. The electric motor that spins the model is outside the tunnel. At the left, a model in a flat-spinning orientation; a normal spin orientation is at the right. This is believed to be the only rotary balance in current (early 1990s) use at Langley. (From Bihrle and Chambers, AGARD AR 265, 1990)

1918) described some of the first tests on autorotating wings. Autorotation is wing negative damping in roll, at angles of attack beyond the stall. Autorotation provides a driving or propelling moment in spins.

Until the coming of the jet airplane and the oscillatory spin, the chief use of rotary balance testing was in finding the steady spin modes of an airplane. That is, would spins be steep, or easily recoverable, or fast and flat, with problematic recoveries? The pioneering rotary balance work of this type was done by P. H. Allwork using the NPL 7-foot wind tunnel in Britain, and Millard Bamber and Charles Zimmerman in the NACA 5-foot vertical wind tunnel.

With simplifying assumptions, the three force equations of the ordinary 6-degree-of-freedom equations of airplane motion reduce to only two equations, which are not simultaneous with the three moment equations. Under steady conditions, angular accelerations drop out. With aerodynamic data from rotary balance coning tests, the remaining three simultaneous moment equations are fairly readily solved for the equilibrium spin.

The groundwork for equilibrium spin analysis was laid in a remarkable 1926 report by Sidney B. Gates and L. W. Bryant. The Gates and Bryant report was far ahead of its time, and quite comprehensive. A modern explanation of the mechanics of the equilibrium spin solution is given in William Bihrle, Jr.'s paper in Sec. 9.1 of AGARD Advisory Report No.265, dated 1990.

9.8   Rotary Balances and the Unsteady Spin

For computer simulations of unsteady spins, incipient spins, and the quasi-spin conditions called post-stall gyrations and departures, data from coning rotary balance tests are helpful but are not sufficient. Thus far, three approaches have been identified to create aerodynamic data bases for calculating unsteady spinning motions, as follows:

1. rotary balance oscillatory coning tests, in which the axis of rotation is misaligned with the tunnel flow, creating a periodic variation in angles of attack and sideslip;
2. combined rotary balance coning or oscillatory coning data and data from forced oscillation tests;
3. orbital or two-axis rotary balance tests.

As an example of the first category, the rotary balance rig at the French ONERA-IMFL 4-meter vertical spin tunnel can be arranged to produce oscillatory coning tests. A remotely controlled mechanism can misalign the spin axis to the tunnel wind direction as much as 20 degrees. This of course makes angle of attack and sideslip periodic instead of constant.

Balance reading time histories under oscillatory coning show results consistent with one's expectations of flow hysteresis. Normal force coefficient variations with angle of attack above the stall of a delta wing form a typical hysteresis loop (Tristrant and Renier, 1985). This means the force coefficient at a given angle of attack is different during angle of attack increases than during decreases. The hysteresis loop shrinks to a normal lift curve for oscillatory coning below the stall angle of attack.

In the second category, the combination of rotary balance coning data with data from forced oscillation tests, a number of investigators have been busy in this challenging work. The well-known theoreticians Murray Tobak and L. B. Schiff at the NASA Ames Research Center propose a set of aerodynamic coordinates that are consistent with data from rotary balances (Tobak and Schiff, 1976). The normal angle of attack of body axes α is replaced by a "total" angle of attack σ of the longitudinal axis to the velocity vector. A sideslip angle is defined by the airplane's roll angle with respect to the plane in which σ is measured. Force and moment coefficients are expanded into series in which each term is identified with a characteristic rotary balance coning motion or ordinary forced oscillation.

Similar schemes have been devised by Juri Kalviste (1978) at Northrop Aircraft and by Martin E. Byers (1995) in Canada. Kalviste projects the airplane's total angular velocity vector onto the coning axis, about which rotary balance data are taken, and the three body axes, for which oscillatory data or estimations are available. A special algorithm is used to reduce the number of components from four to three. The algorithm selects components that are close angularly to the total angular velocity vector. This is intended to avoid using aerodynamic data formed by the differences of large numbers.

The third category of data base formation for computer simulation of unsteady spins, the use of orbital or two-axis rotary balances, is at the time of writing only a concept. In orbital rotary balance testing, coning motions would be superimposed on circular pitching and yawing at a different rate. This would yield small-amplitude angle of attack and sideslip perturbations about large fixed mean values of angle of attack and sideslip in a rotary flow. Practical difficulties appear to be formidable. Two-axis rotary rigs would have to be small enough for wind-tunnel installations and yet have good rigidity.

9.9   Parameter Estimation Methods for Spins

The use of rotary balances of ever-increasing complexity for measuring aero-dynamic forces and moments in spins is avoided if aerodynamic forces and moments can be inferred directly from free–spinning model or airplane tests. Two promising approaches to this application of parameter estimation have been reported.

The first approach (Fremaux, 1995) extends the Gates-Bryant equilibrium or steady-state spin analysis to include the nonequilibrium angular acceleration terms $\dot{p}$, $\dot{q}$, and $\dot{r}$ and the spin acceleration term $\dot{\Omega}$. Calculated aerodynamic moments by this method vary with time if the spin is oscillatory. The calculated moments oscillate about the Gates-Bryant values, which also can be measured independently on a rotary balance. This method requires the investigator to record rapid angular motions in a spin, feasible now with the advent of modern data-acquisition techniques.

A second parameter estimation approach for spins (Jaramillo and Nagati, 1995) appears to have been inspired by the finite-element methods used in structural analysis. A set of control points are established. Aerodynamic force coefficients at these points are correlated with local angles of attack and sideslip during spinning motions. These aerodynamic force (and moment) coefficients are in effect influence coefficients. The influence coefficients are found by minimizing cost functions based on the errors between measured vehicle accelerations and those calculated using forces and moments derived from the influence coefficients. Once the dimensionless influence coefficients are found, the method appears to have predictive capabilities. An improved version (Lee and Nagati, 1999) of the original method reduces the number of unknown parameters to be solved for by using static wind-tunnel test data.

9.10   The Case of the Grumman/American AA-1B

The Grumman/American AA-1B Yankee, its Tr-2 trainer version, and the Tiger four-seat variant are clever, innovative personal airplane designs. Compared with most airplanes of the type, which are built as riveted metal structures, metal-to-metal bonding on these airplanes eliminates drag due to rivets and skin waviness and the points of stress concentration common to riveted structures. More than 2,000 AA-1B's have been built, under several designations. Yet the AA-1B has compiled a sad record of crashes as a result of unrecoverable spins. The AA-1B has a flat spin mode that leads to a high-impact vertical crash with the fuselage level, a crash that has made paraplegics out of student pilots and their instructors.

The airplane's three-view diagram suggests that NACA spin recovery criteria were quite disregarded in the original design. The horizontal tail is mounted low on the fuselage, providing very little tail damping ratio, or TDR (Figure 9.7). By NACA correlation, this would promote a high angle of attack, or flat spin. Also, there appears to be virtually no unshielded rudder area, the NACA URVC factor.

The AA-1B's poor spin and recovery characteristics are recognized in the 1975 version of the owner's manual. In both the operating procedures and operating limitations sections

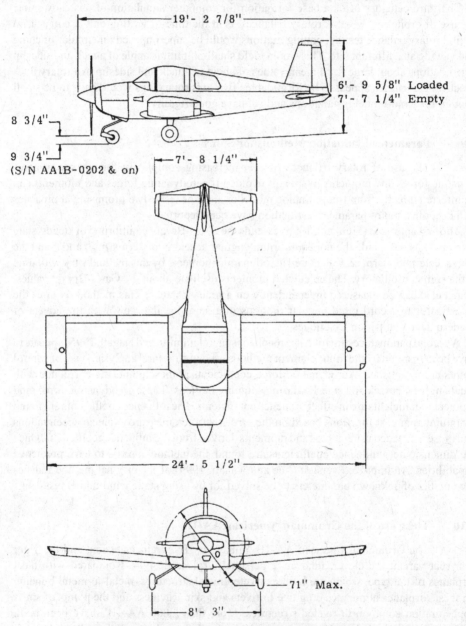

Figure 9.7 Grumman-American AA-1B, known for unrecoverable flat spins. There is very little fuselage side area under the horizontal tail, needed for spin damping according to the NASA TDR criterion. There is also little unshielded rudder area above the horizontal tail. (From AA-1B *Owner's Manual*)

the legend "SPINS ARE PROHIBITED" is displayed. This is followed by the recovery technique in the event of inadvertent spins, and the note:

> If recovery controls are not briskly applied in the first turn, more than one additional turn will be required for recovery. For quick recovery, apply full anti-spin controls as the spin begins, before one turn is completed.

A later version of the owner's manual has wording that reflects actual NASA flight-testing experience:

> There is evidence that permitting the airplane to go beyond one turn without initiating proper recovery procedures can allow a spin mode to develop from which recovery is not possible.

To illustrate this point, the 1/15 December 1990 issue of *Aviation Consumer* reports that American Aviation test pilot Bob Hommel was forced to jump when a modified AA-1A failed to recover during a spin test. The airplane was said to have completed over 100 turns before crashing. There is no reason to think that the AA-1B airplane is unique in having recovery problems in spins that go beyond one turn. James S. Bowman, Jr., writes as follows:

> I think it is important to mention that all normal category airplanes are tested for one-turn spins only and if taken beyond one turn, recovery may be slow, or there may be no recovery at all.

9.11 The Break with the Past

The 1947 NACA tail design requirements for satisfactory spin recovery stood relatively unchallenged until a series of NASA spin tunnel tests and some experiments at the Cessna Company in the late 1970s. Motivated somewhat by the Grumman/American AA-1B Yankee experience, NASA started a broad-based review of light airplane spin recovery. W. H. Phillips credits Joseph R. Chambers with initiating this work. The centerpiece of the program was a flight test fleet of four airplanes: a Cessna 172 Skyhawk, a modified Beech C23 Sundowner, a nonproduction Piper PA-28R T-tail Arrow, and a modified Yankee. Initial results from the review represent a distinct break with past NACA work, in particular, the 1947 TDPF tail design criterion. Nine tail configurations were tested on a model of the Yankee in the 20-foot Langley Spin Tunnel. Six of the nine designs were predicted to have satisfactory spin recovery characteristics according the 1947 TDPF criterion, yet only four showed satisfactory recovery in the spin tunnel (Burk, Bowman, and White, 1977). The investigators concluded:

> On the basis of the results of the present investigation, the tail design criterion for light airplanes, which uses the tail damping factor (TDPF) as a parameter, cannot be used to predict spin recovery characteristics.

According to Burk, Bowman, and White, TDPF was intended to serve only as a conservative guideline for tail design, not as a criterion. Having made this decisive break with 30 years of stability and control design practice, the statement is softened somewhat in words that followed those quoted above, as follows:

> However, certain principles implicit in the criterion are still valid and should be considered when designing a tail configuration for spin recovery. It is important to provide as much damping to the spin as possible (area under the horizontal tail), and it is especially important

to provide as much exposed rudder area at spinning attitudes (unshielded rudder volume coefficient (URVC)) in order to provide a large antispin moment for recovery.

The real thrust of the NASA review of the 1970s lies in the investigation of factors for light-plane spin recovery other than tail design. The NASA and contractor investigators, including H. Paul Stough III, William Bihrle, Jr., James M. Patton, Jr., Steven M. Sliwa, Joseph Chambers, and Billy Barnhart, found that wing and aft fuselage design details affected the results in ways that cannot be ignored. According to John C. Gibson, British spin tests in the 1930s had already disclosed the importance of rear fuselage design.

The evidence on fuselage aft details is not completely clear, because it is bound up in scale effects, or Reynolds number. Side forces, contributing to damping, of square or rectangular fuselage cross-sections appear to be particularly sensitive to Reynolds number. Thus, results from small-scale spin model tests that pin flat, unrecoverable spins to flat-bottomed rear fuselages (Beaurain, 1977) must be considered only tentative. On the other hand, the recent NASA findings on wing design effects on spins are conclusive and important, as detailed in a following section.

Having seen the NASA spin experts make a decisive break with the past, represented by NACA 1946 and 1947 tail design criteria, what advice can one give to designers of new general-aviation airplanes? Well-funded military programs present no problem, since modern spin testing techniques, such as drop models and rotary-balance tests, that are recommended by NASA are available to them. The concern is with light-airplane designers who have been cast adrift, so to speak, with NASA's abandonment of the TDPF design criteria.

The most reasonable course to take for designers of new light airplanes who have no budget for extensive spin model testing probably is as follows:

1. Follow the 1947 TDPF criteria. The evidence is that the criteria deal with the right design details, even if the numerical values are incorrect in some cases because of the influence of other parameters.
2. Avoid the design details that are implicated in flat, unrecoverable spins: flat-bottomed rear fuselages and wings with full-span leading-edge droop.
3. Design the outer wing panels to be able to accommodate a drooped leading edge, if spin problems appear during flight test.
4. Check with NASA on the possibility of doing spin tunnel, rotary balance, or model drop tests for the new design. NASA is able to consider such tests if the results would be of general scientific interest, covering new ground.

9.12   Effects of Wing Design on Spin Entry and Recovery

Modified tail arrangements did not really accomplish much for the Yankee, and NASA turned its attention to modifications of the wing outer panel. In so doing, a line of research was reopened that had been followed at NACA by the versatile Fred E. Weick and Carl J. Wenzinger in the 1930s. There was also the work by R. A. Kroeger of the University of Michigan and T. W. Feistel of NASA Ames Research Center in 1975, in which the concept of a segmented or discontinuous wing leading edge was developed to control stall progression and minimize loss in roll damping at the stall. This type of wing leading edge had been seen previously on the McDonnell F-4 Phantom II.

The best NASA wing leading-edge extensions tested on general-aviation airplanes also have sharp discontinuities at their inboard ends (Figure 9.8). The sharp discontinuities evidently trigger vortices that slow the spread of inboard wing stalling to the outer wing panels.

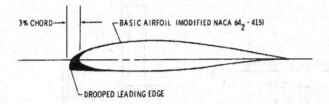

Airfoil modification

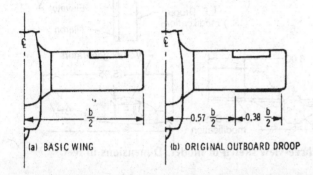

Figure 9.8 Discontinuous wing leading-edge extensions tested by NASA on the modified Grumman/American AA-1B Yankee (*above*) and the modified Piper PA-28R Arrow (*below*). (From DiCarlo AIAA Paper 80-1843 and Jane's *All the World's Aircraft* (1987–1988))

For each installation, there appears to be an optimum location for the droop discontinuity, to delay spin entry and improve spin recovery. Full-span leading-edge droop was actually a detriment on the Yankee, causing easily entered flat spins where there had been none before.

The discontinuous outboard wing leading-edge droop modification provides generally good spin resistance and spin recovery characteristics on a number of different

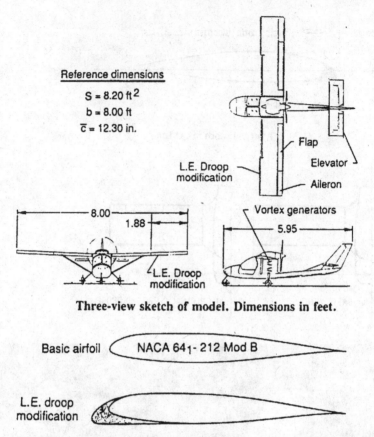

Figure 9.9   Wing leading-edge droop tested by NASA on a 1/4-scale model of the DeVore Aviation trainer. The modification eliminates an abrupt, uncontrollable roll departure at the stall. (From Yip, Ross, and Robelen, *Jour. of Aircraft*, 1992)

configurations. These include the modified Grumman/American AA-1B Yankee; the modified Piper T-Tail Arrow; and models of the Smith Aviation Trainer, the Questair Venture, and the DeVore Aviation Corporation Trainer, an ultralight pusher (Figure 9.9).

9.13   Drop and Radio-Controlled Model Testing

A major disadvantage of free-spinning wind-tunnel model testing is that the Frisbee-throw method of launching the model into the tunnel prevents study of real-life stalls and spin entries. Also, model size and Reynolds number are limited by the wind-tunnel size. Both of these problems with spin tunnels are partially eliminated for drop or radio-controlled models. For spin research, NASA drops models from helicopters and airplanes and operates radio-controlled dynamically scaled models that resemble superficially those flown by hobbyists.

For example, a 1/12-scale drop model of a typical fighter airplane would weigh about 300 pounds. The NASA Langley helicopter drop model work is led by Charles E. Libbey. Both radio-controlled and helicopter drop model testing go on at the NASA Plum Tree test site, near Poquoson, Virginia.

9.14   Remotely Piloted Spin Model Testing

Perhaps the closest that one can get to testing a new airplane configuration for spin entry and recovery conditions in advance of the real thing is the remotely piloted drop model technique originated by Euclid C. Holleman and other NASA staff people at the Dryden Flight Research Center. A test pilot controls the drop model from a ground-based simulator-like isolated cockpit that has instrument displays run by telemetered data. The airplane's flight control system is simulated on a ground computer and commands are up-linked to the model.

Considering that the setup resembles a space vehicle control center, design and operating expenses must be high compared with alternative spin model testing methods. However, the method efficiently produces good flight test data in a short time. Unplanned angles of attack and sideslip reached in remotely piloted tests of a 3/8-scale model of the McDonnell Douglas F-15 fighter might have resulted in loss of control for a less forgiving airplane.

On the negative side, pilots report that the absence of motion cues is a serious drawback for spin testing and might lead to overly conservative results. Holleman notes that before the actual drop tests pilots practice the flight plan on fixed-base ground simulators. Speeding up the pace for the fixed-base training sessions by a factor of 1.4 prepares the pilot for the actual flight.

9.15   Criteria for Departure Resistance

The word "departure" is used for uncontrolled flight following stalls, the first stages of spin entry. Rather violent departures appeared with the advent of swept wings and long, heavily loaded fuselages, the same features that lead to inertial coupling. Pilots reported nose slices, whirling motions, wing rock, and roll reversals and divergences. There was interest almost at once in finding aerodynamic parameters specifically tied to these anomalies.

The initial search led back to stability and control's early days, a time of high interest in Routh's criterion and stability boundaries. Lateral stability boundaries are formed from the lateral characteristic equation

$$\lambda^4 + B\lambda^3 + C\lambda^2 + D\lambda + E = 0.$$

A necessary condition for stability is that the constants B, C, D, and E all be positive in sign. The last two constants, D and E, are associated with real roots, or convergences or divergences, rather than an oscillation.

In the lateral stability boundaries developed by L. W. Bryant in 1932 and by Charles H. Zimmerman in 1937, sign changes in D and E plotted as functions of static lateral and directional stability define the boundaries. Zimmerman specifically associated the constant D with directional divergence.

Two later investigators, Martin T. Moul and John W. Paulson, followed in Bryant's and Zimmerman's footsteps, but associated directional divergence with the constant C rather than D. Robert Weissman's name is also associated with this development (Figure 9.10). Moul and Paulson coined a new term, "dynamic directional stability," or $C_{n_{\beta\mathrm{dyn}}}$ for an approximation to C. This approximation is

$$C_{n_{\beta\mathrm{dyn}}} = C_{n_\beta} \cos\alpha - (Iz/Ix)C_{l_\beta} \sin\alpha.$$

Together with another factor called "lateral control departure parameter" or LCDP, departures and post-stall gyrations, as well as roll reversals and tendencies to spin, are

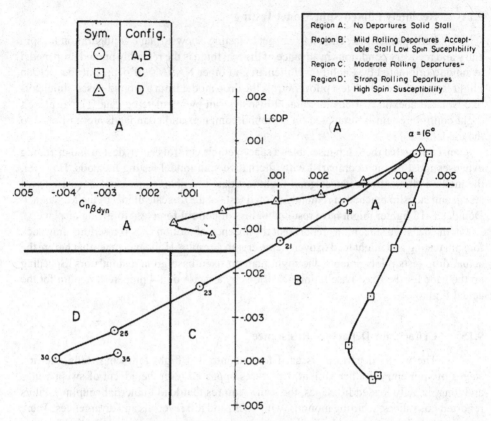

Figure 9.10 The 1972 Weissman spin susceptibility and departure boundaries. The curved lines are parameter values for the McDonnell Douglas F-4J and two variants (C and D) with improved departure resistance. (From Mitchell and Johnson, AFWAL-TR-80-3141, 1980)

correlated with stability derivatives. LCDP in particular, developed by Pinsker (1967), predicts high angle of attack nose slice departures while the pilot attempts to hold the wings level.

The pioneering work of Moul, Paulson, Pinsker, and Weissman in revisiting the Bryant-Zimmerman concept of stability boundaries for departures and post-stall gyrations was followed by several other noted investigators, who developed additional criteria for departure resistance. Figure 9.11 shows another set of departure and roll-reversal boundaries that are based completely on static aerodynamic derivatives (Bihrle and Barnhart, 1978). The advantage of using static derivatives in setting the boundaries is that they can be applied on the spot during conventional low-speed wind-tunnel tests of a complete model.

The Bihrle boundaries are based on digital simulation of a canned full control deflection maneuver involving an initial steep turn followed by full nose-up pitch control, then full opposite roll control. In papers given in 1978 and 1989, Juri Kalviste and Bob Eller proposed coupled static and dynamic stability parameters that are based on separation of the full airplane equations of motion into rotary and translatory sets. These parameters amount to generalizations of the Moul-Paulson $C_{n_{\beta dyn}}$ and LCDP parameters.

The level of sophistication of this work was raised by use of pilot-in-the-loop considerations, which are treated more fully in Chapter 21, "Flying Qualities Research Moves with the Times." In the period 1976 to 1980, David G. Mitchell and Donald E. Johnston correlated

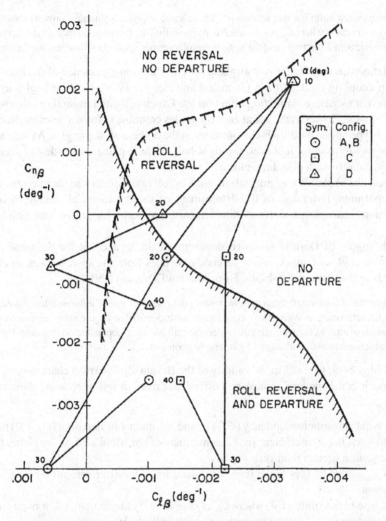

Figure 9.11 The simplified 1978 Bihrle roll reversal and departure boundaries for the case of adverse yawing moment due to aileron deflection. These simplified boundaries involve static parameters only and can be applied during a routine wind-tunnel test. In configuration A the circles apply to the unmodified McDonnell Douglas F-4J and the other symbols are for F-4J variants. (From Mitchell and Johnston, AFWAL-TR-80-3141, 1980)

some departure characteristics with airplane frequency-response parameters involved with pilot loop closures. Negative values of the lateral transfer function numerator term $N\delta_a$ causes roll reversal, for example. Mitchell and Johnston found two additional closed-loop departure parameters for either rudder or aileron maneuvering control that correlate with nose-slice departures.

The vexing question of just how far one can go in using departure criteria based on linearized aerodynamics, or the stability derivatives, is addressed in a paper delivered by Donald Johnston at a 1978 AGARD Symposium on Dynamic Stability Parameters. Although the linearized parameters clearly have some predictive value, Johnston concludes:

> the more common frequency domain linear analysis techniques applied to symmetric, frozen point airframe models may produce totally erroneous answers if the aircraft exhibits

significant coupling due to sideslip. These same analytic techniques provide valid predictions in cases where C_m, C_l, C_n are $f(\alpha, \beta)$ providing the frozen point model represents asymmetric trim conditions and the analytic results are not applied to β excursions through zero.

Calculations on the Vought A-7 airplane provide additional evidence of the importance of sideslip coupling on departures (Johnston and Hogge, 1976). Sideslip angles up to 15 degrees have a moderate stabilizing effect on the Dutch roll and longitudinal short-period modes of motion. However, the spiral and roll modes combine to form a new low-frequency or phugoid-type lateral mode, which is unstable at the higher sideslip angles. A pitch attitude numerator factor, or zero, moves to the right half of the s-plane with sideslip, creating a potential instability with pilot loop closure.

The presence of stability augmentation, such as roll rate feedback to the ailerons, clearly affects departures. Extension of the determinant of the equations of motion to include augmentation feedbacks produces modified departure criteria (Lutze, Durham, and Mason, 1996).

Peter Mangold of Dornier reviewed departure again, to account for the trend toward higher usable angles of attack, where forebody vortices from the nose, strakes, or canards dominate lateral-directional stability. From Mangold's paper (1991):

> For the older aircraft the dynamic [rotary] data were of minor influence and the departure characteristics in Weissman's correlation were dominated by the static derivatives. High angle of attack characteristics of modern aircraft are more dependent on dynamic derivatives which are heavily influenced by forebody geometry.

Mangold goes on to reaffirm the validity of the Bryant-Zimmerman characteristic equation approach in the modern context and offers four rules to follow to avoid departure, as follows:

1. Avoid autorotation tendency ($C_{l_p} < 0$) and maintain yaw damping ($C_{n_r} < 0$) in order to keep the B coefficient [not Zimmerman's D or Moul's C] of the characteristic equation greater than zero.
2. $C_{n_{\beta\text{dyn}}}$ must be kept larger than zero, since this parameter determines the C coefficient.
3. Close to maximum lift, where C_{l_r} is considerably larger than 1.0, it is essential to maintain $C_{n_{\beta\text{dyn}}} > 0$ with negative C_{l_β} and only slightly positive C_{n_β}.
4. Nonlinearities and hysteresis effects versus sideslip have to be avoided.

The interesting reason for the fourth rule is the inaccuracy of sideslip sensors at high angles of attack, which makes it difficult to schedule control laws to cope with the problems of nonlinearities and hysteresis.

Recognizing the ingenuity and skill of the developers of departure parameters, it is still possible to question the place of these parameters in modern design practice. Departure parameters such as $C_{n_{\beta\text{dyn}}}$ and LCDP may remain of interest in preliminary design. However, in a sense, departure parameter research is working behind the curve of modern computer development. Designers responsible for the stability and control of expensive new airplanes will recognize the essential nonlinearity of the departure and spin problems and plan exhaustive digital simulations to explore the full envelope of flight and loading conditions and control inputs. Automated analysis can stack cases on cases and then winnow out those of no interest by algorithms that scan the results.

There is still the major effort required to build aerodynamic data bases for such enterprises. Furthermore, such efforts would presumably evaluate candidate stability or control augmentation schemes, as well. Augmentation schemes for fly-by-wire systems, such as used on the Tornado, can produce care-free airplanes that cannot be made to depart under any pilot action. Departure parameters would be of interest after the fact, to help make sense of the results and as a guide to the future.

9.16   Vortex Effects and Self-Induced Wing Rock

Self-induced wing rock on slender delta wings was observed first at NASA's Langley Research Center in the late 1940s, in the free-flight wind tunnel. Wing rock appeared as a limit cycle, or undamped, roll oscillation at angles of attack below the stall. We know now that wing rock is typically associated with separated flows and time-dependent effects.

Because of interest in both supersonic transports and reentry vehicles, research activities into wing rock continued in both the United States and the United Kingdom (Ross, 1988). Attention turned later to combat airplanes, where wing rock was thought to have contributed to loss of control in high angle of attack maneuvers. Attempts to alleviate wing rock by stability augmentation have been successful, as in the case of the Grumman X-29A research airplane (Clarke, 1996).

However, attempts to correct the problem aerodynamically have been less successful because of the complex flow mechanisms involved. On the X-29A, the driving mechanism for wing rock was determined to be the interaction of vortices from the forebody with other components, as was the case for the Northrop F-5, at low airspeeds. On the other hand, a high-airspeed wing rock of the F-5 was driven by shock-induced separation on the wing. Some measurements indicate asymmetry of the vortices shed from wing leading edges as driving the motion, with vortex breakdown limiting the motion's amplitude (Ericsson, 1993). Dr. Ericsson is a leading expert and prolific author on the effects of unsteady flows on stability and control.

It turns out that self-induced wing rock can also occur on very low-aspect-ratio rectangular wings, caused by vortices shed from the side edges. This is interesting but academic, since rectangular wings of aspect ratio less than 0.5 have never been considered for actual airplanes or missiles. What is decidedly not academic is the role of highly swept wing leading-edge extensions, or LEXs, in wing rock and other undesirable behavior. Wing leading-edge extensions were pioneered on the Northrop YF-17 and F-5 airplanes to increase maximum lift and reduce drag at high lift, by vortex interactions with the main wing surface. Wing leading-edge extensions have gone on to be used on many other modern fighter airplanes, such as the F/A-18. The highly swept side inlets of the Russian MiG-25 airplane act as leading-edge extensions, developing vortices at high angles of attack.

Wing rock is often studied in wind-tunnel tests in which a model is mounted on low-friction roll bearings and is free to roll. Forced roll oscillations can also reveal the wing rock tendency by regions of negative roll damping or positive signs of the rotary derivative C_{l_p}. A comparison between wing rock amplitude measurements on a free-to-roll model F-18 in a wind tunnel and on the F/A-18 HARV (High Angle of Attack Research Vehicle) in flight shows good agreement (Nelson and Arena, 1992). There was also a fair correlation for (reduced) frequency between wind-tunnel and flight testing. This supports the notion that flow conditions during flight wing rock are close to the single-degree-of-freedom wind-tunnel conditions.

Contradicting the reasonable correlation of F/A-18 HARV single-degree-of-freedom wind-tunnel tests and flight test measurements of wing rock is the finding on the F-4, Tornado, and RAE HIRM (High Incidence Research Model) that wing rock occurs at frequencies close to those of the classical Dutch roll. This is unusual since wing rock is a nonlinear limit cycle, while the Dutch roll is consistent with linearized equations of motion and requires roll, sideslip, and yaw degrees of freedom.

In addition to the wing rock phenomenon, vortex bursting at high angles of attack has undesirable effects, such as loss in lift and negative dihedral effects. Vortex bursting is associated with leading-edge sweep angles less than about 75 degrees. The interactions of wing leading-edge vortices with other airplane components of modern fighter airplanes is covered in a comprehensive paper by Andrew M. Skow and G. E. Ericson (1982). A more recent review was provided in 1992 by John E. Lamar, published in AGARD Report 783.

Lamar finds that the leading-edge suction analogy, first proposed by Edward C. Polhamus in 1966, provides a powerful tool for estimating aerodynamic forces and moments for sharply swept wings with vortex flows. It appears that the leading-edge suction force in attached flow is reoriented in the direction of the rotating vortex, when the vortex forms. Typically, lift and moment terms using the analogy are added to linear aerodynamic and vortex lattice computer codes.

9.17   Bifurcation Theory

Bifurcation theory is a classical analysis method in the study of nonlinear differential equations. Bifurcations are said to occur when nonlinear dynamic systems undergo changes in qualitative behavior. In bifurcation analysis, system steady states, or solutions with all time derivatives set equal to zero, and the stability about those steady states are calculated. The steady states are continuous functions of control surface angles. A bifurcation occurs when stability changes from one steady state to the next as a system parameter, such as control surface angle, is varied (Figure 9.12). A particular type of bifurcation known as the Hopf can lead to periodic motions such as wing rock.

A number of investigators, led initially in 1982 by J. V. Carroll and R. K. Mehra, have used bifurcation theory in the study of nonlinear airplane motions, including wing rock and spins (Jahnke and Culick, 1994). P. Guicheteau in France extended the wing rock application to include unsteady aerodynamic effects, and ONERA ran German-French Alpha Jet flight tests to compare with his theory. Drs. J. B. Planeaux, Jahnke, and Culick have studied bifurcations in the United States. Also, a nonlinear analogy to linear indicial response methods has been proposed for understanding the response singularities that appear at large angles of attack and sideslip, and large rolling velocities (Tobak, Chapman, and Schiff, 1984).

Bifurcation analysis, in conjunction with piloted simulation, has been recognized as a potential aid in flight test planning (Lowenberg and Patel, 2000). This approach was experimented with using the aerodynamic and mass characteristics of the R.A.E. High Incidence Research [drop] Model, or HIRM, in bifurcation analysis and simulation in the DERA Advanced Flight Simulator, or AFS. The experimenters concluded that simulation validated the nonlinear characteristics predicted by bifurcation analysis. Thus, bifurcation analysis may be used to good effect in planning simulation and flight test programs.

9.18   Departures in Modern Fighters

In spite of the best efforts of designers to apply the departure research lessons of Moul, Paulson, Pinsker, Weissman, and others, fighter airplanes of the F-14, F-15, F-16, and F-18 generation have departure problems. The situation was summarized by the high angle of attack researcher and test pilots, NASA veterans Seth B. Anderson and Einar K. Enevoldson, and the young NASA Langley engineer, Luat T. Nguyen (1983). This is what they reported for specific airplanes:

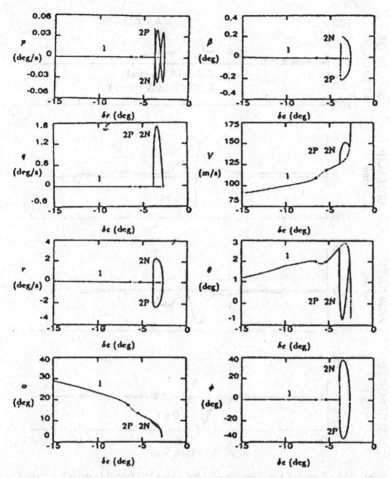

Figure 9.12 System steady states plotted as a function of stabilizer angle for the Grumman F-14A Tomcat. The solid curves, for stabilizer angles more negative (trailing edge up) than −7 degrees, indicate stable trim conditions. Dashed curves for stabilizer angles between −5.4 and −6.7 degrees represent unstable trim points and those between two Hopf bifurcations are represented by the small dots. (From Jahnke and Culick, *Jour. of Aircraft*, 1994)

Grumman F-14A Mild directional divergence and roll reversal start at an angle of attack of 15 degrees. Divergent wing rock and yaw excursions occur at an angle of attack of 28 degrees in the takeoff and landing configurations. A snap roll series can occur if the airplane is rolled at a high angle of attack. The pilot is located some 22 feet ahead of the airplane's center of gravity. As a result, if yaw excursions are allowed to build up, cockpit lateral and longitudinal accelerations are high enough to interfere with the pilot's ability to apply recovery control. A cure for F-14A departures was found by the NASA team, in a switch of roll control at high angles of attack from differential deflection of the horizontal tails to rudder deflection. This feature and foldout canards on the fuselage forebody are used on advanced, digitally controlled F-14As (Chambers, 2000).

General Dynamics F-16A and F-16B Yaw and roll departures are effectively prevented by a system that detects yaw rate above a threshold and automatically applies spin recovery control: ailerons with and rudder against the yaw. However,

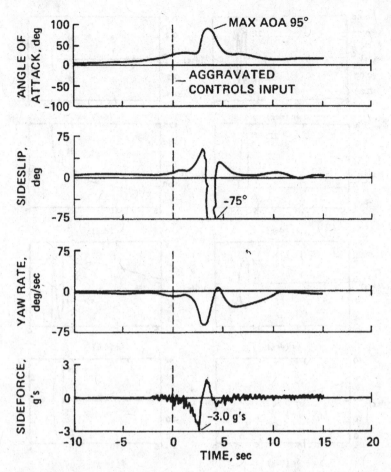

Figure 9.13 Transonic yaw departure obtained with a McDonnell Douglas F/A-18A by using pro-spin controls, or aileron against, and rudder with a turn at a medium angle of attack. (From Anderson, Enevoldson, and Nguyen, AGARD CP-347, 1983)

an angle of attack limiting system can be defeated in a number of ways, leading to excessive angles of attack.

McDonnell Douglas F/A-18A   An automatic spin recovery mode, providing full control authority when yaw departure is sensed, can be defeated if the airplane goes into a spin mode in which the yaw rate is relatively low. Although not a departure problem, the F/A-18A has an odd falling leaf spin mode, involving large sideslip, roll rate, and pitch rate oscillations. Response to pitch recovery control is slow. Finally, yaw departures are triggered by pro-spin control applications at medium angles of attack and high subsonic Mach numbers (Figure 9.13).

Boeing F/A-18E/F   An automatic spin detection and recovery mode has been added relative to older F/A-18 models. The falling-leaf characteristic of the F/A-18A/C is still present with the bare airframe. However, a new $\dot{\beta}$ feedback eliminates the falling-leaf mode (Heller et al., 2001).

Grumman EA-6B   This airplane was not included in the AGARD survey paper by Anderson, Enevoldsen, and Nguyen. However, its nose-slice behavior at the stall is documented in AIAA Paper 87-2361 by Frank L. Jordan, David E. Hahne,

Matthew F. Masiello, and William Gato. Approaching the stall, the EA-6B first experiences a rolloff, followed by a nose slice. Numerous EA-6B accidents in fleet service attributed to departures led to a NASA research program on the problem (Chambers, 2000). An EA-6B fitted with NASA modifications (higher vertical tail, inboard wing leading-edge droop, etc.) had departure-free performance, but budgetary constraints prevented their application to service airplanes.

McDonnell Douglas F-15E This airplane is likewise not included in the AGARD survey. Its departure characteristics are described in Sitz, Nelson, and Carpenter (1997). Control laws are modified for yaw rates above 42 degrees per second, to increase differential tail power in recovery. Departures are found with lateral loading asymmetry. With the left wing loaded, at angles of attack above 30 degrees, rolls to the right are rapid and hard to counter.

Rockwell/MBB X-31 Departures of this research canard fighter at angles of attack of 60 degrees were corrected by a fuselage nose strake (Chambers, 2000).

The conclusion to be drawn from this survey is that departures can still be obtained on modern fighter airplanes. Designers should concentrate on understanding and controlling the vortex flows that often underlie departures, on simple warning cues and recovery procedures, and on crew restraint systems that permit functioning in the face of strong accelerations.

CHAPTER 10

Tactical Airplane Maneuverability

Tactical airplanes have always had special stability and control problems because of the extreme maneuvers required of them. The rapid aileron roll, the sharp pullup, and the rapid turn entry all present special problems. Some examples are the level of rolling velocity actually required, overcontrol in pullups, and badly coordinated turn entries. Finally, controlled flight at angles of attack beyond the stall is a new field of required maneuvers for tactical airplanes.

10.1   How Fast Should Fighter Airplanes Roll?

Fighter roll capability became a crucial question during the early days of World War II. Many allied fighter airplanes carried gun cameras into combat. Gun cameras are movie cameras pointed in the direction of the ship's fixed-wing guns. Movies are taken as long as the firing trigger was pressed, witnessing hits (or misses) on enemy airplanes or missiles. Gun cameras carried on Curtiss P-40s and North American P-51s witnessed interesting moments in dogfights and bore out pilots' accounts of loss in combat advantages due to relatively low rates of roll on the U.S. aircraft.

Some Axis aircraft, particularly the Mitsubishi Zero at low airspeeds, would feint a roll in one direction and then roll rapidly in the other direction. The horizon or cloud background in the gun camera pictures would show the Allied airplane following the feint, a bit slower perhaps, then be left behind as the Zero did a rapid roll in the opposite direction and disappeared from the gun camera's view.

Clearly, high-rolling velocity performance was needed at dogfight airspeeds in order not to lose firing opportunities when in the favorable trailing position. At the low end of the fighter airspeed range the Gilruth/NACA criterion $pb/2V = 0.07$ was a reasonable guide, although higher levels, up to 0.10, were considered. Higher $pb/2V$ levels could be attained with extra-large ailerons. But in early World War II days, before hydraulically powered controls, the wide-chord, long-span ailerons that provided high $pb/2V$ values meant high stick forces, restricting rolling velocities at high airspeeds.

In other words, an airplane could be designed for fast rolling performance at either low airspeeds, say below 200 knots, or high airspeeds, but not both. The Curtiss P-40 was typical in that its maximum rolling velocity of 95 degrees per second occurred at an airspeed of 270 miles per hour. At 400 miles per hour (not shown in Fig. 10.1) maximum available rolling velocity dropped to 65 degrees per second, limited by a nominal 30-pound stick force.

Restricted maneuverability due to high stick forces started an intense research program on both sides of the Atlantic. The British seemed to have had the innovative edge, coming up with two significant stick force reduction schemes: the spring tab, ultimately used on the Hawker Tempest, and the beveled-edge control surface. The history of these devices is given in Chapter 5, "Managing Control Forces." Beveled-edge ailerons worked quite well for the P-51 Mustang, almost doubling the available rate of roll.

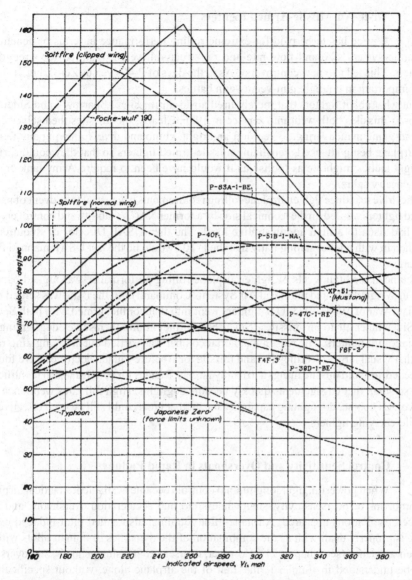

Figure 10.1 Rolling velocities obtainable with 50 pounds of stick force for a number of World War II fighter airplanes, all at an altitude of 10,000 feet. These data were heavily classified during the war. (From Toll, NACA Rept. 868, 1947)

Hydraulic power assistance came into the picture for fighter-type airplanes only at the very end of World War II, on the ailerons of the late version Lockheed P-38J Lightning. However, once power controls became common, in about 1950, stick force limitations to rate of roll were overcome. Now the only limits were hydraulic system capacity, control system and wing strength, wing torsional stiffness, and the inertial coupling phenomenon discussed in Chapter 8. The military specification version of that period reflected these new capabilities. Fighter roll rates up to 360 degrees per second were required. A limiting factor in fighter roll maneuverability at high airspeeds and low altitudes is wing twist, treated in Chapter 19, "The Elastic Airplane."

10.2   Air-to-Air Missile-Armed Fighters

A price has to be paid for extreme rolling performance in terms of demands on hydraulic system size and flow rate and on structural weight required for strength and stiffness. This led to a new controversy. As in the days of P-40s versus Zeros, high roll rates were important in dogfight gun-versus-gun battles.

But what about fighters that merely fired air-to-air missiles? Sparrow I and Sidewinder air-to-air missiles both went into service in 1956. Clearly, the missiles themselves can do the end-game maneuvering, to veer left and right, climb and dive, following any feints by the airplane being attacked. Penalizing missile-armed fighters so that they could carry out dogfight tactics might be as foolish as it would have been to require Army tank crews to wear cavalry spurs.

The drive to reduce fighter airplane rolling requirements because of the advent of missile-armed fighters was led on the technical side by a former NACA stability and control engineer who had risen to a high administrative level. The then USAF Director of Requirements weighed in with a letter stating flatly that the F-103 would be the last USAF manned fighter airplane.

The need for high levels of fighter airplane rolling performance was argued back and forth at Wright Field and the Naval Air Systems Command until the issue was settled by the Vietnam War of 1964–1973. U.S. fighters went into that conflict armed with both Sparrow and Sidewinder air-to-air missiles. Nevertheless, they found themselves dogfighting with Russian-built fighters. The reason that aerial combat was carried out at dogfighting ranges was that visual target identification and missile lock-before-launch doctrines were found to be needed, to avoid missile firings at friendly targets. Ranges for positive visual identification were so small that engagements quickly became dogfights. High roll rates were once more in favor. Of course, dogfighting capability meant that guns could still be used effectively on missile-carrying fighters.

10.3   Control Sensitivity and Overshoots in Rapid Pullups

When powerful, light longitudinal controls became available for tactical airplanes, the problems of oversensitivity, sluggishness, normal acceleration overshoots, and pilot-induced oscillations appeared. Airplane-pilot coupling, also called pilot-induced oscillations, is properly dealt with as the combination of the dynamics of human pilots with that of their airplanes (see Chapter 21). However, oversensitivity, sluggishness, and overshoots may be understood in simpler terms, that of the airplane alone, without specifically involving pilot dynamics. A fundamental indicator of airplane-alone pitch response is the pitch rate transfer function for elevator or stabilizer control inputs (Figure 10.2). Under the usual constant-airspeed assumption, this function has a second-order denominator and a first-order numerator. Although a pure delay may be added, only three parameters are involved: the frequency and damping ratio of the second-order term and the time constant of the first-order term. A number of criteria on oversensitivity, sluggishness, and overshoots deal with this airplane-alone transfer function.

10.3.1   *Equivalent Systems Methods*

Equivalent systems or low-order approaches refer to fitting an airplane-alone transfer function to the complex dynamics of actual airplane and flight control systems. Hodgkinson, La Manna, and Hyde (1976) are generally referenced as the origin of the

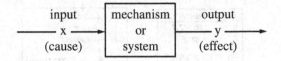

A mechanism or complete system with input x and output y defined by the differential equation

$$\left[\frac{d^{m+n}}{dt^{m+n}} + b_1 \frac{d^{m+n-1}}{dt^{m+n-1}} + \ldots + b_{m+n-1}\frac{d}{dt} + b_{m+n}\right] y(t)$$

$$= K\left[\frac{d^n}{dt^n} + a_1 \frac{d^{n-1}}{dt^{n-1}} + \ldots + a_{n-1}\frac{d}{dt} + a_n\right] x(t)$$

can be represented by the transfer function in the Laplace variable s:

$$\frac{Y(s)}{X(s)} = \frac{K(s^n + a_1 s^{n-1} + \ldots + a_{n-1}s + a_n)}{s^{m+n} + b_1 s^{m+n-1} + \ldots + b_{m+n-1}s + b_{m+n}}.$$

An example is the pitch rate transfer function for elevator or stabilizer inputs, with the airspeed degree of freedom suppressed:

$$\frac{q(s)}{\delta(s)} = \frac{(M_\delta + Z_\delta M_{\dot{w}})s + Z_\delta M_w - M_\delta Z_w}{s^2 - (U_o M_{\dot{w}} + Z_w + M_q)s + M_q Z_w - U_o M_w}.$$

In these equations,

    a,b = constants

    K = gain

    M_δ, Z_w, etc. = control and stability derivatives

    q = pitching velocity

    s = Laplace variable

    U_o = forward speed

    δ = elevator or stabilizer deflection.

Figure 10.2 The transfer function concept. (Adapted from *Aircraft Dynamics and Automatic Control*, by McRuer, Ashkenas, and Graham, Princeton U. Press, 1973)

equivalent systems method. The McRuer, Ashkenas, and Graham approximate factors, with time delay added from variable stability NT-33 tests carried out by Dante DiFranco, were used to match frequency responses of the Neal-Smith data set.

Transfer function criteria, for the airplane alone or the equivalent system, have the authority of a great deal of analysis, simulator, and flight research. Excellent reviews of this field are given by Gibson (1995) and by Hoh and Mitchell (1996). While the original work on transfer-function–based criteria was concerned with tactical airplanes, these criteria were used as well in the flight control designs of modern transport airplanes such as the Boeing 777 (Ward, 1996) and the Airbus series, starting with the A320.

10.3.2 *Criteria Based on Equivalent Systems*

A brief summary of the criteria based on airplane-alone or equivalent system transfer functions is as follows:

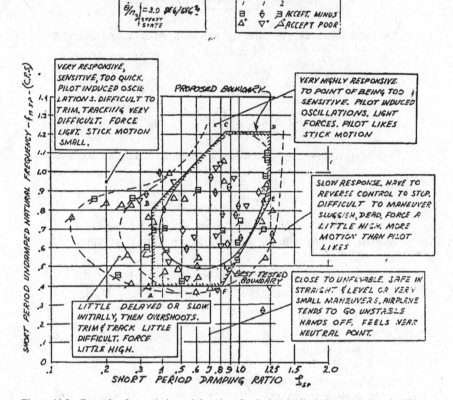

Figure 10.3 Example of an early iso-opinion chart for the longitudinal short-period mode. This one was derived from flight tests of the variable-stability F-94F airplane. (Mazza, Becker, Cohen, and Spector, NADC Report ED-6282, 1963)

Frequency-Damping Boundaries Historically, the earliest findings on pitch sensitivity and sluggishness from variable-stability airplane research was bulls-eye–type pilot opinion contours of the two denominator parameters of the pitch transfer function: natural frequency and damping ratio (Figure 10.3). This work was done by Robert P. Harper and his associates at the Cornell Aeronautical Laboratory in the early 1950s. Gibson (1995) comments that these boundaries ignore the attitude response. He suggests adding quantitative information on attitude response, such as delay, dropback (see subsequent definition), and overshoot.

Numerator Time Constant Requirements The numerator time constant, $T_{\theta2}$, controls the rapidity with which attitude changes result in flight path changes. Shorter values, corresponding to high lift curve slope and light wing loadings, give faster path responses and lower, or better, Cooper-Harper ratings. However, the benefits of low numerator time constants are mainly confined to landing approach control and have little to do with tactical airplane maneuverability.

Bihrle's Control Anticipation Parameter By far the most successful of the criteria based on pitch transfer function parameters is the control anticipation

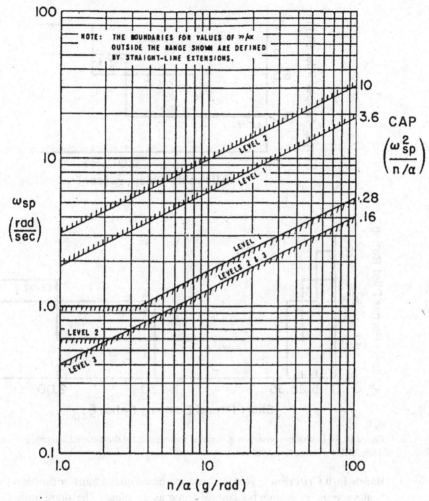

Figure 10.4 MIL-F-8757C short-period mode equivalent natural frequency and CAP (Control Anticipation Parameter) requirements (1980).

parameter, or CAP (Bihrle, 1966). CAP is the ratio of the airplane's initial pitch acceleration in an abrupt pullup to the steady-state normal acceleration produced. The initial pitch acceleration lets the pilot anticipate the final acceleration response. It turned out that CAP also could be expressed as the ratio of the pitch natural frequency to a function of the numerator time constant. In that form CAP appears in MIL-F-8785C (Figure 10.4) and is also referenced in the newer MIL-STD-1797. CAP is augmented by requirements on damping ratio and time delay (Figure 10.5).

Gautrey and Cook's Generic CAP, or GCAP The CAP criterion can be extended to augmented aircraft without recourse to equivalent systems. The generic CAP criterion, or GCAP, uses different parameters than CAP but has the same interpretation. GCAP is neither based on short-period transfer function parameters nor does it require a steady-state normal acceleration, as does CAP. GCAP parameters are well defined even for fully augmented pitch control systems such as are found on the Boeing 777 and Airbus A320–A340 series.

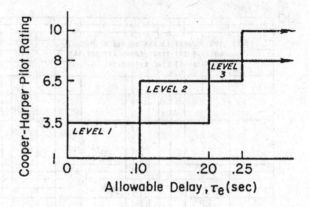

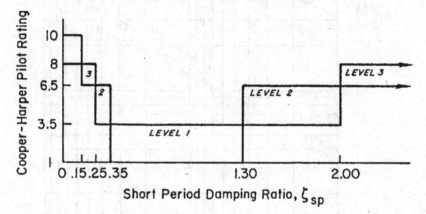

Figure 10.5 Equivalent systems requirements for longitudinal short-period damping and time delay. (From MIL-F-8785C, Nov. 1980)

Bandwidth Criterion   This is a criterion based on the transfer function for pitch attitude as an output for control force as an input. The pitch attitude bandwidth is defined arbitrarily as the lower of two frequencies: the gain bandwidth frequency, at which there is a 6-db gain margin, and the phase bandwidth frequency, at which there is a 45-degree phase margin. An additional factor is the phase delay, which accounts for phase lags introduced by higher frequency components, such as control actuators. A typical bandwidth criterion chart is reproduced in Figure 10.6. The bandwidth criterion is considered significant, although the exact shape of suitable boundaries is still a research matter.

Gibson Nichols Chart Criterion   This criterion defines satisfactory and unsatisfactory flying qualities regions in the Nichols plane of open-loop transfer function gain and phase. An early version of this criterion is shown in Figure 10.7. The concept of attitude dropback appears on the chart, a term defined subsequently.

10.3.3   *Time Domain–Based Criteria*

Time domain response specifications get around the need for equivalent systems. A standard time domain response form was used in the 1987 version of the U.S. flying

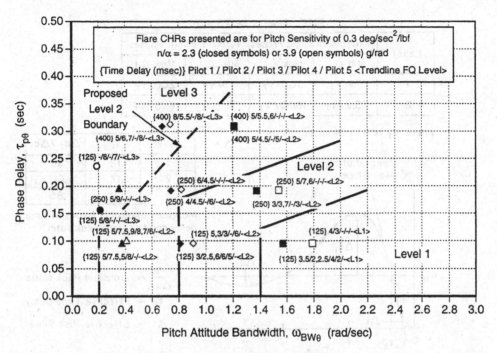

Figure 10.6 Example pitch attitude bandwidth/phase delay criterion, with test results. (From Field and Rossitto, 1999).

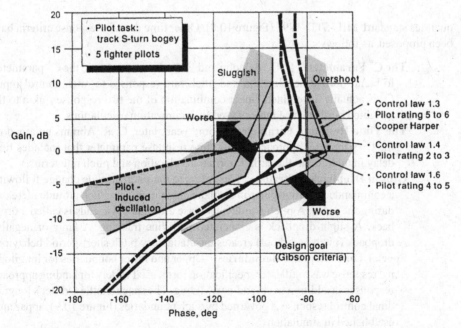

Figure 10.7 Pilot evaluation of pitch response using Gibson Nichols chart template. (From Blight 1996)

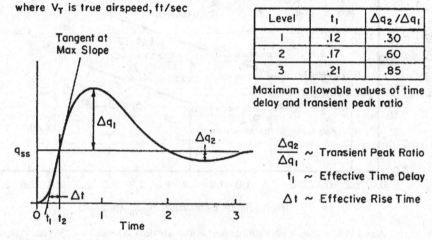

RISE TIME LIMITS

Level	Nonterminal Flight Phases		Terminal Flight Phases	
	Min Δt	Max Δt	Min Δt	Max Δt
1	$9/V_T$	$500/V_T$	$9/V_T$	$200/V_T$
2	$3.2/V_T$	$1600/V_T$	$3.2/V_T$	$645/V_T$

where V_T is true airspeed, ft/sec

Level	t_1	$\Delta q_2 / \Delta q_1$
1	.12	.30
2	.17	.60
3	.21	.85

Maximum allowable values of time delay and transient peak ratio

$\dfrac{\Delta q_2}{\Delta q_1}$ ~ Transient Peak Ratio

t_1 ~ Effective Time Delay

Δt ~ Effective Rise Time

Figure 10.8 Generic pitch rate response to abrupt control input. This type of transient response description has the advantage of applying to high-order stability-augmented as well as unaugmented airplanes. (From Mil Standard MIL-STD-1797, 1987)

qualities standard, MIL-STD-1797 (Figure 10.8). Other time domain response criteria have been proposed, as follows:

The C^* Parameter   L. G. Malcolm and H. N. Tobie originated the C^* parameter, to blend normal acceleration and pitch rate responses to pitch control input. C^* is actually a weighted, linear combination of the two responses, akin to the weighted performance indices used in optimization calculations.

The Time Response Parameter   Some years later, C. R. Abrams enlarged on the C^* parameter approach with a time response parameter that includes time delay in addition to the earlier normal acceleration and pitch rate terms.

Gibson Dropback Criterion   This refers to the pitch attitude change following a commanded positive pulse in airplane angle of attack. Pitch attitude increases during the pulse. A pitch attitude decrease after the pulse ends is called a dropback. A slight dropback is associated with fine tracking. A large or negative dropback (pitch overshoot) creates unsatisfactory pitch short-period behavior.

Special Time Response Boundaries   Upper and lower boundaries for longitudinal response was a still later specification form, used widely for landing approach responses in addition to up-and-away flying. The space shuttle Orbiter's longitudinal control response is governed by such boundaries (Figure 10.9), apparently established in simulation.

Gibson (2000) comments that the upper boundary in particular severely limits rapid acquisition of angle of attack change in response to pitch demand and was responsible for space shuttle touchdown problems. He says further:

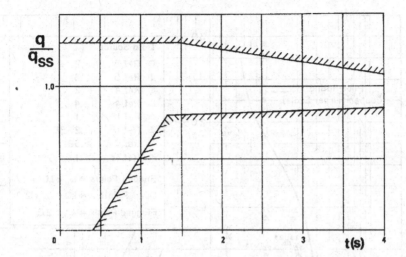

Figure 10.9   An example of a time response boundary. The pitch rate response to a step-type manipulator input must lie between the boundaries. Pitch rate response q is normalized by the steady-state value q_{ss}. This particular time response boundary applies to the space shuttle Orbiter. (From Mooij, AGARD LS 157, 1988)

The UK HOTOL project (a horizontal take off Shuttle equivalent) was studied at Warton... By designing to optimum piloted pitch response dynamics, i.e., with a rapid flight path response and hence considerable pitch rate overshoot, accurate automatic touchdown was easily achieved in simulation.

Further progress in understanding and improving longitudinal maneuverability has made use of closed-loop studies using the human pilot model (see Chapter 21).

10.4   Rapid Rolls to Steep Turns

Effective use of ailerons for rapid rolls to steep turns requires not only good roll response but also coordination, or the suppression of adverse yaw. The airplane's lift vector should remain close to the airplane's plane of symmetry during the roll and turn entry. The ball of the turn and slip indicator (see Chapter 15, Sec. 10.1) will then remain close to center, and the maneuver will be called coordinated. An alternate coordination condition is suppression of sideslip, which puts the velocity vector in the airplane's plane of symmetry.

Starting with the 1943 Gilruth requirements for satisfactory flying qualities, coordination requirements were examined in rapid aileron rolls with the rudder held fixed at the initial trim position. The peak sideslip excursion and the phase angle of the Dutch roll component of the excursion were correlated with pilots' ratings and used as the basis of U.S. Air Force coordination requirements.

More recent studies of tactical airplane roll response and steep turn entries have focused instead on the use of the rudder for coordination. Airplane transfer function theory has been applied, as in the case just described for pitch maneuvers. As in Figure 10.10, pilot ratings are compared with parameters derived from the roll and sideslip due to aileron and rudder transfer functions (Hoh and Ashkenas, 1977). Rudder deflection is assumed to be used in a coordinated fashion to hold the sideslip angle to zero in abrupt aileron rolls, as pilots are trained to do. The essence of the Hoh and Ashkenas method is a solution for the precise rudder cross-feed that accomplishes this, using linearized transfer functions.

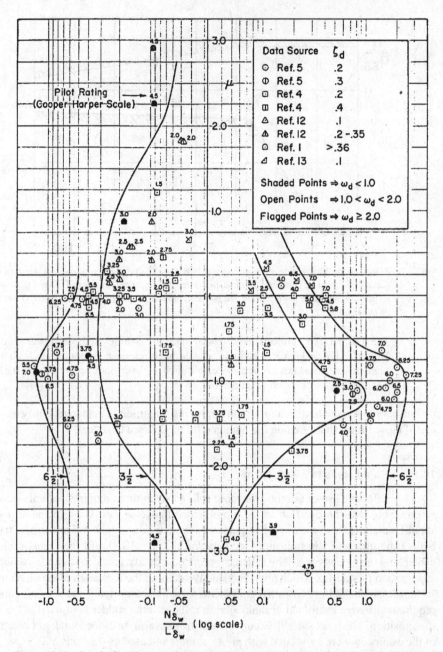

Figure 10.10 Required rudder cross-feed to coordinate turn entry, a significant factor for airplanes with good Dutch roll characteristics. The amount and sense of rudder required is plotted on the abscissa. The ordinate μ shows the required phasing of the rudder input. Rudder angle is sustained after initial input for positive values of μ and reversed for negative values of μ. The greatest pilot tolerance for required cross-feed occurs with $\mu = -1.0$, for which cross-feed fades to zero after the turn is established. (From Hoh and Ashkenas, *Jour. of Aircraft*, Feb. 1977)

The solution is in two parts, magnitude and phasing. The phase dependence means that, depending on the details of the airplane's lateral-directional dynamics, the required rudder deflections for coordination, or cross-feed, may increase or decrease after the initial rudder application.

The end result of the analysis shows a strong favorable effect for a particular required rudder cross-feed phasing. Pilots tolerate the largest amount of rudder angle cross-feed for the case in which the required rudder angle tapers off toward zero as the turn is established. Conversely, if the required rudder angle cross-feed either increases beyond the initial value or changes sign during the turn, pilot ratings suffer and smaller cross-feed levels are tolerated.

The cross-feed phasing parameter μ that expresses all of this is derived from the ratios of the transfer function numerators of rudder to sideslip and aileron to sideslip. Excluding low (gravity) and high (direct force) frequency terms, the parameter μ expresses the separation between simple zeros in these numerators. Positive values of μ correspond to increasing rudder requirements during the turn and negative values to decreasing rudder requirements. The optimum case, in which the steady-state value in a turn goes to zero, corresponds to $\mu = -1.0$.

10.5   Supermaneuverability, High Angles of Attack

Until the 1970s, fighter air-to-air combat followed the pattern set during World War I. Fighter pilots maneuver behind opposing fighters to bring fixed guns to bear long enough for a burst. The tactics are much the same for narrow-field-of-view guided missiles, such as the AIM-9 Sidewinder. In the missile case, a tail position is held long enough for an acquisition tone; then the missile is launched.

Hawker-Siddeley in Britain came up with the thrust-vector–controlled Taildog missile concept in the late 1960s, making an off-boresight launch a possibility. Combined with a helmet-mounted sight, a Taildog-type missile can be launched at target airplanes at almost any position where the pilot can follow the target with his eyes. However, even with off-boresight missile lockons and launches now possible, there is still interest in gunnery for air-to-air combat. Furthermore, there is interest in gun bearing at high angles of attack, increasing firing opportunities in a dogfight.

Supermaneuverability is defined as controlled, or partially controlled, flight in the stalled regime. It takes two forms: first, a dynamic maneuver to a high angle of attack, beyond any equilibrium or trim point. Pitching angular momentum carries the airplane to a momentary peak angle of attack. The second form of supermaneuverability is flight to a sustainable trim equilibrium beyond the stall. Supermaneuverability is seen as a way to get into the tail chase position, by a feint, tricking a pursuing airplane into overrunning one's position. Supermaneuverability adds to a dogfighting airplane's options.

The Cobra maneuver, demonstrated with a Sukhoi Su-27 airplane by the Russian pilot Viktor Pugatchov at Le Bourget in 1989, is in the first category. After Pugatchov's demonstration in the Su-27, the same maneuver was performed in a MiG-29. The Cobra is started from unstalled flight with a rapid application of full nose-up control, which is held up to the maximum angle of attack point, about 90 degrees. Control is neutralized for the recovery, assuming that the airplane has a negative or nose-down pitching moment at that point.

The entire maneuver takes about 5 seconds. There is a small altitude gain but a huge loss in airspeed and kinetic energy. Ordinarily, during air combat, one tries to maximize airspeed and total (potential plus kinetic) energy as a reserve for further maneuvers. Thus, U.S. Major Michael A. Gerzanics, project test pilot for a vectored-thrust F-16, has stated that supermaneuverability is not beneficial in all tactical situations, but is rather something that he would like to have available for close combat with a strong adversary. Clearly, any uncontrolled yawing and rolling moments that develop in the 5-second period beyond the stall must be small. The Cobra maneuver has been elaborated with a sidewise variant, called the Hook.

10.6   Unsteady Aerodynamics in the Supermaneuverability Regime

Mathematical modeling in the supermaneuverability regime has to account for unsteady aerodynamic effects above the stall (Zagainov, 1993). Zagainov describes a state variable mathematical model, developed by M. G. Goman and A. N. Khrabrov, for coefficients such as C_z and C_m. The model has a first-order state equation that defines time dependence (Figure 10.11). The typical hysteresis loop found in forced oscillation tests into the stalled regime can be modeled in this way. Zagainov also discusses the strong rolling and yawing moments that appear in the angle of attack range where vortices are shed from inboard strakes and extended forebodies. These vortex-generated rolling and yawing moments not only appear to exceed values measured in steady wind-tunnel tests, but they are also time-dependent, exhibiting hysteresis loops.

Additional light on the complex, unsteady air flows in the supermaneuverability regime has been shed by a combined wind-tunnel test and flow visualization program (Ericsson and Byers, 1997). A major factor is a coupling between vehicle motion and asymmetric cross-flow separation on a slender forebody. Wing leading-edge extensions or LEX, such as found on the F-16 and F-18 airplanes, change the nature of the cross-flow separation, apparently in a beneficial direction.

10.6.1   *The Transfer Function Model for Unsteady Flow*

Aerodynamicists familiar with the classical Bryan formulation of the perturbation equations of airplane motion expect to find aerodynamic forces and moments expanded in Taylor series. As an example, the yawing moment coefficient C_n is expanded as $C_n = C_{n\beta} \times \beta + C_{np} \times \text{pb}/2\text{V} + C_{nr} \times \text{rb}/2\text{V} + C_{n\delta} \times \delta + \cdots$. The series uses the first derivative only of the function (C_n) with respect to the independent variables, which are the vehicle's state variables β, p, r, δ, etc. With this background, it is natural to treat unsteady flow effects by adding higher derivative terms to the expansion, such as $C_{n\dot{\beta}} \times \dot{\beta}$.

Although mathematically sound, this approach has a serious flaw (Greenwell, 1998). Numerical values of higher order derivatives such as $C_{n\dot{\beta}}$ can be correct at only one oscillation frequency. Numerical values obtained in oscillating wind-tunnel rigs are correct at the frequency tested, but are in general invalid for the free or controlled angular motions of an airplane.

The solution of this problem is readily apparent to engineers trained in servomechanism theory. That is, treat aerodynamic force and moment as the result of dynamic processes much as hydraulic actuators and electrical networks are treated. The transfer function concept shown in Figure 10.2 is ideal for this application. Other modeling methods, such as Fourier function analysis, can produce equally valid results, but as Greenwell points out, the transfer function approach has the great advantage of being easily integrated into flight simulation computer codes (Abzug, 1997). Greenwell further proposes parallel transfer functions for applications at angles of attack that lead to separated flows and vortex bursting, each with its characteristic model. Transfer functions are not limited to first-order lag forms, but these have dominated the field so far. A first-order lag form adds one additional state to a state space aerodynamic model, as in the Goman and Khrabov example of Figure 10.11.

$$C_z = C_{z_{st}}(\alpha, x) + C_{z_q}\frac{qc}{v}$$

$$C_m = C_{m_{st}}(\alpha, x) + C_{m_q}\frac{qc}{v} + C_{m_{\delta_e}}\delta_e$$

$$\tau_1\frac{dx}{dt} + x = x_0(\alpha - \tau_2\dot{\alpha}), \quad (|x| \leq 1)$$

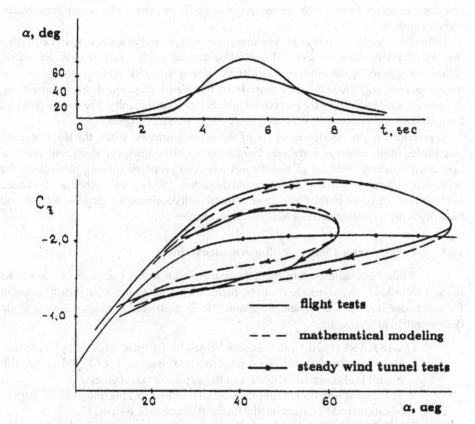

Figure 10.11 Time-dependent mathematical model for aerodynamic force and moments proposed by M. G. Goman and A. N. Khrabov for the fully stalled regime encountered in supermaneuvers. *Below*, lift coefficient variation with angle of attack, using this model. (From Zgainov, AIAA Paper 93-4737, 1993)

The transfer function concept applied to modeling unsteady aerodynamics in simulations is typical of many developments in that it is difficult to establish priority. Greenwell credits Dr. Bernard Etkin as the originator of the concept, with his publication of a 1956 University of Toronto paper (UTIA Report 42). Early work on the concept also was done by Kenneth Rogers, Thomas Burkhart, J. Roy Richardson, Moti Karpel, William P. Rodden, and R. Vepa.

A 100-state aeroservoelastic model of the Grumman X-29A forward-swept wing research airplane uses the transfer function model for unsteady aerodynamics. The transfer function model was also used with success at the DLR to model lift hysteresis at the stall for the Fairchild/Dornier Do 328 (Fischenberg, 1999) (see Chapter 14, Sec. 8.4).

10.7   The Inverse Problem

A requisite for linear analysis is a reference motion about which small perturbations occur. Generating reference motions in the case of fighter airplane supermaneuvers can be a particular problem. A planar reference maneuver, such as a Cobra-type snapup, may be generated with no particular difficulty from flight path equations (short-period dynamics suppressed) for a specific airplane. One would apply full nose-up aerodynamic or thruster control until the desired peak attitude or angle of attack is reached, followed by full nose-down control. This open-loop maneuver would yield a time history of airspeed, attitude, and angle of attack from which operating points could be selected for small-perturbation stability analysis.

Difficulties can be expected only if a maneuver path in inertial space is specified, rather than an open-loop time sequence of control or thruster angles. In that case, an inverse solution is required to determine the airplane's velocity along the path and to be sure that the maneuver is possible in the sense that control limits are not exceeded. Again, specifying a planar reference trajectory for inverse solution presents no difficulty. The case is different for nonplanar maneuvers in that the geometry could become complex.

In principle, spatial sequences of six of the normal airframe states, the three position coordinates of the center of body axes and the three Euler angles, can define any airplane maneuver. A natural path set of coordinates has been proposed instead, particularly for nonplanar maneuvers (Myers, McRuer, and Johnston, 1987). The method is illustrated with the familiar yo-yo tactical maneuver. Natural path coordinates – tangent, normal, and binormal – are a familiar concept in classical mechanics.

10.8   Thrust-Vector Control for Supermaneuvering

While supermaneuvering flight maneuvers such as the Cobra can evidently be made with ordinary aerodynamic controls, there is a growing interest in thrust vectoring for supermaneuvering (Gal-Or and Baumann, 1993). Four recent thrust-vectoring flight demonstration programs are

> **F/A-18 HARV** (High Alpha Research Vehicle)   Thrust is deflected by three vanes per engine for pitch and yaw control to a maximum angle of 12.5 degrees. Roll control is also available because of the airplane's two engines.
>
> **X-31**   Thrust is deflected for pitch and yaw control to a maximum of 15 degrees by carbon paddles, integrated with the flight control system.
>
> **F-16D MATV**   This airplane's thrust-vectoring system is integrated into the engine, with maximum yaw and pitch deflection angles of 17 degrees (Figure 3.13).
>
> **YF-22 Prototype**   Engine nozzles are deflected in pitch at angles of attack above 12 degrees and airspeeds under 200 knots, blended with horizontal tail deflections. The airplane is controllable at an angle of attack of 60 degrees (Barham, 1994).

10.9   Forebody Controls for Supermaneuvering

Alternatives to thrust-vector controls at the high angles of attack for supermaneuvering are the blowing or strake controls that act on the vortex systems shed by tactical airplanes' forebodies. There is an extensive literature on the effects of vortices shed by slender body noses on airplane forces and moments. The intent of blowing and strakes or tabs is to modify these vortices for control purposes, particularly in the supermaneuvering high angle of attack regime.

Pedriero et al. (1998) demonstrated both the promise and problems of forebody blowing. Rolling and yawing moment coefficients as large as 0.02 and 0.4, respectively, are available with blowing to one side, for a cone-cylinder body with a 70-degree delta wing. However, moment linearity with jet mass flow is too poor for closed-loop control purposes. Adding controlled amounts of blowing to the opposite side improves linearity to the point where closed-loop control is possible, with no sacrifice in available control moment. In tests of forebody blowing for a model with a chine at the body's widest point, control linearity with mass flow appears to be improved without resorting to blowing to the opposite side (Arena, Nelson, and Schiff, 1995).

The F/A-18 HARV was used to experiment with deflectable foldout strakes on the forward forebody for roll control at high angles of attack, with successful results (Chambers, 2000).

10.10   Longitudinal Control for Recovery

Tactical airplanes are able to reach supermaneuvering angles of attack by low or even negative static longitudinal stability. Full nose-up control starts the pitchup; unstable or nose-up pitching moment keeps it going. Recovery requires a nose-down pitching moment that will overcome the unstable pitching moment and leave a margin for nose-downward angular acceleration.

A rule of thumb for recovery nose-down pitching moment has been proposed, based on simulation studies and practical fighter design (Mangold, 1991). A pitching acceleration of 0.3 radians per second squared is said to be adequate. This leaves a margin for inertial coupling due to rolling during the pitching maneuver. A related problem is the amount of longitudinal control power required for very unstable airplanes, not necessarily during supermaneuvers. For that problem, Mangold correlates required pitching acceleration control with time to double amplitude.

The recovery control problem also has been attacked using the classical Gilruth approach (Nguyen and Foster, 1990). Satisfactory and unsatisfactory recovery flight characteristics are used to draw a criterion line in a plot of minimum available pitching moment coefficients with full-down control versus a moment of inertia and airplane size parameter. With only five flight data points, Nguyen and Foster call their criterion preliminary.

10.11   Concluding Remarks

Current tactical airplane maneuverability research spans all aspects of the stability and control field, from linearized transfer functions to unsteady aerodynamics and the complex, vortex-imbedded flows found at very high angles of attack. Further advances and new theories appear likely with the advent of thrust-vectoring and direct side and normal force control.

CHAPTER 11

High Mach Number Difficulties

As airplanes approach and exceed the speed of sound, 761 miles per hour at sea level, the air's compressibility changes the nature of the flow. The Mach number, the ratio of airspeed to the speed of sound, is how we keep track of these flow changes and their effects on airplane stability and control. First encountered in flight in the early 1940s, compressibility effects are still a consideration for designers of high-speed airplanes.

11.1   A Slow Buildup

Our understanding of the compressibility effects on airplane stability and control grew slowly by a buildup of theory and wind-tunnel data, with no counterpart in flight test experience for many years. The buildup in the theory started as far back as 1916, with work by Lord Rayleigh, followed by G. H. Bryan in 1918. Wind-tunnel studies started, too, but it was not for more than 20 years, or in the early years of World War II, that compressibility suddenly appeared as a stability and control problem in flight.

A key early theoretical result came from the ubiquitous Hermann Glauert, around 1927. This was the Prandtl-Glauert rule, applying to the variation of pressure coefficient with Mach number. The rule gives the pressure coefficient at any Mach number as the incompressible value, increased by a simple function of the Mach number. The Prandtl-Glauert rule was developed by the theory of small perturbations. A similar rule was developed around 1941 by Theodore Von Kármán and H. S. Tsien. Their formulation is called the Kármán-Tsien rule.

Early high-speed wind-tunnel tests were made in very small wind tunnels, compared with the larger sizes available for low-speed testing. Drs. Hugh L. Dryden and Lyman J. Briggs tested airfoils in a small supersonic jet in the 1920s. At NACA's Langley Laboratory in the early 1930s, John Stack built small high Mach number wind tunnels as adjuncts to an existing pressurized low-speed wind tunnel. High-pressure air from the big tunnel was vented into a small vertical wind tunnel, downstream from the vertical tunnel's test section.

At first, Stack and his group limited their tests to airfoils used in propellers, since at that time only propeller tips had experienced compressibility effects. By the end of the 1930s the work had broadened to include other airfoils. Pressure distributions showed a distinct upper surface discontinuity or jump, which Stack called the compressibility burble (Figure 11.1). Burble occurs at a critical airspeed at which the local surface velocity reaches the speed of sound. The local surface velocity at any point on an airfoil is the sum of the airspeed and the velocity induced by the airfoil shape. Stack reasoned that increases in the critical airspeed or Mach number could be attained through the development of airfoils that had minimum induced velocity for any given lift coefficient and thickness. This insight was the genesis of the first airfoils designed specifically for high Mach number flight.

11.2   The First Dive Pullout Problems

The Lockheed P-38 Lightning, a powerful and effective fighter airplane of World War II, is believed to have been the first airplane to have experienced adverse compressibility

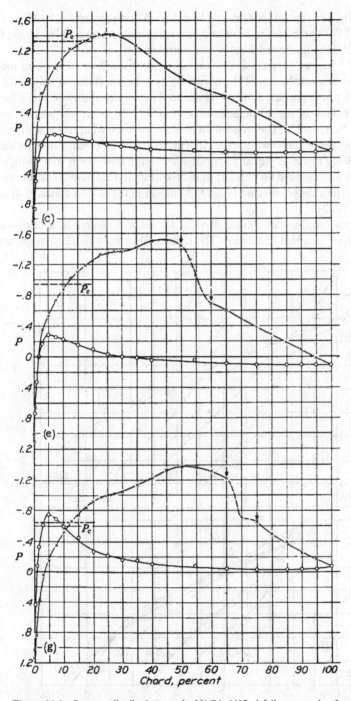

Figure 11.1 Pressure distributions on the NACA 4412 airfoil at an angle of attack of 1.9 degrees. From top to bottom, the Mach numbers are 0.60, 0.66, and 0.74. Separation, or the compressibility burble, starts at a Mach number of 0.66 on the upper surface areas between the arrows. (From Stack, Lindsey, and Littell, NACA Rept. 646, 1938)

effects on stability and control. A May 1942 Lockheed Company report by C. L. (Kelly) Johnson reported P-38 problems in dives and dive recoveries. These problems were typical of those found later in high-speed dives of other airplanes.

An important effect of compressibility on the P-38 was a great increase in the static longitudinal stability contribution of the horizontal tail. This was due to loss in lift curve slope of the thick center portion of the wing due to the compressibility burble. The P-38's wing center section was the 15-percent-thick NACA 23015; the tip section was the 12-percent-thick NACA 4412. The reduced lift curve slope of the wing center section also reduced the rate of change of downwash at the tail with angle of attack, accounting for the static longitudinal stability increase. Equally serious was a large nose-down trim change at the positive lift coefficients needed for pullouts from dives. Figure 11.2 shows both the increased stability and the trim change measured on a P-38 model in the NACA Ames 16-foot wind tunnel (Ericson, 1942).

Ericson and the NACA staff looked at fixes for these problems in a later series of 16-foot wind-tunnel tests that same year. The rather stubby P-38 fuselage aggravated the

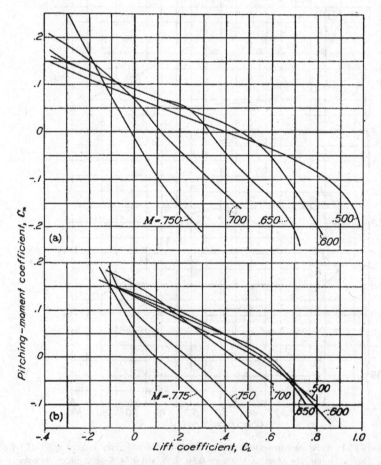

Figure 11.2 Variation of pitching moment coefficient with lift coefficient for two models tested in the NACA Ames 16-foot high-speed wind tunnel. Static longitudinal stability increases greatly at the higher Mach numbers and the trim lift coefficient (at zero pitching moment) is reduced. The upper data are for the Lockheed P-38, the lower for the Douglas DC-4 airplane. (From Hood and Allen, NACA Rept. 767, 1943)

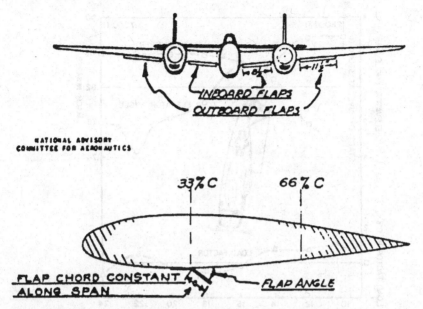

Figure 11.3 Dive-recovery flaps fitted to the Lockheed P-38 in the NACA 16-foot high-speed wind tunnel and later used on the P-38J-LO. (From Erickson, NACA WR-A-66, 1943).

problem by inducing high velocities over the already too thick wing center section, and some improvement was made when the model's fuselage lines were straightened, by lengthening it. However, the most significant gain was made with lower surface auxiliary split-wing flaps, located 1/3 of the wing chord behind the leading edge (Figure 11.3). The auxiliary flaps, later to be called dive recovery flaps, increased the P-38's trim lift coefficient by 0.55 at a Mach number of 0.725 (Ericson, 1943).

Best of all, auxiliary flap effectiveness decreases as the Mach number decreases. The importance of decreased effectiveness is that overcontrol is avoided during the dive recovery. Overcontrol in a dive recovery was a serious problem without dive recovery flaps, when pilots had only elevator control power to rely on to produce positive or nose-up pitching moments. In contrast to dive recovery flaps, pitching moment coefficient per degree of up-elevator travel actually increases as the Mach number decreases. Thus, if the pilot exerts a heavy pull force and nose-up trim through trim tabs to eke out a dive recovery against strong longitudinal stability and a nose-down trim change, when the Mach number drops and the latter two factors go away the airplane will respond to the heavy pull force and nose-up trim tab setting by making an excessively rapid pullout. If the normal acceleration or pullout g is high enough, the pilot blacks out and the wings fold upward.

The −25 block of the P-38J-LO series was fitted with dive recovery flaps. This same block of airplanes got power-boosted ailerons, further improving high-Mach–number controllability. Dive recovery flaps also were installed on P-47, A-26, P-59, and P-80 airplanes.

11.3   P-47 Dive Tests at Wright Field

The Republic P-47 Thunderbolt was, like the P-38 Lightning, an important fighter airplane of World War II. Like the P-38, the P-47 had a supercharged engine and could climb to altitudes of about 35,000 feet and reach high enough airspeeds in dives to have compressibility effects on stability and control. The P-47 experience was sufficiently different

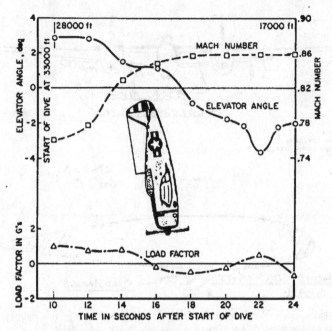

Figure 11.4 A Republic P-47 Thunderbolt fails to respond to 4 degrees of up-elevator in a dive at a Mach number of 0.86. (From Perkins, *Jour. of Aircraft*, July–Aug. 1970)

from the P-38's to merit retelling (Perkins, 1970). Following split S entries to vertical dives at 35,000 feet, the P-47's nose would go down beyond vertical. No recovery seemed possible even with full-back stick and nose-up tab (Figure 11.4). At 15,000 feet high normal acceleration would suddenly come on, and airplanes would recover at 20,000 feet, with bent wings.

Three possible reasons for this were examined in a 1943 conference held at the NACA Langley Laboratory:

1. ice formation on the elevator hinges at 35,000 feet;
2. elevator hinge binding due to loads;
3. Mach number effects on stability and control.

As Perkins recalls, it was Theodore Theodorsen, the eminent NACA mathematician and flutter theorist who championed ice on the hinges as a possible cause for the problem. The NACA structures researcher Richard V. Rhode proposed elevator hinge binding as the cause. Robert Gilruth and the forceful John Stack claimed correctly that it was all transonic aerodynamics. Quoting from the Perkins paper:

> It became obvious that one simple test would resolve the major difference in the theories. When the pilot pulled on the stick, did the elevator go up or didn't it? If it didn't go up, then one of the first two theories would be correct; but if the elevator did go up and the airplane did not respond as it should, then the third theory would be the most likely answer.
>
> The U.S. Air Corps at Wright Field agreed to run these tests and attempts were made to sign up a test pilot to perform the experiment. None of the contract test pilots were very anxious to do this and would have agreed only at very high fees. The problem was resolved when one of the Air Corps's strongest and ablest test pilots, [Capt] P. [Perry] Ritchie [Figure 11.5] said he would perform the tests for nothing. He performed some thirty dive tests on an instrumented P-47 and his reward was an Air Medal.

Figure 11.5 Capt. Perry Ritchie (1918–1944), the courageous U.S. Air Corps test pilot who made 30 test dives in the P-47 Thunderbolt. (USAF photo)

It was found at once that the elevator did go up to the predicted angle. However, while at that high airspeed the measured amount of elevator angle should have produced 20 to 30 g, the actual response was about 0.5 g, which appeared to the pilot as no response at all. This behavior was also found later by Republic Corporation test pilots. The P-47 was clearly experiencing the same Mach number phenomena as did the P-38. Compressibility burble on the inboard wing sections led to lift curve slope reductions and reductions in rate of change in downwash over the horizontal tail. This caused an increase in longitudinal static stability and a nose-down trim shift.

11.4 P-51 and P-39 Dive Difficulties

North American P-51 Mustang compressibility dive tests were made at Wright Field in July 1944 in response to fighter pilot reports from combat theaters. Captains Emil L. Sorenson and Wallace A. Lien and Major Fred Borsodi were the pilots in these tests (Chilstrom and Leary, 1993). The P-51 was climbed to an altitude of 35,000 feet, then power-dived to reach Mach numbers where compressibility effects on stability and control were found. Using a newly developed Mach number meter, the onset was found to be at a Mach number of 0.75. The tests were carried out to a Mach number of 0.83.

Longitudinal trim changes and heavy stick forces were encountered, but for the P-51 Mach number increases beyond 0.83 were limited by heavy buffeting. So many rivets were shaken loose from the structure that the airplane was declared unsafe, and the tests were concluded. It was on this series of dive tests that Major Borsodi saw the normal shock wave as a shimmering line of light and shadow extending spanwise from the root on the upper surface of the wing. Skeptics were silenced only when photos taken by a gun sight camera on later flights showed the same thing.

The Bell P-39 Airacobra was dive tested a few years later at the NACA Ames Laboratory. L. A. Clousing was the pilot, a flyer who had a strong interest in stability and control theory. The P-39 had a fairly thick wing; the NACA 0015 at the root, tapering to the NACA 23009 at the tip. Nose-down trim changes and increased stability were encountered in dives up to a Mach number of 0.78. Compressibility effects were a bit obscured by fabric distortion on the airplane's elevator.

11.5   Transonic Aerodynamic Testing

Aerodynamics engineers, including stability and control designers, were baffled in their attempts to get reliable wind-tunnel measurements at transonic speeds, near a Mach number of 1.0. High-speed wind tunnels suffered from the choking phenomenon, in which normal shocks originating on models under test spread across the test section as speed was increased, preventing further increases.

W. Hewitt Phillips credits Robert R. Gilruth with the invention of one method to circumvent this problem, the wing flow method. Figure 11.6 shows how small wing or complete configuration models are mounted normal to the wing upper surface of an airplane, in a

Figure 11.6   A sweptback-wing half-model mounted on the upper wing surface of a North American P-51, for wing flow testing during dives. The model is transonic, while the airplane is not. (From Phillips, *Jour. of the Amer. Avia. Histor. Soc.*, 1992)

region where the local Mach number is much higher than the airplane's flight speed. Phillips describes the method as follows:

> A special glove is built on the wing to give a more uniform flow region. As the airplane [P-51 Mustang] goes through its dive and pullout, the model is oscillated back and forth at a frequency of about one cycle per second, to vary either the angle of attack or flap deflection. The forces on the model are continuously recorded with a strain gage balance and a recording oscillograph. The dive lasts about 30 seconds and in this period the Mach number at the model increases from about 0.7 to 1.2.... A vacuum-operated windshield wiper motor was usually used to oscillate the model (Phillips, 1992).

The wing flow method and data from small drop models were both effectively obsoleted with the invention of the porous or slotted-throat transonic wind tunnel by Ray H. Wright, of the NACA Langley laboratory, around 1948. The slotted-throat wind tunnel allows measurements to be made through a Mach number of 1.0.

11.6 . Invention of the Sweptback Wing

The story of the independent invention of the sweptback wing in the United States by Robert T. Jones and in Germany by Adolph Busemann has been told many times. But some accounts of the early work on the stability and control effects of wing sweep belong to this history.

The dive pullout problems of thick, straight-wing airplanes such as the Lockheed P-38 were mainly due to a large increase in static longitudinal stability at high Mach numbers. Thus, an early theoretical result (Jones, 1946) seemed too good to be true. Jones showed that the static longitudinal stability or aerodynamic center location of sharply swept delta wings is invariant with Mach number, from zero to supersonic speeds. A test of a triangular wing of aspect ratio 0.75, with leading-edge sweep of 79 degrees, confirmed the theory. The catch turned out to be that wings of that low aspect ratio are impractical for airplanes that operate out of normal airports.

More practical swept wings for airplanes have higher aspect ratios. In moderately high-aspect-ratio–swept wings there is an outboard shift in additional span loading (Figure 11.7). The outboard shift in additional span load leads to wing tip stall at low angles of attack for moderate- to high-aspect-ratio–swept wings. Outflow of the boundary layer adds to this tendency. Wing tip stall causes an unstable break in the wing pitching moment at the stall (Figure 11.8). That is, loss of lift behind the center of gravity causes the wing (and airplane) to pitch nose-up at the stall, driving the airplane deeper into stall. On the other hand, a stable pitching moment break or nose-down pitch leads to stall recovery, provided that the elevator is moved to trim at a lower angle of attack. Tip stall also leads to an undesirable wing drop and reversal or positive signs for the roll damping derivative C_{l_p}, making spins easier to enter and sustain. A condition for autorotation in spins is a positive C_{l_p}.

The situation changes for low-aspect-ratio–swept wings, where leading-edge vortex flow acts to create diving, or stabilizing, pitching moments at the stall. A striking correlation was produced showing the combinations of wing sweep and aspect ratio that produce either stable or unstable pitching moment breaks at the stall (Shortal and Maggin, 1946). Figure 11.9 shows an extended version that includes taper ratio effects (Furlong and McHugh, 1957). The stable region was shown to be broadened for sharply tapered wings.

The McDonnell-Douglas F-4 Phantom's wing, with aspect ratio 2.0 and quarter-chord sweep of 45 degrees, is precisely on the Shortal-Maggin stability boundary, signifying a neutral pitching moment break at the stall. High-aspect-ratio–swept wings, typical of transport

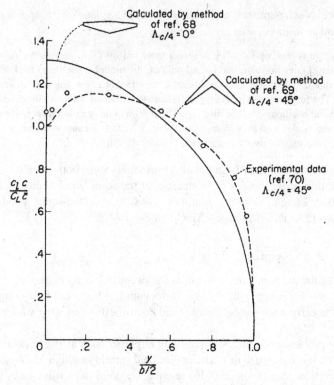

Figure 11.7 The outward shift in additional span load distribution caused by using wing sweepback. Load increases at the tip at high angles of attack, leading to tip stalling. (From Furlong and McHugh, NACA Rept. 1339, 1957)

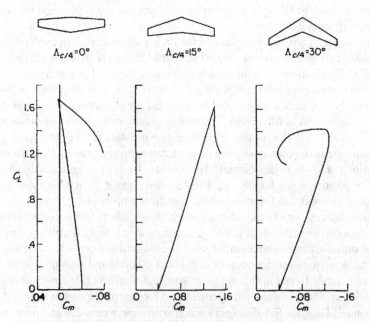

Figure 11.8 The effect of sweepback on the pitching moment coefficient break at the stall. The straight and 15-degree swept wings are stable beyond the stall; the 30-degree swept wing is unstable (noses up). (From Furlong and McHugh, NACA Rept. 1339, 1957)

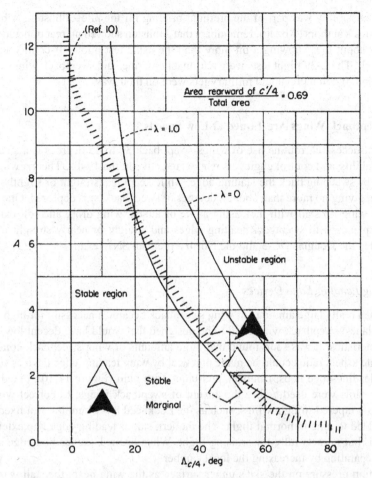

Figure 11.9 Shortal and Maggin's celebrated empirical longitudinal stability boundary for sweptback wings, extended to include the effect of taper ratio. (From Furlong and McHugh, NACA Rept. 1339, 1957)

airplanes, are unstable at the stall without auxiliary devices. For example, the Lockheed 1011's wing, with aspect ratio 6.95 and quarter-chord sweep of 35 degrees, is in the unstable zone at the stall. An early attempt to evaluate in-flight the low-speed stability and control characteristics of moderate-aspect-ratio–swept wing airplanes was made by simply removing the wings of a Bell P-63 Kingcobra and reattaching them to the fuselage at a sweep angle of 35 degrees. The tail length was increased at the same time by adding a constant cross-section plug to the fuselage aft of the wing trailing edge. NACA called this early research airplane the L-39. The L-39's first flight was made by A. M. (Tex) Johnston. He was to become famous a few years later as the test pilot for Boeing's prototype 707 jet airliner.

Wind-tunnel tests of the L-39 showed the usual increase in dihedral effect with increasing angle of attack. That is, the rolling moment coefficient in sideslip becomes quite high in the stable direction at attitudes near the stall. There was real concern on the L-39 that if sideslip angles occurred during liftoff or the landing flare, as a result of gusts or rudder use for cross-wind corrections, the rolling moment from dihedral effects would quite overpower the ailerons and the airplane would roll out of control.

This dire possibility was part of the preflight briefing for the pilot Johnston. A briefer, one of this book's authors (Abzug), remembers that Johnston showed no reaction and asked no questions about this, showing a bit more than the usual test pilot self-confidence. All turned out well. The L-39 flight tests were reasonably routine, and sweptback wings for the next generation of commercial and military jets were on their way.

11.7   Sweptback Wings Are Tamed at Low Speeds

The successful, routine use of wings swept back 30 to 45 degrees is a source of wonder to stability and control engineers who were active in the 1940s. Then, a wing that was tapered by sweeping back the leading edge, while keeping a straight or slightly swept trailing edge, giving no more than about 5 degrees of sweepback, was deplored. One could expect early wing tip stall with increasing angle of attack, wing drop, and roll damping reversal. Airplanes with sweptback leading edges and straight or nearly straight trailing edges included the Douglas DC-3 and the North American SNJ Texan.

11.7.1   *Wing Leading-Edge Devices*

When really large amounts of wing sweepback became a necessity for high Mach number airplanes, sweptback wings had to be designed that would have decent low-speed stalling chacteristics. It was soon found that large amounts of wing sweepback combined with moderate aspect ratios could be made practical by wing leading-edge devices such as slots, slats, leading-edge flaps, cambers, and blunt-nose radii (Figure 11.10). Fixed-wing leading-edge slots were used before the advent of sweptback wings, to correct wing tip stall on heavily tapered straight wings, such as the Lockheed PV-1 Ventura. But fixed slots obviously added to drag in normal flight. The modern slat or leading-edge flap extends at low airspeeds but retracts fully for cruising flight. When opened, the leading-edge slat or flap delays separation by increasing the local camber.

High suction pressure on the slat's upper surface as the wing nears the stall was used successfully to open wing slats on the North American F-86 and the Douglas A3D and A4D airplanes, avoiding hydraulic opening and closing systems. These self-opening slats could prove nerve-racking to pilots. When angle of attack is increased one wing slat tends to bang open a bit earlier than the other. However, at the angles of attack considerably below the wing stall point where air loads open the slats the opened slats have little effect on wing lift. While unsymmetrical slats look dangerous, at angles of attack well below the stall the airplane has little or no tendency to roll.

With the coming of better hydraulic systems wing slats are now universally powered. Wing slats are undesirable in some applications, as in stealth airplanes, because of the special treatments needed to avoid radar returns from the slat-wing seams. Where slats are impractical designers have learned to use wing camber and nose radius changes as substitutes. However, it may be doubted whether these measures can be effective as slats in preventing early wing tip stall. Other wing devices to improve stalling characteristics are considered next.

11.7.2   *Fences and Wing Engine Pylons*

Wing fences are streamwise panels on the wing's upper surfaces. They are intended to interrupt and shed the wing's low-energy boundary layer outflow (toward the wing tips) that would otherwise accumulate and cause flow separation and tip stall. Fences are found

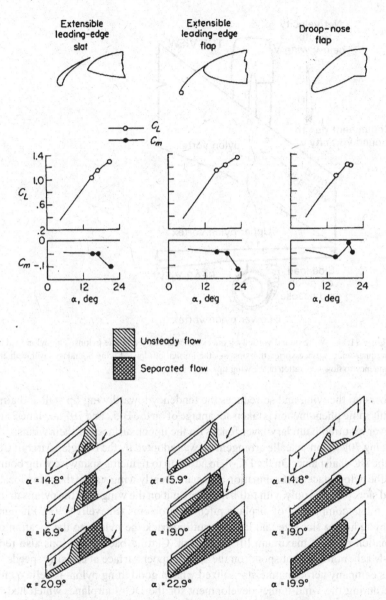

Figure 11.10 Stall patterns on a sweptback wing equipped with slat, leading-edge extension, and drooped-nose leading-edge flap. Initial tip stall is prevented in all three cases. (From Furlong and McHugh, NACA Rept. 1339, 1957)

on some early swept-wing jets, such as the Comet 1, Sud Caravelle, Tupolev Tu-54M, and Gulfstream II.

The pylons for underwing jet engines can be a substitute for wing fences on high-aspect-ratio–swept wings. This was discovered by the Boeing Company, probably during the wind-tunnel test program for the B-47 airplane. Figure 11.11 shows how bound vorticity of a lifting wing induces sidewash at the nacelle–pylon combination, which in turn causes a sideload. The side-loaded pylon–nacelle combination creates a tip or edge vortex over the top of the wing, opposing the normal outward wing boundary layer flow, which tends to

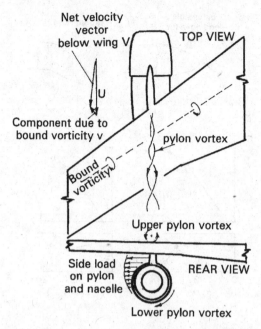

Figure 11.11   Wing bound vorticity induces sidewash on jet engine pylons. The pylon load creates an upper wing surface vortex that opposes the normal outflow of wing boundary layer, reducing the tendency to flow separation at the wing tips.

follow isobars on the wing and so reduces the tendency toward wing tip stall and airplane pitchup. This same phenomenon is taken advantage of on the B-52 and 707 airplanes, neither of which ever required boundary layer fences on the upper surfaces of their wings.

The Boeing 707 pylon–nacelle arrangement was adopted by the Douglas Aircraft Corporation for the DC-8 airplane. On the DC-8, in addition to reducing spanwise wing boundary layer flow, the pylon–nacelle combination also caused early wing stall at the pylon locations. When fixed slots (opened only with full flaps) were put on the wing near the pylons to inhibit local stall, pilots complained of airplane pitch-up problems (Shevell, 1992). Pitchup was remedied by reducing slot size and later by cutting back the pylons to the location of the wing stagnation point at maximum lift coefficient. Cutting back the pylons also reduced local high-Mach–number "hot spots" on the wing's upper surface at cruising speeds.

Douglas company aerodynamicists realized what a good thing pylon–nacelle combinations were during the wind-tunnel development for the DC-9 airplane, which had none. Fitting two pylon–nacelle combinations from the DC-8 model to the DC-9 model cured its tip stalling problems; removing the nacelles from the pylons worked, too. Finally, the pylons were reduced in size, streamlined, renamed "vortillons," and patented. The spanwise vortillon location was chosen to produce desirable vortex flow at the tail, which up till then had been insufficient to recover from a deep stall.

The DC-10 airplane has large vortex generators, or strakes, on the sides of its nacelles to alter nacelle and wing stall behavior at high angles of attack. Nacelle strakes also are found on some Boeing airplanes, but on one side of the nacelle only. David A. Lednicer writes:

> The story I have heard is that McDonnell Douglas [held] the patent for the use of strakes on both sides of the nacelle, so Boeing circumvented the patent by putting them on only one side.

11.8   Trim Changes Due to Compressibility

One compressibility effect on airplane stability and control that was not done away with by wing sweepback or thin wings was the longitudinal trim change when accelerating or decelerating through the speed of sound, or Mach 1. This was a much more severe problem for the Lockheed P-38 and its contemporaries. But along with swept wings designers began to have the hardware available for electronic trim change compensation. Ideally, this is done either on a different longitudinal control surface than the one hooked up to the pilot's cockpit controller or by a series connection for the compensation system between the artificial feel system in the cockpit and the actuator at the control surface.

The nose-down longitudinal trim change or "tuck under" near Mach 1 was a particular problem for the Douglas F4D-1 Skyray, used by the U.S. Navy in 1953 for an assault on the world's speed record. The F4D-1, later called the F-6, had fully powered elevon controls but no Mach trim change compensation.

The speed record flights were made by test pilot Robert O. Rahn at very low altitudes over a measured course at Edwards Air Force Base in California. The low altitudes at which the compressibility trim change occurred exaggerated its effect. At Mach 1, sea level, the F4D-1 changed load factor or g by about 1.5 for each degree change in angle of attack. At the highest speed attained, Rahn used a pull force to overcome the nose-down trim change. At the end of the runs, turning to return to the course, speed dropped off and a push force was required. This of course was contrary to the usual pull control forces required in turns.

When the F4D-1 was fitted with a higher powered engine, the J57-P-2, Rahn flew the new version to maximum speed at low altitude. The airplane reached a Mach number of 0.98, 500 feet over the ocean. This time Rahn used the F4D's trim surfaces in the nose-up direction to overcome the diving tendency near Mach 1. This provided more precise path control at that tremendous speed close to the water. However, when Rahn cut off the afterburner to decelerate the airplane, the nose-up trim setting produced an uncontrollable pullup to 9.1 g. The airplane was overstressed and badly buckled but landable.

Flight tests of a well-instrumented North American F-86 Sabre provide an unusually good look at the transonic trim change problem (Anderson and Bray, 1955). The measurements show marked increases in longitudinal static stability and decreases in elevator control power as the Mach number increases from 0.94 to 0.97. The record of a dive pullout (Figure 11.12) shows a trim change when the F-86 traversed the same Mach number range, slowing down in a dive pullout.

The transonic trim change problem also was experienced with the North American F-100 Super Sabre, although less dramatically. In unpublished correspondence Paul H. Anderson recalls these events:

> The first complaint was that the airplane could not be trimmed at cruise speed. Much time and effort was spent redesigning and flight testing modifications to the trim system (with no improvement) until we finally recognized what was happening. The answer, of course, was to feed back Mach number to the flight control system. . . .
>
> Prior to that, the slope of stabilizer position versus speed was called static longitudinal stability. When the aerodynamic center shift [with Mach number] was encountered some people said that the airplane was statically unstable, when it was actually more stable than before. We finally changed the name of [the] stabilizer position versus speed to Speed Stability and the problem went away.

Mach trim compensators as separate systems continued to be features of transonic airplanes for many years, up to the advent of integrated fly-by-wire control systems. As an

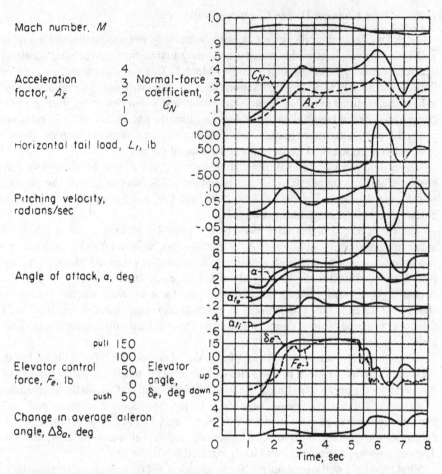

Figure 11.12 Transonic trim change of the North American F-86 Sabre. At time equal to 5 seconds, normal acceleration continues to increase although the elevator is moving down. This is because the Mach number is reducing from 0.97 to 0.94 and the airplane is becoming less stable. (From Anderson and Bray, NACA Rept. 1237, 1955)

example of one of the older separate Mach trim compensators, the Boeing 707 transport has an automatic Mach trimmer that puts in 2 degrees of nose-up stabilizer trim starting at Mach number 0.82. In modern integrated fly-by-wire control systems Mach trim change compensation is just one of the many stability augmentation programs in a flight control computer.

11.9   Transonic Pitchup

Transonic high angle of attack pitchup is an instability caused by reversal of the normal, stable wing–fuselage pitching moment variation with angle of attack. In the normal, stable angle of attack range, increases in angle of attack cause negative, or nose-down, pitching moments. At high angles of attack on sweptback wings at transonic speeds flow separation on the outboard panels reverses the wing–fuselage pitching moment from negative to positive, or nose-up. The airplane will continue to rotate in a nose-up direction, increasing the angle of attack. This can happen even against pilot application of nose-down control.

This phenomenon was first seen in flight in August 1949, on the Douglas D-558-II Skystreak research airplane being flown by Robert Champine. While pulling 4 g at a Mach number of 0.6, the airplane suddenly pitched up to 6 g. This was not much of a surprise since wind-tunnel tests had shown wing–fuselage pitching moment reversal at high angles of attack. Outer panel vortex generators delayed slightly the onset of pitchup on the D-558-II, by about 0.05 in Mach number.

Seth B. Anderson and Richard Bray (1955) had the opportunity to analyze in detail the same transonic pitchup phenomenon as it occurred on the North American F-86 Sabre, in flight tests at Ames Aeronautical Laboratory. The key was strain gage measurements of horizontal tail load, which separate wing–fuselage instability from changes in downwash at the horizontal tail. In windup turns at a constant Mach number, load factor or g increased during the pitchup, although the tail carried an increasing, stabilizing up-load as the pitchup continued (Figure 11.13). Thus, F-86 pitchup is caused by wing instability, independently of the tail.

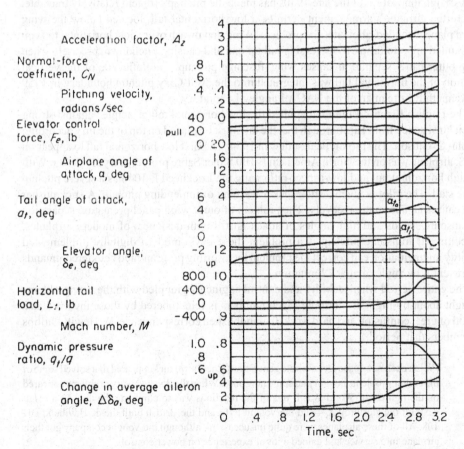

Figure 11.13 Pitchup on the North American F-86 Sabre in a turn. Normal acceleration increases after 1.5 seconds, although the elevator angle is constant or moving down. By measuring the horizontal tail load, NACA verified that the pitchup was caused by wing–fuselage instability, just as predicted in the wind-tunnel test. (From Anderson and Bray, NACA Rept. 1237, 1955)

A refinement of the Shortal-Maggin instability boundaries (Figure 11.9) was made by Joseph Weil and W. H. Gray in 1953. They showed that transonic speeds shift the boundary toward lower values of wing sweepback. That is, for the same wing aspect ratio, less sweepback is allowable at transonic speeds.

The Boeing B-47 has an interesting transonic pitchup case history. With a wing aspect ratio of 6.0, a quarter-chord sweepback angle of 35 degrees, and a taper ratio of 0.23, the B-47 wing falls close to the Furlong and McHugh pitchup boundary, which applies at low speeds. With the Weil-Gray shift in pitchup boundary toward less allowable wing sweep at transonic speeds, the B-47 would be expected to have transonic pitchup, and indeed it does (Cook, 1991).

As with the Douglas D-558-II, outer wing panel vortex generators reduced the pitchup instability. Vortex generators made no improvement in B-47 wind-tunnel tests, possibly because of the small scale of the generators, but Cook reports that two rows of generators actually eliminated the problem in-flight (Figure 11.14). The B-47 installation is apparently better than the one that created only a slight delay in pitchup to a higher Mach number for the D-558-II.

A design innovation of the late 1940s has made the pitchup problem relatively tractable. Where the airplane's arrangement permits, a low horizontal tail, located below the wing chord plane extended, alleviates the problem. Wing downwash over a low horizontal tail can be counted upon to drop off, increasing tail upload and causing pitchdown, precisely when wing outer panel separation causes wing–fuselage pitchup. The influence of the vertical position of the horizontal tail was integrated into the Weil-Gray pitchup boundary in 1959 by Kenneth P. Spreeman of the NACA Langley Laboratory.

The English Electric Lightning applied the low horizontal tail principle "against strong official insistence from Farnborough that the tail must be placed on top of the fin," according to John C. Gibson. The prototype first flew in 1954. The first low horizontal tail to appear on a U.S. airplane was on the North American F-100 Super Sabre prototype a year earlier. With its high horizontal tail and low-aspect-ratio wing the Lockheed F-104 has a severe pitchup at the stall. The pilot is given a stick shaker warning of impending pitchup. A stick pusher then causes an automatic recovery. The problem of outer wing panel premature separation at transonic speeds and high angles of attack is still with designers of modern airplanes. Of course, if transonic pitchup is a problem these days, modern digitally implemented stability augmentation can correct the problem by inserting programmed countercommands or prevent it by angle of attack limitation.

The entire period when stability and control engineers grappled with the new problems brought about by flight at transonic Mach numbers is remembered by those involved as a period of great confusion and hard work. In unpublished correspondence, W. Hewitt Phillips remembers that period:

> [W]ith each new discovery a whole generation of new airplanes appeared that solved some of the earlier problems but got into some new ones. When the dive recovery problems appeared on the fighters of World War II one of the reactions was to eliminate the tail. As a result we built the Northrop X-4 and the Vought F7U, and the British built the de Havilland DH 108. All of these airplanes were quite unsuccessful, although the Vought company got their airplane into service and gained a lot of experience on power controls.
>
> After that came the studies of the effects of sweep and aspect ratio, and companies built planes that looked like the earlier unswept versions but with swept wings. These include airplanes like the F9F-6 and the F-84F. About this time came the discovery of notched wings, low tails, etc., and many of the pitchup problems were cured. This generation includes the F8U and the F-100.

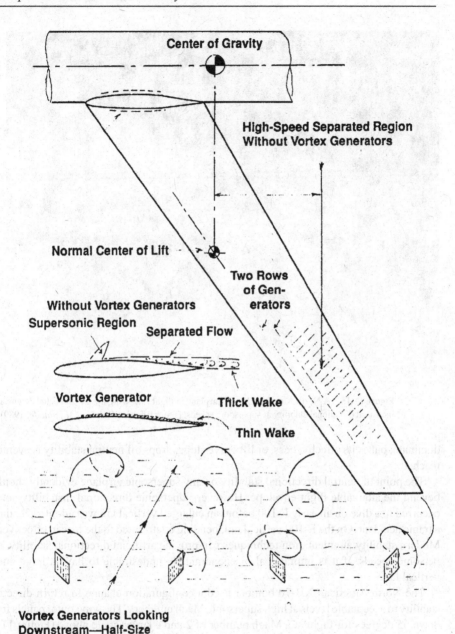

Figure 11.14 Boeing B-47 wing showing the locations of the two rows of vortex generators that eliminate transonic pitchup. (From Cook, *The Road to the 707*, 1991)

11.10 Supersonic Directional Instability

A rather simple static directional instability problem first appeared in a test flight of the North American F-100 Super Sabre. It is simple because the problem has one well-known cause, the loss in lifting surface effectiveness as Mach number increases beyond 1. The instability of bodies of revolution, on the other hand, remains essentially invariant with Mach number. Static directional stability is to a first order the balance between the unstable fuselage and the stabilizing vertical tail. The vertical tail is supposed to

Figure 11.15   A North American XB-70 airplane in flight. The wing tips are deflected downward for increased directional stability at supersonic speeds. (From Bilstein, *Orders of Magnitude*, 1989)

dominate, but as its effectiveness, or lift curve slope, drops off neutral stability is eventually reached.

The point of neutral directional stability on any supersonic airplane evidently should be beyond the attainable flight envelope. However, supersonic directional instability actually occurred in a dive on an early F-100 before an enlarged vertical tail was adopted, leading to a tragic accident. On the F-100 vertical tail, bending contributed to the loss in effectiveness. Modern stability augmentation techniques can provide artificial directional stability at supersonic speeds, if it is impractical or economically undesirable to have a large enough vertical tail.

The North American XB-70 bomber used a configuration change to return directional stability to acceptable levels at high supersonic Mach numbers. The wing outer panels folded down 65 degrees for flight at a Mach number of 2 and a larger angle above (Figure 11.15). Unfortunately, this made the dihedral effect negative, resulting in poor flying qualities. This was corrected on the second XB-70 prototype by a triangular wedge welded between the fuselage and wing, producing 5 degrees of geometric dihedral.

There was concern that if the XB-70's wing tips ever stuck down in the folded position, the airplane could not be landed because of lack of ground clearance. Fortunately, this never happened. An additional benefit of the folded-down wing tips was reduction in excess static longitudinal stability at supersonic speeds, due to the change in planform. Also, compression lift was generated at supersonic speeds by shock waves from the folded tips producing positive pressures on the bottom of the wing and fuselage.

The British Aircraft Corporation's TSR-2, designed for a Mach number of 2.0, had neutral directional stability at a Mach number of 1.7. The vertical fin was made small to reduce tail loads in high-speed flight at low altitudes. The airplane was canceled for other reasons before a directional stability augmenter could be installed for flight faster than a Mach number of 1.7.

11.11 Principal Axis Inclination Instability

Lateral-directional dynamic instability due to nose-down inclination of the principal axis is not strictly a high Mach number or compressibility phenomenon. However, this type of instability is linked to high-speed flight, and so it is included in this chapter.

The symmetric principal axis is defined as that airplane body axis in the plane of symmetry for which the product of inertia I_{xz} vanishes. Mathematically, $I_{xz} = \int xz \, dm$, where x and z are the X- and Z-axis coordinates of each elementary mass particle dm. Weights high on the vertical tail, such as a T-tail, cause the principal axis to be inclined nose-downward with respect to normal body axes.

A nose-down inclination of the principal axis with respect to the flight path destabilizes the lateral-directional or Dutch roll oscillation (Sternfield, 1947). Actual lateral-directional dynamic instability due to a nose-down inclination was encountered dramatically in May 1951 by the NACA test pilot Bill Bridgeman. This was in a series of flight tests of the Douglas D-558-2 Skyrocket research airplane. In tests reaching a Mach number of 1.79 serious rolling instability occurred during pushovers after rocket-powered steep climbs. The principal axis inclination to the flight path becomes quite nose-down during pushovers.

The test team evidently failed to connect the rolling instability with the principal axis effect and concluded that even higher speeds could be reached safely. Bridgeman was asked to nose over from the climb to a very low factor of 0.25, in an effort to reach a Mach number of 2.0. According to Richard Hallion (1981):

> the Skyrocket rolled violently, dipping its wings as much as 75 degrees. He cut power, but the motions, if anything, became even more severe. Finally he hauled back on the control column, for the Skyrocket was in a steep dive and getting farther and farther away from the lakebed. The plane abruptly nosed up and regained its smooth flying characteristics, and he brought it back to Muroc.

However, concerns about Dutch roll instability due to principal axis nose-down inclination have been eliminated by the almost universal use of yaw damping stability augmentation on high-speed airplanes.

11.12 High-Altitude Stall Buffet

A Mach number–altitude operating envelope is a useful concept for jet airplanes. With Mach number plotted as the abscissa, the right-hand boundary gives the maximum operating Mach number at each altitude. This is ordinarily established by buffet for commercial airplanes. Military airplanes have higher design load factors, and structural strength or controllability ordinarily sets the right-hand maximum operating Mach number. When buffet, structural, or controllability limits do not apply, maximum Mach number is established by the highest speed attainable in dives from maximum altitude.

The left-hand boundary gives the minimum operating Mach number at each altitude. Ordinary stall generally defines the low-altitude minimum operating Mach number. Airframe buffeting due to flow separation can define the minimum operating Mach number at high altitude, since that Mach number can be high enough for compressibility-induced flow separation. There is a significant stability and control involvement when stall buffet is responsible for the minimum operating Mach number at high altitude. This is because when a high-altitude stall buffet boundary is breached when an airplane is upset by turbulence, the pilot may use the controls too vigorously to recover.

This seems to have happened in 1992 with a McDonnell Douglas MD-11 airliner cruising at a Mach number of 0.70 at an altitude of 33,000 feet. The airplane was upset by turbulence. In recovering from the upset, it slowed to a Mach number of 0.50 while making four successive penetrations of the stall buffet boundary.

11.13   Supersonic Altitude Stability

A somewhat strange lack of stability cropped up when airplanes began to operate at supersonic speeds above about Mach 2 at quite high altitudes. This showed up as an inability to control altitude and airspeed precisely in flights of the North American XB-70 Valkyrie and the Lockheed SR-71A. The Concorde SST with a maximum speed around Mach 2 is believed to have difficulties of this sort, as well. According to Glenn B. Gilyard and John W. Smith (1978), on the SR-71A:

> Decreased aircraft stability, low static pressures, and the presence of atmospheric distur-
> bances are all factors that contribute to this degraded control. The combination of high
> altitude and high speed also contributes to an unfavorable balance between kinetic and po-
> tential energy, thereby requiring large altitude changes to correct for small Mach number
> errors when flying a Mach hold mode using the elevator control.

In simulating SR-71A supersonic altitude and airspeed control problems NASA found it necessary to add to the normal equations of aircraft motion inlet geometry effects on airplane motion, inlet operating characteristics up to the unstart boundary, and the afterburning equations for the two engines. While with these additions simulation presents no unusual difficulties, attempts to find suitable theoretical models are another matter.

The applicable body of theory begins with Lanchester's 1897 analysis of phugoid motion. Lanchester's model, and the Bryan and Williams analyses that followed, neglected atmospheric density changes as an airplane's height changes during a longitudinal oscillation. F. N. Scheubel added density gradient to the mathematical model in 1942. In 1950, Stefan Neumark added the effects of thrust and sound speed variations with altitude to the equations.

While the classical Bryan-Williams model leads to a fourth-degree characteristic equation, with both short- and long-period longitudinal oscillations, density gradient increases the characteristic equation degree to 5. A new aperiodic height mode appears, typically a very slow divergence (Figure 11.16). The height mode was first identified, or rather predicted, by Neumark. The supersonic altitude stability problems thus far encountered probably involve both the phugoid or long-period mode and the height mode.

Thrust effects are significant on both the phugoid and height modes (Stengel, 1970; Sachs, 1990). Aside from the effects of possible thrust offsets from the airplane's center of gravity, the throttle-fixed variation in thrust with airspeed affects both the phugoid and height modes. Both modes are stabilized when thrust decreases with increasing airspeed, and vice versa.

Lanchester's original analysis (Durand, 1934) assumes an airplane whose lift is always at right angles to the flight path and numerically proportional to the square of the airspeed. These simple assumptions and small-angle approximations lead to Lanchester's phugoid motion, an undamped oscillation of period $\sqrt{2}\pi V/g$, where V is the flight velocity and g is the acceleration of gravity.

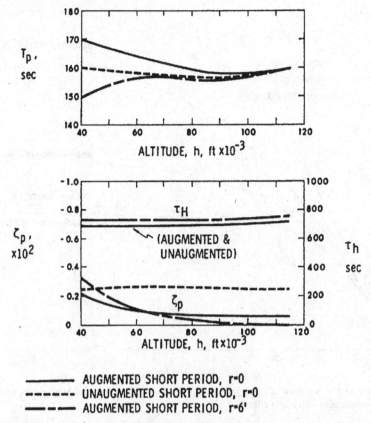

——————— AUGMENTED SHORT PERIOD, r=0
– – – – – – UNAUGMENTED SHORT PERIOD, r=0
—— – —— AUGMENTED SHORT PERIOD, r=6'

Figure 11.16   Effect of altitude on the phugoid and height modes of a hypothetical SST, cruising at a Mach number of 3.0. (From Stengel, *Jour. of Aircraft*, Sept.–Oct. 1970)

However, the linear increase of period with airspeed predicted by Lanchester does not occur at high airspeeds. The reason is that the density gradient effect that Scheubel wrote about in 1942 becomes very important at high airspeeds. The phugoid period is shortened compared with the Lanchester case. In effect, as the airplane noses down, picking up speed and giving up potential energy for kinetic energy, higher density at the lower altitude increases lift, bending the path upward again. Higher density at lower altitudes acts as an extra spring, shortening the period.

A simplified model developed in 1965 by Lockheed's John R. McMaster predicts drastic reductions in the phugoid period relative to the classic Lanchester values at high airspeeds. McMaster's predicted period at a Mach number of 3.0 is about 150 seconds, compared with the Lanchester value of 401 seconds. Calculations by Stengel in 1970 for an SST configuration at a Mach number of 3.0 give a phugoid period of about 160 seconds, close to the McMaster value. A later simplified model (Etkin, 1972) shows a reduction from the Lanchester values, but not quite so large.

A formulation by Regan (1993) also corrects the Lanchester approximation for density gradient. Regan's approximation may be derived from the small-perturbation longitudinal equations of motion of Figure 18.4 by adding a height degree of freedom and the height derivative $\partial Z / \partial h$, where Z is the Z-axis aerodynamic force and h is altitude perturbation from equilibrium flight. The Regan approximation for phugoid period is

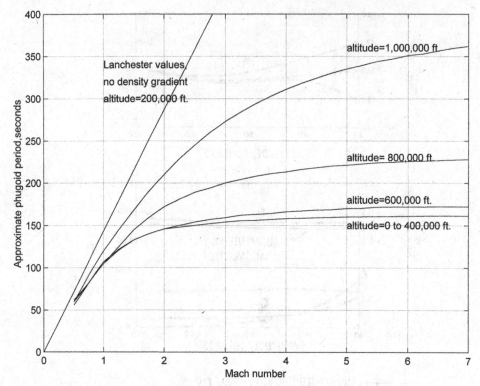

Figure 11.17 Variation of approximate phugoid period with altitude and Mach number, including density gradient with height effects.

$$\text{Period} = 2\pi(V_0/g)/\left(2 - (1/\rho)(\partial\rho/\partial h)\left(V_0^2/g\right)\right)^{1/2},$$

where V_0 = equilibrium flight speed,

ρ = equilibrium air density,

g = acceleration of gravity at the equilibrium altitude,

$\partial\rho/\partial h$ = density gradient with altitude at the equilibrium altitude.

This relationship is used to show the general trend of phugoid period with Mach number and altitude in Figure 11.17. Omitted are possible thrust effects. Figure 11.17 shows that density gradient causes the phugoid period to reach asymptotic values as airspeed increases indefinitely, which is at odds with the classical Lanchester approximation. Neglect of density gradient incorrectly doubles the approximated phugoid period at an altitude of 200,000 feet and Mach number = 2.

Figure 11.17 predicts a phugoid period of 154 seconds at Mach 3 at altitudes below 400,000 feet. This is close to McMaster's predicted value for Mach number 3 and also to Stengel's results in Figure 11.16. Eigenvalue calculations for the NASA GHAME hypersonic vehicle give remarkably close results to the Regan approximation for the phugoid period at Mach numbers above 2.

Flight test data that would support one simplified phugoid model or another seem not to be available, for good reason. Typically, airplanes are flown at those very high Mach numbers and altitudes with either or both altitude and Mach hold control loops active, as a direct result of the altitude stability problem. Loop closures modify the basic phugoid

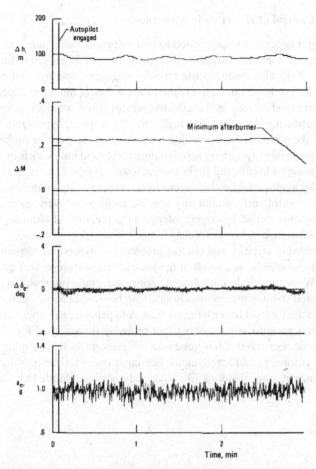

Figure 11.18 YF-12A altitude hold performance with an optimized autopilot, at Mach 3, altitude 77,500 feet. Altitude oscillation about the desired value is held to within plus or minus 25 feet. (From Gilyard and Smith, NASA CP 2054, 1978)

motion to the point where its period and damping would be difficult to detect, even if tests could be devised to measure periods as long as 160 seconds.

NASA flight tests of the YF-12A are encouraging in that properly designed and compensated altitude and Mach hold systems, working through the pitch and thrust controls, respectively, seem to be able to hold reasonably stable cruise conditions at Mach 3 (Figure 11.18). One physical limitation is instability in the atmosphere itself, notably temperature shears that change the indicated Mach number even though the true airspeed and altitude have not changed. Gilyard and Smith noted that baseline YF-12A altitude hold mode operation varied from day to day. "Occasionally altitude could be held reasonably constant; at other times, it diverged in an unacceptable manner."

A related phenomenon was found in flight testing the XB-70 around a Mach number of 3. Indicated altitude changes of 1,000 feet were seen in 2 or 3 seconds, quite evidently the result of atmospheric temperature gradients, since the airplane could not have possibly changed altitude so quickly. To avoid having altitude and Mach hold systems chase after atmospheric instabilities, it may be necessary to smooth atmospheric data with inertial data or position measurements derived from satellites.

11.14   Stability and Control of Hypersonic Airplanes

Hypersonic flight is generally understood to mean flight at Mach numbers above 5.0. The only experience with manned airplanes at hypersonic speeds has come from the North American/NASA X-15 and space shuttle Orbiter programs. Stability and control phenomena at hypersonic speeds are qualitatively no different than at moderate supersonic speeds. There is the same relative loss in the effectiveness of lifting stabilizing surfaces relative to fuselage-destabilizing moments. The high altitudes at which hypersonic flights are carried out lengthen the periods of uncontrolled motions, always a piloting problem.

The influence of the propulsion system on aerodynamic forces and moments is expected to be more extreme in powered hypersonic flight than at lower speeds. The reverse is also true in that slight sideslip angles could cause severe inlet problems, depending on details of the design. Sufficient control surface authority may be required to overcome yawing, pitching, and rolling moments caused by engine inlet unstarts precisely at altitudes where control moments are low because of low air density.

However, the most pressing stability and control problems of hypersonic airplanes are probably encountered at low speeds, as a result of the unique design features that go along with hypersonic flight. Wing slats or drooped leading edges that could improve low-speed longitudinal and directional stability are apparently ruled out because of aerodynamic heating problems at seams in the forward lower wing surface. A hypersonic passenger airplane for long over-ocean flights remains an interesting, but probably distant, goal for aviation planners. The aerodynamic research that has gone into this concept so far has quite properly dealt mostly with performance and aerodynamic heating. Conceptual designs that have been published show configurations that look like stretched-out space shuttle Orbiters.

CHAPTER 12

Naval Aircraft Problems

Airplanes operating from aircraft carriers have stability and control problems not present in land-based airplanes. Some problems arise from the size constraint, to allow airplanes to fit on the elevators of as many carriers as possible. For stability and control engineers this translates into restrictions on tail length, since wings can be folded. Good pilot visibility over the nose is needed for nose-high landing approaches, affecting the airplane's design at many points. Waveoffs or missed approaches must be made starting from more adverse airspeed and attitude conditions than from field landings. This means positive, safe control near the stall and careful integration with the airplane's performance design.

Finally, there is the matter of carrier landings. From the moment of starting a final approach to either field or carrier landings an airplane's path and airspeed must be controlled. Path control is needed to make a touchdown in the correct area, with a reasonable vertical velocity. Airspeed control is needed to keep the touchdown speed within limits. Depending on the on-board avionic equipment, weather conditions, and pilot training and preferences, path and airspeed control for field landings use a variety of visual cues and instrument readings. The important point is that touching down at a precise point is seldom required for field or airport runway landings.

In contrast to the airport runway case, touchdown point precision to within a very few feet is necessary for successful landings on aircraft carriers. Carrier landings are made without flare. Thus, low approach speeds are desirable to reduce touchdown vertical velocity and landing gear loads. There is little tolerance for errors in touchdown airspeed between stalling and excessive speed, leading to hard landings. As a result, carrier landing accidents, mainly due to hard landings and undershoots, are statistically more common than airport landing accidents.

12.1   Standard Carrier Approaches

Naval aviators have developed a distinctive landing approach procedure to make touchdown point precision a routine matter. U.S. carrier-based airplanes turn onto a short final approach path in a steep left turn, avoiding as much as possible the ship's turbulent wake or burble. Carrier final approaches are typically less than 3/4 mile long, taking some 15 to 20 seconds to complete (Craig, Ringland, and Ashkenas, 1971).

In manually flown approaches the pilot relies on an optical projection device for a vertical reference, rather than the view of the ship's landing area. The optical device, mounted on the carrier's deck, is gimbaled to project a stable glide slope. The pilot sees a projection of a solid circle and short horizontal datum bar. Below the glide slope the ball appears below the bar, and vice versa. Radar-controlled automatic carrier landings have also been developed, using an SPN-42 tracking radar mounted on the ship. Aircraft attitude changes are sent as commands to the airplane's pitch attitude autopilot, to correct perceived height errors relative to the ideal glide path.

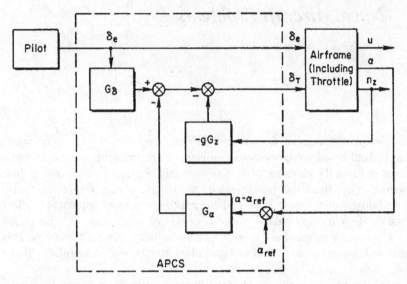

Figure 12.1   Block diagram for the AN/ASN-54 Approach Power Compensation System. This was designed for carrier airplanes to hold the angle of attack constant using thrust variations. However, path control was unsatisfactory using this system. (From Craig, Ringland, and Ashkenas, Syst. Tech., Inc. Rept. 197-1, 1971)

In order to control airspeed closely, the final approach is made at a constant angle of attack. Precise control of angle of attack as a means of controlling airspeed is considered so important that a special throttle control system – the Approach Power Compensation System, or APCS – was developed for that purpose. There were many experiments with different feedbacks; the final APCS design uses angle of attack and normal acceleration feedbacks to the throttle and some pilot stick feedforward (Figure 12.1).

12.2   Aerodynamic and Thrust Considerations

It has been known for some time that landing approach path control by elevator or pitch adjustments does not work for low-aspect-ratio (stubby) straight or sweptback wings. This is due to the variation of drag with airspeed when the lift is equal to the gross weight in level flight conditions. We normally expect level-flight drag to increase rapidly with increasing airspeed, and so it does at cruising airspeeds. At cruising airspeeds height control by pitch attitude changes using the elevator is stable and effective. The throttle can be left fixed.

However, the level-flight drag for any airplane increases with decreasing airspeed near the stall, as a result of induced drag increases and flow separation at high angles of attack. As airspeed is reduced from cruising values level-flight drag reaches a minimum and then actually increases again as the airspeed is reduced still further. The airspeed at which level-flight drag, and thrust required to hold level flight, reach minimums was given the name "minimum drag speed" by Stefan Neumark in Britain (1953).

The increase in level-flight drag near the stall is accentuated for airplanes with low-aspect-ratio wings, leading to increases in minimum drag speed. The minimum drag speed for an airplane with a low-aspect-ratio wing can be well above the low approach airspeed desired for carrier landings. Thus, if an airplane with a low-aspect-ratio wing is on a stabilized descent at a low landing approach speed typically used for aircraft carriers and the pilot retrims the airplane to a higher angle of attack, reducing airspeed, the airplane will rise at first relative to the original path and then settle even faster. The flight path will become steeper, a counterintuitive result.

For landing approaches below the minimum drag speed, where increasing thrust is required for decreasing airspeed in level flight, sometimes called "the back side of the thrust required curve," pitch attitude control by the elevator is unsatisfactory, even with the throttle used to control height. Thrust control by the pilot or an automatic system (the Navy's APCS) to hold constant airspeed or angle of attack has been used to artificially create the normal variation of thrust required for level flight.

"Backside" carrier-based approach problems were first recognized about 1950 (Shields and Phelan, 1953). Pilots needed to use higher approach speeds for the XF-88A and XF3H-1 airplanes than the standard rules of thumb based on stalling speed. Shields and Phelan proposed a fixed-throttle pitch-up test maneuver that is similar to a popup maneuver later adopted as one criterion for minimum carrier-approach speed. The first large-scale organized set of data on minimum approach airspeed behavior for jet airplanes was taken at the NACA Ames Aeronautical Laboratory (White, Schlaff, and Drinkwater, 1957). Carrier-type landing approaches were made with seven straight- and swept-wing jet airplanes, the FJ3, F7U-3, F9F-6, F4D, F-100A, F-94C, and the F-84F. The objectives of the 1957 Ames tests was to find the minimum "comfortable" approach airspeeds for carrier-type landings for these representative jet airplanes.

The reason most frequently given by the NACA Ames pilots for minimum approach airspeeds was inability to control precisely altitude or flight path at lower speeds. However, there was a surprising lack of correlation between the minimum comfortable approach airspeed and the Neumark minimum drag speed. For example, Ames pilots set the minimum comfortable approach airspeed for the Douglas F4D-1 Skyray at 121 knots, while the minimum drag speed is 152 knots. Similar results appeared with the North American F-100A Super Sabre, where a minimum approach airspeed of 145 knots was selected, as compared with the minimum drag speed of 150 knots (Figure 12.2). Clearly, some other factors than inability of the elevator or stabilizer to control height without reversal were critical.

Another set of carrier-approach tests (Bezanson, 1961) found that flight path control of the Vought F8U and Douglas F4D-1 airplanes at low landing approach speeds required use of the throttle and was not satisfactory by angle of attack or pitch control modulation alone. Bezanson found that with thrust modulation as the primary path controller the dynamic characteristics of the thrust control system became important, including such factors as throttle friction and breakout force, throttle sensitivity (pounds of thrust per inch of throttle movement), and thrust time lag following abrupt throttle movements.

In contrast to pure jet engines, turboprops are operated at high RPMs all the time. Thrust modulation is done by propeller pitch changes, with very small time lags. The poor engine dynamic behavior of pure jet engines, particularly engine thrust time lag at low power levels (Figure 12.3), kept U.S. Navy interest alive in turboprop combat airplanes long after the U.S. Air Force had switched to pure jets. For example, the Douglas/Navy turboprop A2D-1 Skyshark made its first flight in 1950, the same year as the start of production on the Boeing/Air Force B-47A six-jet bomber.

12.3   Theoretical Studies

The carrier-approach problem for naval aircraft received a great deal of attention from leading aeronautical research organizations, starting in the late 1950s. We note

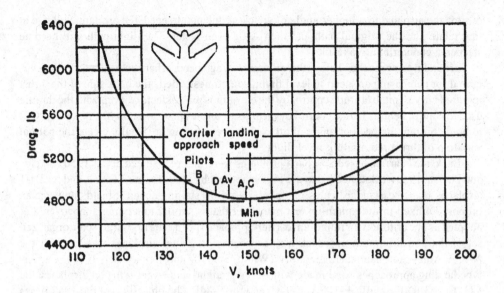

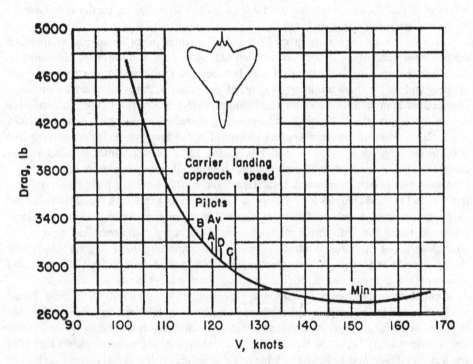

Figure 12.2   Carrier landing approach airspeeds chosen by pilots are below the airspeed for minimum drag for both the North American F-100A (*top*) and the Douglas F4D-1 (*bottom*). (From White, Schlaff, and Drinkwater, NACA RM A57L11, 1957)

particularly contributions to the theory by groups at the NACA Ames Aeronautical Laboratory, the Royal Aeronautical Establishment, and Systems Technology, Inc. Two main lines of investigation were prediction of the minimum acceptable carrier-approach airspeed for any airplane and the physics of optimum vertical path control during approaches.

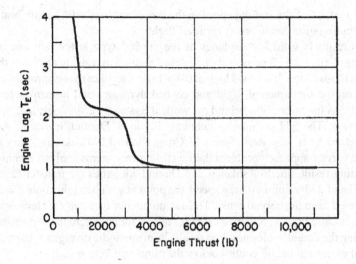

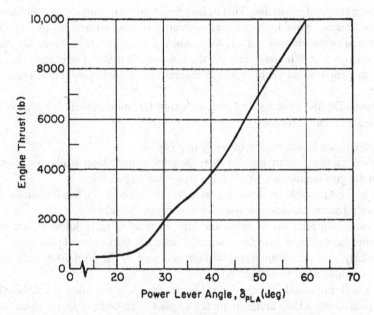

Figure 12.3 Ling-Temco-Vought A-7E engine response characteristics. Lag in developing engine thrust is large at low power settings, creating path control problems in carrier approach. (From Craig, Ringland, and Ashkenas, Syst. Tech., Inc. Rept. 197-1, 1971)

The NACA Ames group examined some five candidate predictors for minimum approach airspeeds (Drinkwater and Cooper, 1958). As found earlier, minimum drag speed correlated poorly with minimum acceptable carrier-approach airspeed. Other performance-related criteria were no better. Two that failed to correlate were the minimum airspeeds at which a given rate of change of flight path angle or a 50-foot climb could be obtained.

In the end, the Ames researchers concluded that a simple criterion based on stalling speed correlated best with the data. The minimum comfortable carrier-approach airspeed agreed best with 115 percent of the stalling speed in the power approach (PA) configuration, that is, flaps and landing gear down, power for level flight.

The Ames result is valid for airplanes of the general type tested, but one might be concerned at applying the 115-percent stalling speed prediction result to airplanes that differ radically from those tested. It seemed logical to try to develop a carrier-approach flight path model based on the fundamental flight and control dynamics and human factors of the problem. That was the motivation behind the work at Systems Technology, Inc., sponsored by the U.S. Navy. The STI engineers, including Tulvio S. Durand, Irving L. Ashkenas, Robert F. Ringland, C. H. Cromwell, Samuel J. Craig, Richard J. Wasiko, and Gary L. Teper, brought to the carrier-approach problem their well-known systems analysis techniques.

An interesting result, due to Ashkenas and Durand, identifies the transfer function parameter associated with minimum drag speed, the point at which height control by elevator becomes reversed from the normal sense. This is a numerator factor in the elevator to height transfer function called $1/T_{h1}$. Negative values of this factor put a zero in the right half of the s-plane. Closing the height to elevator loop results in an aperiodic divergence corresponding to the reversal of normal height control below the minimum drag speed.

Around 1962, Ashkenas came up with the first systems analysis basis for minimum carrier-approach airspeed prediction. That is, his prediction for minimum approach airspeed was based on assumed pilot loop closures with an airframe defined by arbitrary mass, aerodynamic, and thrust characteristics (Ashkenas and Durand, 1963). Since the systems analysis approach does not merely correlate the behavior of existing airplanes, the results should apply to airplanes not yet built whose characteristics are beyond the range of those tested so far.

The Ashkenas-Durand systems analysis prediction for minimum carrier approach airspeeds can be explained as follows:

1. The approach is assumed to be made in gusty air.
2. In gusty air the pilot attempts to close the pitch attitude loop at a higher frequency than the gust bandwidth, or as high a frequency as possible.
3. The highest possible pitch attitude loop bandwidth occurs when the pilot's gain is so high that the closed-loop system is just neutrally stable.
4. By excluding pilot model leads and lags, or treating the pilot as a pure gain, a definite gain value is associated with the neutrally stable closed pitch loop.
5. Similarly, an outer altitude control loop is closed by the pilot using pure gain, or thrust proportional to altitude error.
6. With both pitch and altitude control loops closed by the pilot, the sensitivity of the pitch control loop break frequency to pilot pitch control gain is calculated, as a partial derivative.
7. The sign of this partial derivative of pitch loop bandwidth to pilot pitch control gain, called the *reversal parameter*, is taken as an indication of carrier-approach performance. Positive reversal parameter values mean that increasing pilot gain improves bandwidth and performance.
8. The lowest airspeed at which the reversal parameter is positive is taken as a prediction of minimum carrier-approach airspeed.

The reversal parameter was refined in subsequent studies by Ashkenas and Durand in 1963 and by Wasicko in 1966. An interesting consideration was the finding in 1964 by Durand and Teper that the carrier-approach piloting technique as the airplane nears the carrier ramp changes from that assumed in the reversal parameter model. However, the approach airspeed would have already been set in the early part of the approach.

A later (1967) study of the carrier-approach problem by Durand and Wasicko went into the problem in greater detail, including the dynamics of the optical projection device that pilots use as a glide slope beam. The $1/T_{\theta 2}$ zero in the pitch attitude to elevator transfer function turned up as a primary factor, both in simulation and in landing accident rates. Unfortunately, this zero is dominated by airplane lift curve slope and airplane wing loading. Lift curve slope in turn is fixed by wing aspect ratio and sweep.

Wing loading, aspect ratio, and sweep are among the most fundamental of all design parameters for an airplane, affecting its flight performance. When a new carrier-based airplane is being laid out and wing loading, aspect ratio, and sweep are being selected to maximize such vital factors as range and flight speed, it is hard to imagine that a statistical connection with landing accident rate will be prominent in the trade-off.

Systems analysis methods were applied again to the carrier-approach problem in 1990 by Robert K. Heffley of Los Altos, California. Heffley studied the factors that control the carrier-approach outer loop involving flight path angle and airspeed. The higher-frequency pitch attitude inner loop was suppressed in the analysis, assumed to be tightly regulated by the pilot.

Heffley closed the outer loop under three different strategies, depending on whether the airplane was on the front side or back side of the drag required curve. The results give interesting insights into factors affecting the approach (Heffley, 1990). Another study in this series is an application of the Hess Structural Pilot Model (discussed in Chapter 21) to the carrier-approach problem, using a highly simplified pilot–airframe dynamic model.

The current U.S. Navy criterion for minimum carrier-approach speed, as exemplified by the system specification for the F/A-18 Hornet, gives no fewer than six possible limiting airspeeds, such as the lowest speed at which a 5-foot per second squared longitudinal acceleration can be attained 2.5 seconds after throttle movement and speed brake retraction. Heffley concludes that two additional criteria might be needed. One is a refinement of existing lag metrics to one that combines coordinated pitch attitude and thrust inputs. The other is an extension of the popup maneuver dealing with the end game, when the airplane is quite near the carrier's ramp.

12.4   Direct Lift Control

Direct lift control, in which airplane lift is modulated to correct flight path errors without changing airplane angle of attack, seems to be a natural solution to carrier-approach path control problems. According to William Koven, the first proposal for direct lift control on carrier airplanes came from Douglas E. Drake, a Douglas Aircraft engineer and former Navy pilot. Professor Edward Seckel directed follow-up studies at Princeton University. Direct lift control was first tested on a carrier-based airplane in 1964. This was a Ling-Temco-Vought F-8C, modified for the ailerons to act as variable flaps. That work was done by J. D. Etheridge and C. E. Mattlage.

The production Lockheed S-3A Viking has direct lift control, to aid in carrier landing approaches. On the S-3A, quite rapid flight path corrections are made by moving both wing spoilers, changing wing lift without changing the angle of attack (Figure 12.4). With direct lift control there is no need to wait for the airplane to respond in pitch to elevator motion, a response that takes place at the short-period pitching frequency of the airplane. A button on the S-3A pilot's yoke commands symmetric spoiler deflection.

In contrast to the relatively crude button-operated S-3A direct lift control system, a sophisticated, integrated, direct lift system is used on the Lockheed 1011 Tristar, which is fitted with the Collins FCS-240 digital automatic pilot. This Collins system, incorporating

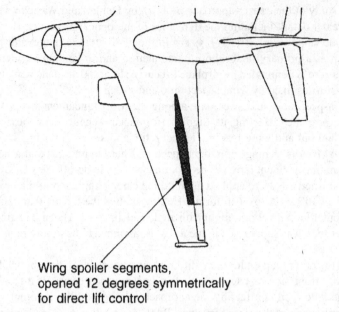

Wing spoiler segments,
opened 12 degrees symmetrically
for direct lift control

Figure 12.4   Direct lift control using wing spoilers provide satisfactory path control for carrier landings on the Lockheed S-3A Viking. (From Jane's *All the World's Aircraft*)

automatic landings or autoland, was originally developed for the L-1011's European market,
where the winter months require frequent low-visibility landings. The FAA certified the
L-1011 for Category IIIA (ceiling zero, visibility 700 feet) landings in 1981.

When the pilot selects landing flaps the four inboard wing spoiler segments are rigged up
to an 8-degree position. They are then modulated upward and downward from the up-rigged
position to obtain direct lift control. Spoiler angle changes from the up-rigged position are
commanded by the cockpit control column moved either by the pilot or the autopilot's
autoland mode, in the normal sense. That is, back control column motion closes the spoilers
and the airplane gains altitude.

If a control column adjustment is sustained for several seconds, the spoiler segment
deflections from their up-rigged positions are gradually washed out and a corresponding
adjustment is made to the horizontal tail angle. In the case of a sustained rearward control
column motion, the washout moves the spoilers back up to their up-rigged positions of
8 degrees and the horizontal tail angle is increased in the airplane nose-up sense. The
Tristar's period in the short-period pitching mode is a full 8 seconds at landing approach
airspeeds. Pitch and path corrections by horizontal tail control alone would take place at
that modal period, or rather slowly. Direct lift control provides a faster path response for
demanding all-weather manual or automatic landings. In contrast to the S-3A case where
the pilot controls vertical path with the direct lift button as well as control column motion,
in the integrated L-1011 case there is only one pilot controller, the control column.

What is missing in considering direct lift control for any airplane, carrier- or land-based,
is a method that determines when that feature is needed. What combinations of pitch and
thrust responses are such that direct lift control is required to meet specific vertical path
control requirements?

12.5   The T-45A Goshawk

Thrust system dynamics reappeared as major problems many years after the Patuxent tests of the F8U and F4D-1 and the NACA, British RAE, and Systems Technology studies. Two carrier-based airplanes, the McDonnell Douglas/British Aerospace T-45A Goshawk and the Lockheed S-3A Viking, had similar problems (Wilson, 1992).

As a land-based trainer, the British Aerospace Hawk has a large speed brake or airbrake under the rear of the fuselage, aft of the wings, in the so-called ventral position. Extended, the speed brake would hit the ground when landing. The Hawk's speed brake is thus designed to retract into the fuselage automatically when the landing gear is lowered. In common with many subsonic jet airplanes, the Hawk's jet engine is the bypass variety. Bypass jet engines provide good low-airspeed performance, with high thrust levels, and they are fuel-efficient. However, high thrust at low airspeeds means that landing approaches are normally made at idle thrust settings, or low-engine RPM. Thus, if a go-around is required, a bit more time is needed to increase RPM to maximum than for an engine without bypass.

According to George Wilson's account, U.S. Navy test pilot Captain George J. Webb, Jr., as a carrier suitability expert, flew the original Hawk airplane in November 1983 to evaluate its behavior in simulated carrier approaches. Quoting from a draft of a memorandum from Webb to Rear Admiral E. J. Hogan, Jr., commander of the Naval Air Test Center:

> Glide slope tracking [with speed brake not extended] was difficult, and corrections from high, low, and off speed conditions often resulted in numerous glide slope overshoots. Use of the ventral speed brake improved glide slope tracking and made any necessary corrections easier to accomplish.
>
> Aircraft attitude changes associated with speed corrections were very small and difficult to discern. The combined effect made it difficult for the pilot to recognize an underpowered, decelerating situation sufficiently early to make timely corrections. Consequently, student pilots will occasionally land hard or short of the runway during syllabus flights not monitored by an LSO [Landing Signal Officer].

It was thought that the T-45A would be a straightforward conversion to naval use of a simple, existing training airplane. Thus, full-scale engineering development leading to production of 300 airplanes was launched in 1984, concurrently with U.S. Navy flight tests of the Hawk, rather than after these had been completed. Some four years later, a first interim report on the McDonnell Douglas version, the T-45A, reported that the British Hawk landing approach deficiencies spotted by Captain Webb and others had been built right into the new U.S. Navy airplane.

In the end, the Hawk's single-speed brake under the fuselage, where it cannot be used in landings, was replaced on the T-45A by a pair of fuselage side brakes. These are just ahead of and under the horizontal tail (Figure 12.5). Carrier landings are made with speed brakes extended and at a high thrust level to overcome speed brake drag. At the higher thrust levels, modulation is rapid and effective in controlling the flight path. Also, just to be sure that student pilots stay out of path control trouble, the T-45A's flight idle RPM was increased from 55 to 78 percent of maximum, by adding an approach idle stop to the throttle mechanism.

There were other changes made to the T-45A, relative to the Hawk, based on Navy flight tests. This all took place after full-scale engineering development had been started back in 1984. Hydraulically operated wing slats were added to increase wing maximum lift coefficient, a higher-thrust Rolls Royce Adour engine was installed to increase forward acceleration and reduce altitude loss when a waveoff was required, the vertical tail span

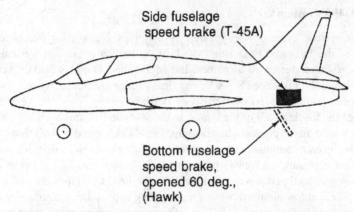

Side fuselage
speed brake (T-45A)

Bottom fuselage
speed brake,
opened 60 deg.,
(Hawk)

Figure 12.5   Path control problems for carrier landings required changes to the McDonnell-Douglas T-45A Goshawk. Speed brakes were moved from the bottom to the side of the fuselage. (From Jane's *All the World's Aircraft*)

was increased, and a yaw damper and rudder-to-aileron interconnect was added to improve lateral-directional (Dutch roll) behavior. Captain Webb had complained about this following his Hawk flights in 1983.

Also, bearing specifically on the thrust lag problem during carrier-landing operations, the Navy installed modified Lucas fuel controls on the T-45A's Adour engines, to minimize thrust lag when power increases are called for. Finally, the speed brakes are interconnected to the horizontal stabilizer to minimize trim changes when the brakes are extended or retracted.

12.6   The Lockheed S-3A Viking

Lockheed S-3A development followed a similar path to that for the McDonnell Douglas T-45A. That is, corrections for deficiencies found in the 1973 flight tests of the fourth S-3A airplane were stretched out over the next ten years. The original S-3A design had the same problem in carrier approaches as did the original T-45A Goshawk. With jet bypass engines delivering enough thrust at low engine rotation speeds to stay on the final approach path, "If all of a sudden you're starting a settle coming into the carrier, you add power to regain altitude but nothing happens because of the delay in getting the engines to respond" (Wilson, 1992).

In the case of the S-3A, the belated fix was the direct lift control system described previously. Another belated stability and control fix to the S-3A for carrier suitability is thrust trim compensation. The S-3A's low-slung engines produce longitudinal trim changes when power is used to adjust the final approach path angle, upsetting the desired constant angle of attack condition. The compensation moves the elevators automatically when the pilot adjusts the throttle position.

12.7   Concluding Remarks

While the special carrier-approach problems for swept-wing jet airplanes have been recognized for over 30 years, there seems to be no clear-cut method for predicting the severity of such problems in preliminary design, much less for adopting solutions at an early stage. The detail specification for one of the U.S. Navy's recent jet airplanes, the McDonnell

Douglas F/A-18, makes that point, listing no fewer than six possible determinants for that airplane's approach speed.

The closed-loop systems analysis approach to the carrier-landing problem would seem to offer the best chance of answering difficult questions, such as whether a new design will need direct lift control and what the upper limit might be for thrust lag following throttle motions. However, the closed-loop systems analysis approach apparently requires additional development before it is ready to be used in this design sense. Systems analysis study reports typically close with a "Need for Further Research" section.

CHAPTER 13

Ultralight and Human-Powered Airplanes

The category of ultralight airplanes ranges from hang gliders to light versions of general-aviation airplanes. They fill a need for experimenters and for pilots who want to fly inexpensively and with little regulation. Ultralight airplanes evolved as did the early flying machines, by much cut-and-try and flight testing. Although these designs have been useful, indications are that many commercial ultralights are deficient in stability and control.

Human-powered airplanes are extreme ultralights, designed not for practicality but to push the engineering and human limits of aviation. Early efforts at human-powered flight were discouraging because of the poor performance and extreme fragility of the machines that were constructed before the first successful one, the Gossamer Condor.

13.1   Apparent Mass Effects

For very light airplanes, not much heavier than the air in which they fly, apparent mass effects must be considered. These effects were first noticed in 1836 by George Green, who found that pendulum masses in a fluid medium were apparently greater than in a vacuum. The apparent mass effect can be described as follows (Gracey, 1941):

> The apparent increase in mass can be attributed to the additional energy required to establish the field of flow about the moving body. Inasmuch as the motion of the body may be defined by considering its mass as equal to the actual mass of the body plus a fictitious mass, the effect of the inertia forces of the fluid may be represented as an apparent additional mass; this additional mass, in turn, may be considered as the product of an imaginary volume and the density of the fluid. The effect of the surrounding fluid has accordingly been called the *additional mass effect*. The magnitude of this effect depends on the density of the fluid and the size and shape of the body normal to the direction of motion.

The primary motivation for Gracey's work was to be able to correct airplane and wind-tunnel model moments of inertia measured by suspending the airplanes or models and swinging them as large pendulums. To the extent that the NACA was involved with equations of motion for the airships of those days, this would have given Gracey yet another motivation to study apparent mass.

The 1941, the NACA apparent mass tests were made by swinging covered frameworks of various shapes as compound pendulums. The test specimens were swung both in air and in a vacuum tank. It is interesting that Gracey started out with balsa wood shapes, but found that their weights varied with air pressure and humidity. Gracey's training in this exacting experimental work must have helped him to appear later on as NASA's expert in airspeed and altitude measurement methods.

Interest in apparent mass effects returned with the advent of the plastic and fiber materials that could be used to build very light airplanes, such as the human-powered Gossamer Condor and the high-altitude, long-duration pilotless airplanes Pathfinder and Helios, all built by AeroVironment, Inc., of Monrovia, California. Apparent mass effects are important as well for lighter-than-air and for underwater vehicles. Mathematical models of these craft for dynamic stability analysis include apparent mass terms, as a matter of course. In the series expansions for aerodynamic forces and moments originated by G. H. Bryan, apparent mass terms appear as derivatives with respect to linear and angular accelerations.

Lacking the vacuum swinging apparatus of Green and Gracey, one can approximate apparent mass terms in the equations of airplane motion by adding cylindrical air masses to the lifting surfaces, with diameter equal to the surface chord for motions normal to the chord and equal to the surface thickness for motions in the chord plane. This approximation yields the following astonishing results for the Gossamer Condor. The apparent masses in lateral and vertical motions are 21 and 170 percent of the actual airplane mass. The apparent moments of inertia in pitch and roll are 140 and 440 percent of the actual moments of inertia.

In addition to measurements on swinging models and the approximations mentioned above, panel computer codes can be used for apparent mass estimation. David A. Lednicer reports that the VSAERO code is used routinely for apparent mass calculations on underwater vehicles.

13.2 Commercial and Kit-Built Ultralight Airplanes

There are three classes of commercial and kit-built ultralight airplanes, each with interesting stability and control characteristics. Using the terminology of the influential *Jane's All the World's Aircraft*, the most simple is the classical modern hang glider, developed from the Rogallo wing (Rogallo et al., 1960). Control is obtained by shifting body weight, as in the nineteenth-century Lilienthal hang gliders. The next level of sophistication is called a parawing. It is a powered ram air parachute. Finally, there is the broad category of microlights. These airplanes range from powered hang gliders to lightly constructed airplanes of conventional layout. Like the hang glider, microlights use fabric-covered light structures.

The FAA in the United States and the CAA in the United Kingdom have each developed certification provisions for ultralight airplanes, FAR Part 103 (1990) and BCAR Section S-CAP 482, respectively. FAR Part 103 applies to unpowered ultralights weighing less than 155 pounds and powered ultralights weighing less than 254 pounds; powered ultralights have top speeds less than 55 knots and stalling speeds less than 24 knots. Part 103 specifically exempts these ultralights from meeting airworthiness standards. Operating rules only are specified. Other countries use the FAA and CAA standards for their own certifications.

A comprehensive review of hang-glider stability and control is presented by Anderson and Ormiston (1994). Longitudinal trim is provided by reflexed airfoil shapes. Directional stability is generally positive because of wing sweep. Geometric dihedral is adjusted for neutral spiral stability at cruise. A surprising finding is low Dutch roll stability at low angles of attack. This has led to pilot-induced oscillations, augmented by inadvertent swing of the body in response to side accelerations. Hang-glider full-scale tests on an outdoor mobile test rig were conducted at Cranfield University (Cook and Kilkenny, 1987).

The stability and control characteristics of powered hang gliders, called flexwing airplanes by the author, were discussed by Brooks (1998). Turn control of these machines is unconventional, as in the case of the Gossamer Condor. To turn right, the pilot's weight is moved to the right by exerting force to the left on the control frame base bar. The weight moment and aeroelastic wing flexing (right wing washout) combine to start a right roll. Adverse yaw causes right sideslip. Anhedral, or negative dihedral effect, increases the right rolling moment, accelerating the turn.

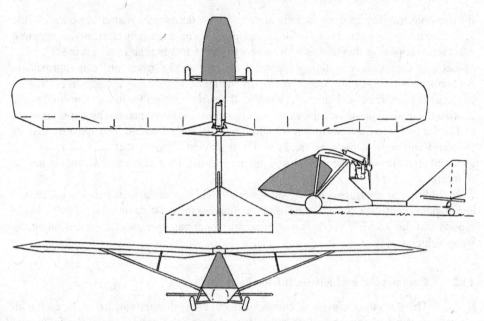

Figure 13.1   Chinook WT-11 ultralight airplane, general arrangement. (From Roderick, 1986)

A brief wind-tunnel test of the fabric-covered wing and tail surfaces of a Chinook WT-11 ultralight airplane (Figure 13.1) conducted in the Canadian NAE 9m by 9m low-speed wind tunnel (Roderick, 1986) showed some unusual characteristics. There was noticeable wing twist at higher dynamic pressures, which decreased wing lift curve slope. With the elevator deflected, the horizontal tail had nonlinear lift curves near its stall. The investigators concluded that a large amplitude pitch down at the stall was a possibility.

Aside from the technical findings of these investigators, experience has shown that inadvertent stalling is a major cause of ultralight accidents. Operators of ultralights under U.S. FAR Part 103 are not required to pass knowledge or experience tests. However, avoidance of inadvertent stalling during the demanding operations of approach and landing requires careful training.

13.3   The Gossamer and MIT Human-Powered Aircraft

A general arrangement drawing of the original Gossamer Condor human powered airplane is shown in Figure 13.2. All of the Gossamer aircraft are canards because of the packaging convenience of the pilot–powerplant combination with its plastic chain-driven pusher propeller. Like the Wright brothers before them, the team found it relatively easy to control pitch attitude (and angle of attack) even though the center of gravity was behind the neutral point. An opposite wing warp and tilted canard method of roll control developed for the Gossamer Condor (Sec. 5) was applied as well to the Gossamer Albatross. Finally, the Gossamer team pioneered in the art of using carbon fiber plastic tubes as bending and compression members in the airframe structure, thereby ensuring a very light weight (Grosser, 1981).

The MIT human-powered Chrysalis biplane and Monarch and Daedalus monoplanes were less radical in design than the Gossamer series, with tractor propellers and aft tails. All-moving tail surfaces and warping wings for lateral control were used at first. The Daedalus machine dispensed with wing warping or ailerons, relying on rudder control and dihedral effect. This proved insufficient.

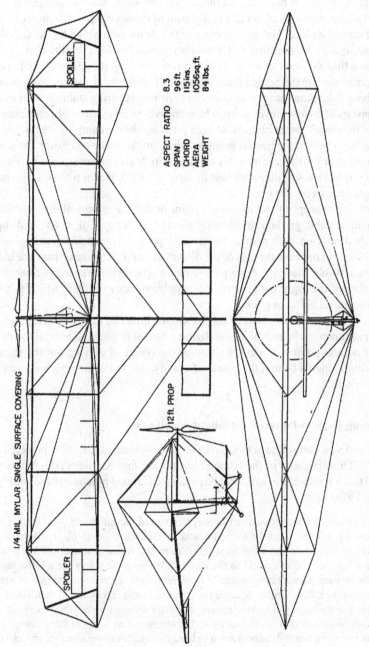

ASPECT RATIO 8.3
SPAN 96 ft.
CHORD 115 ins.
AERA 1056 sq. ft.
WEIGHT 84 lbs.

SPOILER

SPOILER

1/4 MIL MYLAR SINGLE SURFACE COVERING

12 ft. PROP

Figure 13.2 Initial version of the Gossamer Condor human-powered ultralight airplane, with single-surface wings and spoilers for lateral control. The airplane could not be turned with the spoilers. (From Burke, "The Gossamer Condor and Albatross," AIAA Professional Study Series, 1980)

13.4   Ultralight Airplane Pitch Stability

Conventional ideas about the need for longitudinal static stability are misleading in the case of ultralight airplanes. The reason is that, instead of the normal short- and long-period, or phugoid, modes of motion, four unfamiliar first-order modes may appear. For example, the Gossamer Condor's center of gravity is aft of the neutral point, in order to unload somewhat the canard surface. This produces a positive or unstable value for the C_{m_α} derivative. As a result, one of the four first-order modes is unstable. However, the corresponding divergence has a time constant of about 1,000 seconds, making it imperceptible to pilots.

Another way to explain the benign pitch behavior of ultralight airplanes flying at centers of gravity behind the neutral point is to consider their maneuvering stability. Maneuvering stability disappears at the maneuver point. The maneuver point of ultralight airplanes tends to be far aft of the neutral point because of high pitch and heave damping levels. For flight at centers of gravity behind the neutral point but ahead of the maneuver point, the machine would have no tendency to diverge unstably in pitch attitude at constant airspeed. Its unstable behavior would require a simultaneous loss of airspeed and nose-up pitch change in level flight, a process that is very slow.

To illustrate the concept of the maneuver point or maneuvering stability, consider an airplane with an unstable gradient in pitching moment with angle of attack, and suppose it to be disturbed nose-up with respect to its flight path. The unstable pitching moment gradient would tend to increase the size of the disturbance, but at the same time the increase of angle of attack would cause the flight path to curve upward if the speed is constant. The upward curvature of the flight path implies an angular velocity in pitch, which is resisted by the aerodynamic damping in pitch.

In the case of the Gossamer aircraft, the stabilizing effect of the pitch damping due to flight path curvature overwhelms the destabilizing gradient of pitching moment with angle of attack. The neutral point is 5 percent ahead of the center of gravity, but the maneuver point is *four* chord lengths behind the center of gravity, due to the large path curvature for a given angle of attack.

13.5   Turning Human-Powered Ultralight Airplanes

Roll and yaw control have emerged as major problems for human-powered ultralight airplanes. This appeared in the competition for the first Kremer Prize. Winning the prize required that a figure-eight maneuver be performed around pylons one-half mile apart. Henry R. Jex (1979) says:

> Early analyses [of the Gossamer Condor] ... revealed the futility of conventional aileron control for roll, the need for some fin area to enhance C_{y_β} and C_{n_β}; and the seeming paradox that twisting the wing for a leftward rolling moment would quickly produce a yawing velocity and roll angle to the *right*! ... Warping the wings for a left-wing-down rolling moment immediately creates a large nose-right adverse yaw torque of about 15 percent of the rolling torque. Because the effective inertia in yaw is less than 1/5 of that in roll [due to the large roll apparent mass] and the weathervane stability is very small, the airframe immediately starts to yaw right. The strong rolling moment from yawing [due to the outside wing moving faster than the inside wing] quickly overpowers the [control rolling moment due to wing warp], so the airplane starts to roll *right* within about 2 seconds!

The initial Gossamer Condor version could not be turned with spoiler ailerons. A solution was found for this failure of conventional aileron control for ultralight airplanes in the invention of a new control method. Like the Wright brothers, the inventors of the new control method, Dr. Paul B. MacCready, Dr. Peter Lissaman, and James D. Burke, applied for a patent for their scheme. In the new method, pilot roll control tilts the canard and its lift vector by means of tabs at the outboard trailing edge and concurrently warps the wing in the opposite direction. Nose-right tilt of the canard and the adverse yaw of the leftward warped wing pull the nose right. The airplane is rolled right by the strong rolling moment due to right yawing C_{l_r}. The result is a slightly overbanked turn in which inward (right) sideslip develops. Canard tilt is reversed a little later. The final wing and canard settings for a stabilized, coordinated right turn at a 2.0-degree bank angle are 4.1 degrees of left wing warp and about 5 degrees of left canard tilt (Figure 13.3). Because of the ultralight airplane's low flight speed of 16 feet per second, the 2.0-degree bank angle produces a standard rate "needle-width" turn of 180 degrees per minute. The turn radius is 300 feet.

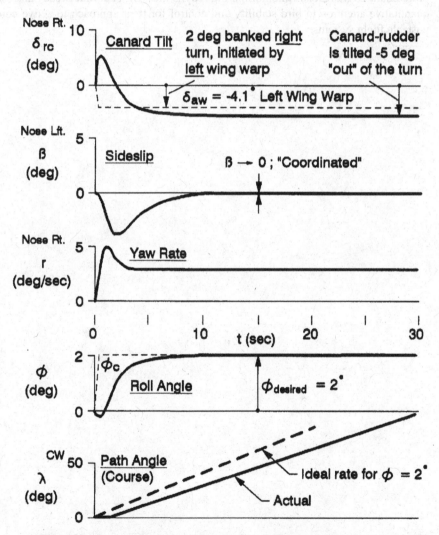

Figure 13.3 Gossamer Condor turn to the right. Coordination requires opposite wing warp (left in a right turn) to overcome higher airspeed of the left or outboard wing. The canard is tilted with a 1-second pulse of aileron tab initially to the right, to start the turn. Canard tilt is then reversed to keep from overbanking the airplane. The resultant 3-degree-per-second yaw rate is sufficient. (From Jex and Mitchell, NASA CR 3627, 1982)

13.6   Concluding Remarks

Ultralight airplanes have satisfied a need for inexpensive, lightly regulated machines. At the same time, microlight versions have proved to be useful in specialized applications such as surveillance and crop dusting. Stability and control deficiencies have surfaced that had no previous history. Inadvertent stalling is a significant cause of accidents.

Human-powered ultralight airplanes have been useful for the original thinking that they have inspired, more than for the utility of the machines produced. But having developed new stability and control concepts for some of these machines, one would like to see the scaling laws that connect these concepts with the characteristics of conventional airplanes. It would be instructive to see how the anomalous longitudinal and turning behavior of human-powered ultralights blends into normal flight dynamics. Dr. Paul MacCready also suggests quantitative attention to bird stability and control for their approach to active control of unstable flight systems.

CHAPTER 14

Fuel Slosh, Deep Stall, and More

In nearly 100 years of controlled flight the stability and control field has seen any number of special gadgets and phenomena that fit into no general category. We recall some of the most interesting of these.

14.1   Fuel Shift and Dynamic Fuel Slosh

The term *fuel shift* refers to long-term motions of the fuel in a partially filled tank, such as a shift to the rear of an airplane's tank caused by a prolonged nose-up attitude, in a climb. The causes and effects of fuel shift are apparent. Aft fuel shift could move the airplane's center of gravity to a dangerously rearward position. This possibility is generally considered by every designer. There is even the possibility of fuel starvation if the tank feeds the airplane's engine from a forward sump.

Dynamic fuel slosh occurs when fuel in a partially filled tank, be it an automobile or airplane tank, sloshes around inside the tank in response to changing vehicle accelerations. As the tank walls contain the sloshing fuel, transient forces are transmitted to the walls, and the vehicle, by the fuel. Dynamic fuel slosh can be a problem in airplane stability and control if the fuel modes of motion couple with the airplane's normal modes of motion. The dynamic fuel slosh problem is worth examining because modern jet airplanes tend to have high ratios of fuel to gross weight and slosh motions could have a considerable effect. Also, on airplanes with relatively thin wings, the main tankage tends to be in the fuselage. Wing tanks are generally interrupted by structural members, which act as baffles to fuel motion, while fuselage tanks can have large uninterrupted volumes.

Although dynamic fuel slosh is an intriguing mathematical and engineering problem, documented cases of dynamic fuel slosh coupling with an airplane's modes of motion are rare. There was a verified case of dynamic fuel slosh coupling with the Dutch roll mode of motion on the Douglas A4D Skyhawk. Partial fuel in a 500-gallon fuselage tank forward of the center of gravity caused undamped roll–yaw oscillations during landing approaches (Figure 14.1). The problem was corrected when fore-and-aft vertical tank baffles were added to the fuselage tank, dividing the tanks into left and right halves. The baffles almost doubled the fuel slosh frequency, decoupling slosh from the Dutch roll.

There was another reported dynamic fuel slosh problem on the Lockheed F-80C airplane. This was during an all-out drive to improve the loitering capability of F-80s in the early months of the Korean War. F-80 units in Korea started carrying unbaffled 265-gallon tip tanks, but soon reported unexplained crashes. At the request of Headquarters, Far East Air Forces, Wright Field flight-tested the 265-gallon tanks on an F-80C.

In a test flight in November 1950, James D. Kelly found no flight problems at takeoff and climb-out, when the wing tip tanks were essentially full. Later, with partial tip tank fuel, an uncontrollable pitching motion started. Kelly could see the wing tips twisting as fuel sloshed fore and aft. He recovered control only after the left tip tank collapsed and fell away and he could jettison the right tank.

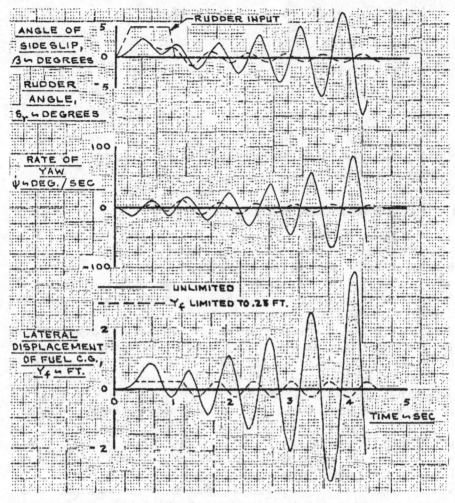

Figure 14.1   Calculated effect of fuselage tank slosh in the Douglas A-4 Skyhawk before installation of a fuselage tank baffle. Fuel motion couples with the Dutch roll mode of motion. With the fuel mass motion limited by the tank sides (dotted curves), a steady limit cycle "snaking" motion results. (From Abzug, Douglas Rept. ES 29551, 1959)

Two additional dynamic fuel slosh cases are known, both documented in U.S. Air Force flight test reports. Both involved fuel slosh coupling with the Dutch roll mode of motion and were excited when the rudder was used to stop the yawing component of the Dutch roll motion. The airplanes with this problem were the Boeing KC-135A and the Cessna T-37A.

An analytical approach to the problem of dynamic fuel slosh coupling with the modes of airplane motion was made possible by a model proposed by Ernest W. Graham (Luskin and Lapin, 1952). Graham used the velocity potential for liquid in a rectangular open-top tank given in Horace Lamb's classic *Hydrodynamics*. Sloshing fuel is modeled as a simple pendulum plus a fixed mass below the pendulum. The pendulum angle from the vertical is taken as the average fuel surface angular displacement in its fundamental mode of motion (Figure 14.2). The fuel's general motion has higher harmonics, of shorter wave lengths, all of which are neglected to define the equivalent pendulum.

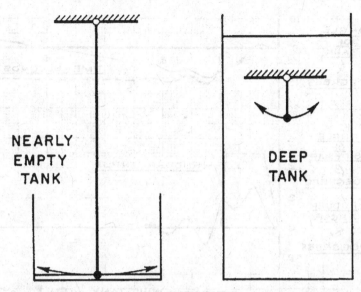

Figure 14.2 Ernest W. Graham's pendulum analogy to sloshing fuel in open-top rectangular tanks, illustrated by two extreme cases. All of the fuel sloshes in the nearly empty tank, but in just the upper portion of the deep tank. (From Luskin and Lapin, *Jour. of the Aeronautical Sciences*, 1952)

The pendulum analogy to dynamic fuel slosh depends mainly on the depth of the fuel in the tank. In the nearly empty case, fuel slosh is simply the wave action in a shallow container. The equivalent pendulum is long, and so is the period of the sloshing motion. In the nearly full or deep tank, the equivalent pendulum is short and so is the slosh period. Graham's model permitted subsequent analysts to treat the problem of fuel slosh by adding the equivalent pendulum to the equations of airplane motion, as an additional degree of freedom for each partially filled tank.

Albert A. Schy, a skilled NACA analyst, set up the fuel slosh problem without using the Graham pendulum model, by assuming fuel is carried in spherical tanks (Schy, 1952). While spherical tanks are never seen in airplanes, Schy's model is altogether equivalent to Graham's model for conventional rectangular or prismatic tanks. Schy's calculations show significant coupling into the Dutch roll mode of an airplane when the sloshing fuel mass is one-fourth the weight of the airplane.

Dynamic fuel slosh coupling with the longitudinal short- and long-period modes of motion is seldom a problem. There is a slight loss in short-period mode damping, but the long-period or phugoid mode cannot couple measurably with fuel in a partially filled tank unless the fore and aft dimension of the tank is impossibly long (Luskin and Lapin, 1952). Recent dynamic fuel slosh studies on a modern swept-wing jet with long fuel tanks running parallel to the wing spars came up dry, in the sense that slosh in any pair of partially filled tanks had negligible effects on the aircraft's modes of motion.

In contrast to the modest effects of dynamic fuel slosh (but not long-term fuel shift) for the airplane case, dynamic fuel slosh has been an ongoing concern for large liquid-fueled boost or launch vehicles, such as NASA's Saturn V. Launch vehicle dynamic fuel slosh problems have included coupling with controlled pitch and yaw modes of motion as well as with elastic body bending modes.

Turning to the fuel shift problem in airplanes, an interesting case occurred on the Douglas A4D Skyhawk during early test flights. The A4D's ultrasimple fuel system has only two

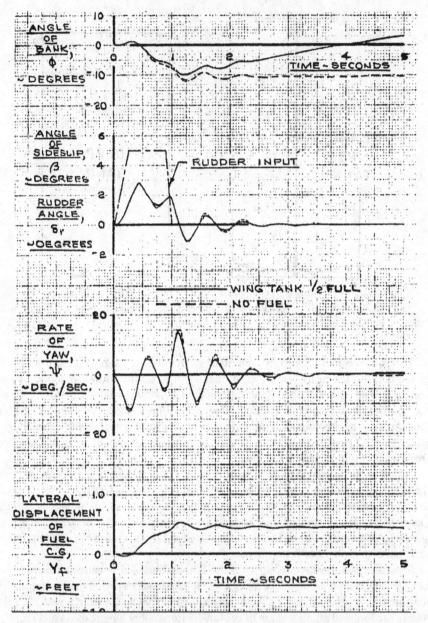

Figure 14.3   Calculated effect of fuel shift in the Douglas A-4 Skyhawk's wing tip–to–wing tip wing tank. With half-wing fuel, a rudder kick shifts the fuel, giving apparent spiral instability. The bank angle increases slowly with time. (From Abzug, Douglas Rept. ES 29551, 1959)

tanks, the fuselage tank, which had sloshing problems before a baffle was installed, and a single integral wing fuel tank, which runs from wing tip to wing tip. The wing ribs provide excellent slosh baffling, but prolonged lateral acceleration can transfer partial wing fuel across the airplane's centerline.

A4D wing fuel shift shows up as spiral instability, easily corrected by the ailerons (Figure 14.3). However, the early A4D airplanes had a single, or nonredundant, aileron hydraulic system. When aileron hydraulics malfunctioned at a high Mach number during an early test flight the airplane and pilot James Verdin were lost. The painful lesson was learned. Production A4D (now A-4) airplanes retain the single integral wing fuel tank, but dual, independent aileron power systems now guard against loss in lateral control due to fuel shift.

Another fuel shift incident from the same time period occurred at Wright Field in a North American YF-100 being flown by Captain H. Z. Hopkins. He took off for a short functional check of the 275-gallon external fuel tanks. Only 50 gallons were loaded into each tank; takeoff acceleration shifted this load aft. Fuel was being burned from the forward fuselage tank, adding to the aft shift in cg. The cg apparently shifted aft behind the maneuver point, the point at which pull stick forces are required for positive-load factors. The airplane went through a rapid sequence of positive and negative maneuvers. The external tank fuel somehow shifted forward, and the structurally damaged airplane was brought back under control and landed.

14.2   Deep Stall

Deep stall requires a stable longitudinal trim point beyond the stall. If, in a stall, the combination of longitudinal trim and control is insufficient to nose the airplane down to an unstalled attitude, the condition is called a deep stall or locked-in deep stall. In some cases short of a locked-in stall, control power is so marginal that recovery takes place slowly or requires unusual measures, such as rolling or sideslipping the airplane or rocking the airplane in pitch. Deep stall was first encountered in-flight on a Handley Page Victor bomber in 1962. Hawker Siddeley's Trident 1C, British Aircraft Corporation's BAC 1-11, and the Soviet Tu 134 subsequently experienced deep stalls. The details of a BAC 1-11 deep-stall crash were widely disseminated, leading to a new series of wind-tunnel tests of airplanes then under development, such as the McDonnell Douglas DC-9. There were subsequent accidents in which deep stall was suspected, notably on the Boeing 727, a jet that resembles the BAC 1-11. The Canadair (Bombardier) Challenger CL600 also resembles the BAC 1-11. A CL600 and a variant were lost in deep-stall accidents.

There is an underlying cause for deep stall in airplanes with horizontal tails mounted on top of the vertical tail, in the T position. The wing trailing vortex system is normally rolled up into concentrated vortices by the time it reaches the horizontal tail. In unstalled flight the rolled-up vortices are generally behind the wing tips, quite a bit outboard of the horizontal tail span. This weakens the downwash at the horizontal tail, a source of nose-up pitching moment.

In airplanes prone to deep stall the outboard wing panels stall first when the angle of attack is increased. In effect, new wing tips are formed at the tips of the unstalled wing portion. The rolled-up trailing vortices are now quite close in span to the horizontal tail. If the horizontal tail is in the T location, at the top of the vertical tail, the rolled-up vortices at high angles of attack are roughly in the same plane as the horizontal tail, the position to exert a maximum downwash and nose-up pitching moment. The aerodynamic flow conditions for a deep stall on an airplane with a T-tail are illustrated in Figure 14.4.

NASA Ames 40- by 80-Foot Wind Tunnel tests of a full-scale Learjet, Inc., Model 23 executive jet provide a clear example of a deep stall at an aft center of gravity position (Soderman and Aiken, 1971). The airplane has a T-tail and a moderate wing sweep of 13 degrees. At the aft center of gravity position of 31.5 percent MAC, the data show a stable trim point at an angle of attack of 39 degrees, well beyond the stall. This is with full airplane nose-down stabilizer trim of 0.4 degree. With full 15-degree down-elevator added for recovery, the diving moment is insufficient to recover unstalled flight (Figure 14.5).

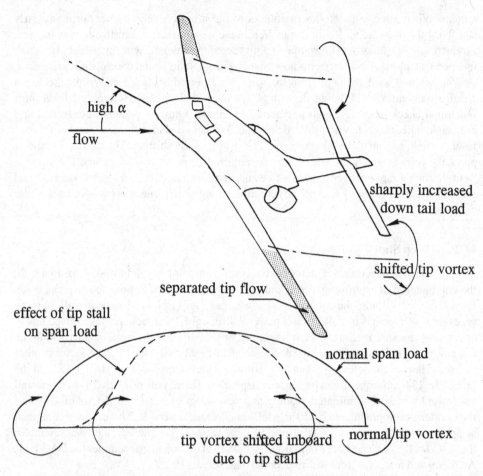

high α

flow

sharply increased
down tail load

shifted tip vortex

separated tip flow

effect of tip stall
on span load

normal span load

tip vortex shifted inboard
due to tip stall

normal tip vortex

Figure 14.4 Aerodynamic flows for deep stall for an airplane with a T-tail. Separated wing tip flow shifts the tip vortices inboard. Cores of the shifted tip vortices are in line with the T-tail, giving maximum downflow and a sharply increased down tail load.

The General Dynamics F-16 can also get into a deep stall (Figure 14.6). The F-16 has a special manual pitch override stick switch giving the pilot full tail travel, canceling roll and stability augmentation functions. This is to allow the pilot to rock out of a deep stall, in phase with residual pitch oscillations (Anderson, Enevoldson, and Nguyen, 1983). They report:

> The instructions [flight manual] note that if no increase in attitude is discerned (with full [nose-up] pitch control), the pilot should wait 3 seconds and apply full reverse control. If the nose does not continue down with full forward stick, but reverses and starts up, full back stick must be applied to continue rocking the aircraft. The pitch oscillation has a period of approximately 3 seconds and the pilot is warned that rapid cycling of the control will be ineffective.

As many as four cycles are needed to break the F-16's deep stall. Proper rock phasing is difficult if the airplane is in a roll oscillation, since severe rolling masks the airplane's pitching motions. The ultimate fix for F-16 deep stall was a 25-percent increase in horizontal tail size, incorporated in all production airplanes (Chambers, 2000).

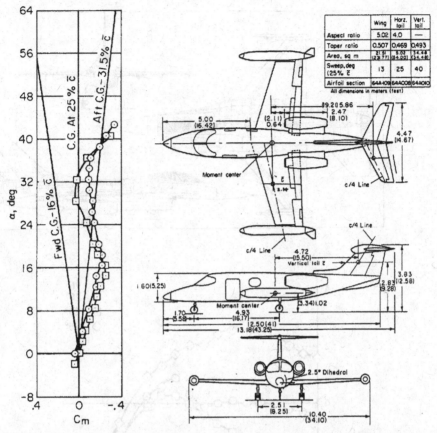

Figure 14.5 Pitching moment coefficient variation with angle of attack for a full-scale Learjet Model 23, as tested in the NASA Ames 40- by 80-foot wind tunnel. At an aft center of gravity position of 31.5 percent MAC there is a stable trim point at an angle of attack of 39 degrees. This is with full nose-down trim. Circle symbols are flaps up, squares are flaps down. Addition of full down-elevator angle is insufficient to regain unstalled flight. (From Soderman and Aiken, NASA TN D-6573, 1971)

The McDonnell Douglas C-17 military cargo airplane also has a locked-in deep stall potential because of its T-tail. However, in contrast to the F-16 case, a sophisticated angle of attack limiter scheme prevents deep stalls from occurring. A control column pusher can prevent deep stall by overcoming the pilot pull force that is leading to a stall with an opposing push force. This approach was rejected because of reliability, particularly the possibility of a single point failure (Iloputaife, 1997).

The deep stall prevention system selected for the C-17 relies on a measured angle of attack providing an initial aural warning and shaking of the control column at a soft limit angle of attack. If the angle of attack continues to increase, or if airspeed drops at an excessive rate, the attack limiter system switches on, an angle of attack command system having a hard limit.

Interesting design features of the C-17 angle of attack limiter system are the array of six fuselage-mounted vane flow direction sensors and the provisions to switch the limiter in and out without causing undesirable transients or secondary stalls. The sensor array permits the system to operate correctly under sideslip conditions, which affect individual vane readings, and with redundancy in the face of vane failure or damage.

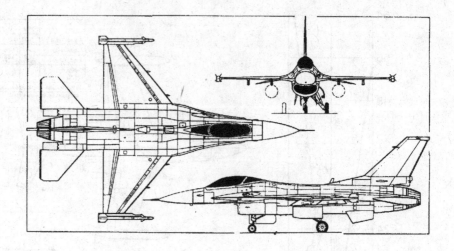

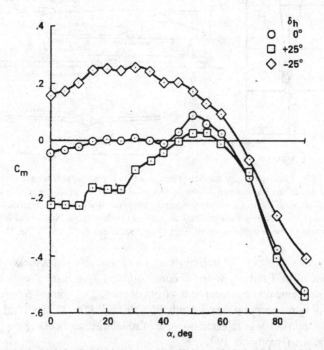

Figure 14.6 Variation in pitching moment coefficient with angle of attack for the General Dynamics F-16A, with zero and maximum tail surface angles. Even with full nose-down control (+25 degrees) there is a stable or locked-in trim point at an angle of attack of 60 degrees. (From Nguyen, Ogburn, Gilbert, Kibler, Brown, and Deal, NASA TP-1538, 1979)

14.3   Ground Effect

The fact that the close approach of an airplane to the ground is accompanied by substantial changes in its aerodynamic characteristics has been known for some time.

This is the opening statement in an NACA report by J. W. Wetmore and L. I. Turner, Jr. (1940). Ground effect theory dates from 1922 to 1924, with work by C. Wieselsberger in

Germany and H. Glauert in Britain. The stability and control effects of the close approach of an airplane to the ground primarily are in longitudinal control and trim. A conventional or tail-last airplane requires more nose-up longitudinal control to hold particular angles of attack, or to stall, near the ground. Likewise, the presence of the ground increases the amount of nose-up control required to lift the nose wheel during takeoff ground runs.

It is interesting that the aerodynamics of an airplane model under test in a closed test section wind tunnel and that of a full-scale airplane near the ground are treated by similar theoretical methods. The solid flow boundaries caused by the wind-tunnel walls and by the ground under the airplane are represented in theory by images of the model or airplane whose downwash and sidewash flows just cancel the flow velocities that would ordinarily cross the solid boundaries.

The ground effect case is altogether simple, since a single image system does the cancellation exactly. The required image is a mirror image of the airplane, the ground itself taken as the mirror's surface. On the other hand, for closed test section wind-tunnel walls an infinite series of images is required. Of course the most distant images are neglected for practical calculations.

The mirror image method of representing ground effect carries over into the modern application of computational aerodynamics to stability and control. That is, vortex lattice models give ground effect on longitudinal control and trim when an image vortex lattice system is added to the basic airframe lattice. This application was used in estimating ground effect for a recent tailless airplane design.

In contrast to the generally well-understood ground effect theory and mathematical modeling, ground effect measurements in the wind tunnel and in-flight testing are usually less than satisfactory. Low-speed wind-tunnel model tests of new designs, if they are reasonably well funded, generally include a few ground board tests. A board spanning the wind tunnel and extending some distance ahead of and behind the model is supported at a distance representing the ground. Pitch runs with ground board in place and removed give the desired ground effect increments in lift, drag, and pitching moment.

There are two main problems with ground effect wind-tunnel model tests, aside from the normal scale or Reynolds number differences between model test and the full-scale airplane. The ground board installation itself creates flow blockage in the wind tunnel, with unwanted cross-flow and buoyancy influences on the measurements. Also, a boundary layer stands on the ground board, starting at the leading edge of the board. There is no analog to the ground board boundary layer in real landings or takeoffs. The ground board boundary layer can be minimized by bleeds or by using a moving belt in place of the ground board. However, blockage effects must somehow be subtracted from the data.

Ground effect flight tests are equally problematic, if one wants to measure actual increments in lift and control angles for trim rather than just whether the airplane has satisfactory control in the presence of the ground. The difficulty is of course maintaining stabilized flight at different angles of attack while flying with wheels inches from the ground and making measurements.

The Wetmore and Turner 1940 work that is quoted at the start of this section was a model of clever flight test design. The problem of getting stabilized flight near the ground was solved by taking ground effect measurements on a glider towed behind a car (Figure 14.7). Car buffs will enjoy the photo of NACA's tow car. It appears to be a Chevrolet faired to resemble a Chrysler Airflow sedan. The glider pilot maintained the prescribed height above the ground by sighting targets mounted on the car.

The Franklin PS-2 glider, towed behind a car for ground effect tests.

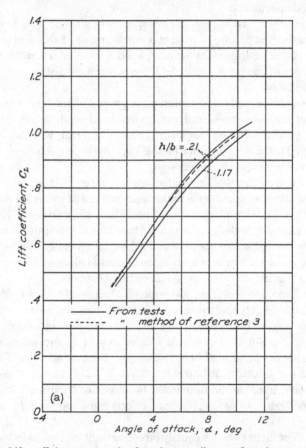

Figure 14.7 Lift coefficient versus angle of attack at two distances from the ground, as measured with a Franklin PS-2 glider towed behind a car. (From Wetmore and Turner, NACA Rept. 695, 1940)

Stabilized flight near the ground could perhaps be maintained for modern ground effect tests using autopilot loops closed around height-finding signals, such as given by radar altimeters. Likewise, NASA has a model-launching rig at Langley Field that can obtain ground effect data without wind-tunnel ground board problems. The Langley rig also simulates any unsteady aerodynamic effects that may be significant. That is, real landings are dynamic affairs in that the airplane comes near the ground and lands in some time that may be short compared with the time required for stabilized flows to be established. Limited vortex lattice calculations show that dynamic effects cause an increase in lift and pitching moment due to ground effect, as compared with steady-state conditions.

14.4 Directional Stability and Control in Ground Rolls

The modern light plane tricycle landing gear has main wheels behind the center of gravity and a steerable castering, or freely swivelling, nose wheel ahead of the center of gravity. This arrangement, invented and applied by Fred C. Weick (1936), put an end to the ground loop. The ground loop is a rapid yaw from the runway heading and a swerve off the runway. It is a problem for tail wheel landing gears, which were still used by some designers for many years after Weick's invention.

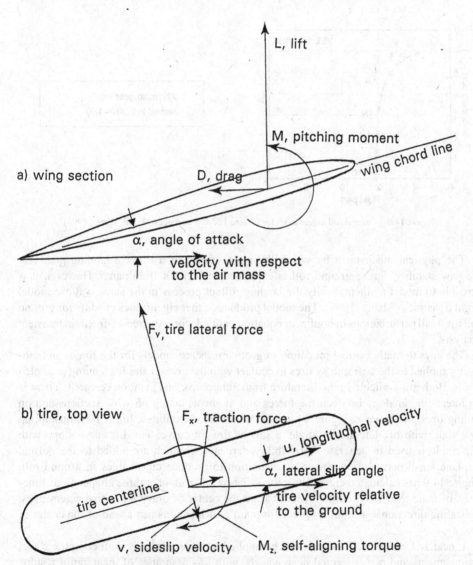

Figure 14.8 Forces and moments acting in wing sections and rolling tires. (a) Wing section. (b) Tire, top view. (From Abzug, 1999).

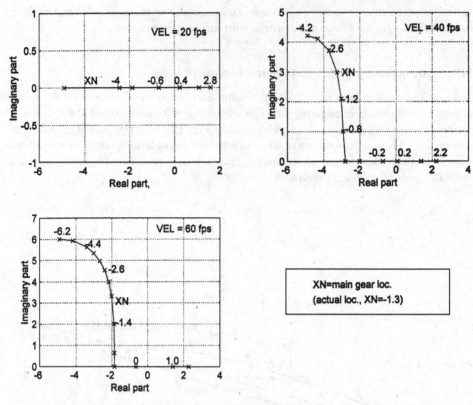

Figure 14.9   Ground roll eigenvalues for Cessna 182 at 3 airspeeds. (From Abzug, 1999)

The physical mechanism by which the main wheels of a tricycle landing gear create yaw stability during ground roll is explained in Weick's 1936 paper. However, it is possible to model mathematically the landing rollout process in the same way we model flight dynamics (Abzug, 1999). The model produces either eigenvalues or roots for ground rollout small perturbations or nonlinear equations suitable for 6-degree-of-freedom transient analysis.

The keys to mathematical modeling of ground rolls are models for the forces and moments applied to the airframe by tires in contact with the ground and for landing gear oleo struts. Both are available in the literature from automotive and aviation research. There is an interesting analogy between the forces and moments acting on wing sections and on rolling tires, as shown in Figure 14.8. Tire lateral force exhibits a linear relationship, up to a stall, with tire lateral slip angle, a sort of tire lift curve. Tire lift curve slope with slip angle is used to generate tire stability derivatives, which are added to the normal airplane small-perturbation equations of motion to produce eigenvalues in ground roll. Figure 14.9 has calculated eigenvalues for a Cessna 182 rollout at three airspeeds, as functions of main gear longitudinal distance from the center of gravity. Positive eigenvalues, indicating directional instability, occur for main gear locations just ahead of the center of gravity.

Linearized ground roll analysis can be applied to large airplanes with complex wheel arrangements and power-steered nose wheels, with less assurance of meaningful results. In those cases, linearized analysis may show ground handling trends, but one should plan on full nonlinear 6-degree-of-freedom analysis, including tire forces. An extended analysis and simulation of ground roll was made for the Navy/Boeing T-45 trainer by the NASA Langley Research Center (Chambers, 2000). With proper tire dynamic models and inclusion of aircraft roll attitude, a pilot-induced yaw oscillation was reproduced.

14.5   Vee- or Butterfly Tails

A vee-tail or butterfly tail is a wing with a large amount of positive dihedral. If the vee-tail is built with negative dihedral, or anhedral, it is called an inverted vee-tail. Vee-tails or inverted vee-tails are used instead of the usual horizontal and vertical tails, replacing three surfaces with two. The Beech Model 35 Bonanza is the vee-tailed airplane most people have seen. Over 12,000 of them have been built. A very early vee-tail installation was made by Rudlicki in 1931 (Figure 14.10).

Why vee-tails? An evident advantage is that there is one fewer tail surface to build, assuming that horizontal tails are built in two halves. There is also less interference drag because there is one fewer tail–fuselage junction. Other possible advantages are dependent on the configuration. The inverted vee-tail of the General Atomics Gnat 750 protects the pusher propeller, serving as a bumper for tail-down landings (Figure 14.11).

Entire vee-tail surfaces, or flaps on their trailing edges, are deflected symmetrically for pitch control and asymmetrically, that is, with equal and opposite angles, for yaw control. There is a small rolling moment generated in the asymmetric case. For vee-tails with positive dihedral the rolling moment is adverse, in the following sense. Right rudder and yawing moment are generally applied by a pilot or autopilot during right rolls, to overcome adverse yaw. However, with a positive dihedral vee-tail, a left rolling moment is generated by the

Figure 14.10   Polish engineer Georges Rudlicki (on the left) with test pilot W. Szubczewski in front of the Hanriot 14 biplane fitted with Rudlicki's V-tail design in 1932. This configuration was successfully tested by Szubczewski and two other pilots. (From *Aircraft Engineering*, March 1932)

Figure 14.11   An inverted vee-tail airplane, the Amber, predecessor of the General Atomics Gnat 750 UAV (Unmanned Aerial Vehicle), used in 1994 for U.S. reconnaissance over Bosnia. (From Leading Systems, Inc.)

right rudder deflection, opposing the ailerons. Of course, this effect is reversed for inverted vee-tails, favorable rolling moments accompanying rudder deflection in rolls.

Negative tail dihedral plays an important role in avoiding pitchup for swept-wing airplanes at high subsonic Mach numbers. Though strictly not inverted vee-tails, negative tail dihedral is found on such airplanes as the McDonnell Douglas F-4 Phantom and the Dassault/Breguet/Dornier Alpha Jet.

Vee-tail surface stalling has been a continuing stability and control concern. A dangerous, abrupt pitch-down due to tail stall during landing approaches is possible. This could happen at forward center of gravity positions, with landing flaps full down, requiring down tail load for trim. If at the same time the landing approach is being made in rough air, appreciable sideslip angles can develop while the pilot fights the Dutch roll mode of motion. The combination of large down load and sideslip excursions could bring momentarily one vee-tail panel beyond its stall, relieving part of the down load and causing an abrupt diving moment.

Professor Ronald O. Stearman has looked into vee-tail stalling in sideslips at forward centers of gravity as a possible cause of the relatively high accident rate of the vee-tailed Beech Model 35. The Model 35's accident rate is seven times as high as the Beech Model

33, which is essentially the same airplane equipped with a conventional tail. Stearman's (1986) tests show that a vee-tail panel would stall, causing the Model 35 to nose over at a sideslip angle of 10 to 12 degrees. Flight tests conducted for Beech failed to reproduce Stearman's results. Vee-tail stalling remains a potential problem, which is an argument for making vee-tails somewhat oversized. Oversized tail surfaces would decrease the panel angles of attack required to produce given tail loads.

Estimation of required vee-tail size and dihedral angles was dealt with definitively by Paul E. Purser and John P. Campbell (1945). A naive approach to the problem would merely take the vee-tail's projected areas in plan and side views as effective horizontal and vertical areas, respectively. However, the Purser-Campbell method deals with the actual panel loads, considers interference between the vee-tail panels, and works out the trigonometry for panel angle of attack and force resolution. Purser and Campbell provide convincing experimental verification of their vee-tail theory, using wind-tunnel test data (Figure 14.12).

14.6   Control Surface Buzz

Control surface buzz is properly a single-degree-of-freedom flutter phenomenon, and thus not in the stability and control domain, but its cures affect controllability. Buzz was first encountered in dives on the Lockheed P-80 Shooting Star. Pilots reported strong, high-frequency lateral shaking of the control stick. Ordinary flutter was ruled out because calculations showed a reasonably high coupled wing-aileron flutter speed. The P-80's wing is stiff, with 13-percent-thick NACA 65-series airfoils, and the ailerons are mass balanced. Moreover, the vibration frequency was higher than would be expected for structural modes.

The cause of the shaking turned out to be an aileron-alone mode, not involving wing bending or torsion at all. At Mach numbers where wing normal shocks perpendicular to the fuselage are in the vicinity of the aileron hinge line, shock position couples with aileron deflection to produce the motion. Trailing-edge-upward deflection of the aileron causes flow field changes that move the upper wing surface normal shock forward and the lower wing surface normal shock aft. This causes relative trailing-edge-down hinge moment, starting the aileron down and eventually reversing the shock wave upper and lower surface positions. This completes the cycle.

Clearly, control system flexibility plays a role in aileron buzz, permitting symmetric or in-phase aileron deflections. Restraining the aileron from symmetric rotations prevents buzz. The F-80's control surface boost cylinder is on the airplane's centerline. Control system flexibility from the centerline out to the ailerons is sufficient for buzz to occur.

The buzz fix for the P-80 and later airplanes with centerline aileron actuators was rotary hydraulic dampers installed at each aileron hinge. Damping is approximately proportional to control surface angular velocity, which is very high during high-frequency buzz but much lower for normal stick motions. Properly sized buzz dampers do not interfere with lateral controllability. Modern jet airplanes normally have irreversible hydraulic actuators located at each aileron. Aileron buzz is not a possibility for this arrangement.

Control surface buzz was a problem for the rudders of two subsonic jet airplanes of the 1950s, a late-model North American F-86 Sabre and the Douglas A4D Skyhawk. Both airplanes used splitter-plate or "tadpole" rudders to overcome buzz. Splitter-plate control surfaces are unskinned from about the midchord to the trailing edge, with loads borne by a central or splitter plate. The normal shock, whose fore and aft motion is central to control surface buzz, is stabilized at the abrupt surface break where the splitter starts. Another

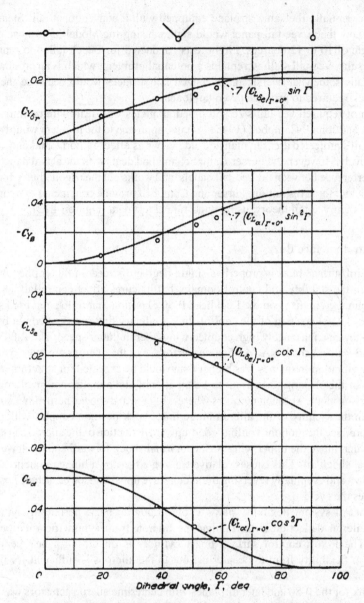

Figure 14.12 Experimental verification (small circles) for the Purser/Campbell vee-tail theory. Isolated tail stability and control contributions are plotted against tail dihedral angle. (From Purser and Campbell, NACA Rept. 823, 1945)

aerodynamic fix for control surface buzz is reported to be the addition of vortex generators just ahead of the control surface hinge line.

14.7   Rudder Lock and Dorsal Fins

Rudder lock occurs at a large angle of sideslip when reversed rudder aerodynamic hinge moments peg the rudder to its stop. The airplane will continue to fly sideslipped, rudder pedals free, until the pilot forces the rudder back to center or rolls out of the sideslip with the ailerons. Aerodynamic hinge moments can peg the rudder against its stops so securely as to defy the pilot's efforts at centering. In that case recovery by rolling or pulling up to reduce airspeed are the only options.

Two things must happen before an airplane is a candidate for rudder lock. Directional stability must be low at large sideslip angles and rudder control power must be high. The relative size of the fuselage and vertical tail determines the general level of directional stability. Directional stability is reduced at large sideslip angles when the fin stalls. The sideslip angle or fin angle of attack (considering sidewash) at which the fin stalls depends on the fin aspect ratio. Unfortunately, tall, efficient, high-aspect-ratio fins stall at low fin angles of attack. As a general rule, fin stall occurs at sideslip angles of about 15 degrees.

Unlike normal wings, whose lift is proportional to angle of attack until near the stall, the lift of very low-aspect-ratio rectangular wings is proportional to the square of the angle of attack (Bollay, 1937). There is very little lift generated in the low angle of attack range. However, the angle of attack for stall is increased greatly, reaching angles as high as 45 degrees.

What this means is that a two-part vertical tail is an efficient way to avoid loss in directional stability at large sideslip angles and rudder lock. One part is a high-aspect-ratio vertical tail, which can provide directional stiffness in the normal flight regime of low sideslip angles and give good Dutch roll damping and suppression of aileron adverse yaw. The other part is a low-aspect-ratio dorsal fin, with a reasonably sharp edge, which will carry very little lifting load in the normal flight regime. However, at a sideslip angle where the high-aspect-ratio fin component stalls, the dorsal fin can become a strong lifting surface, maintaining directional stability.

Returning to the role of the rudder, large rudder areas and control power are needed for two-engine airplanes with wing-mounted engines, for the condition of single-engine failure at low airspeeds. This is especially true for propeller-powered airplanes, since full-throttle propeller thrust is highest at low airspeeds, and wing-mounted engines tend to be further outboard than for jets, to provide propeller–fuselage clearance.

Although a four-engine rather than a two-engine airplane, rudder lock was experienced on the Boeing Model 307 Stratoliner, with its original vertical tail. This occurred during an inadvertant spin. From William H. Cook (1991):

> On a demonstration flight for KLM and TWA, the KLM pilot applied rudder at low speed. The rudder locked full over in the spin, and the control forces on the rudder were too high [to center it]. Wind tunnel tests showed that a long dorsal fin would prevent the rudder locking over. A hydraulic servo on the rudder was also added.

The addition of a dorsal fin to the Stratoliner and a reduction in rudder area corrected the problem (Figure 14.13) (Schairer, 1941). George Schairer recently commented that he was unaware of the true inventor of dorsal fins, but that a member of the GALCIT 10-foot tunnel staff might have installed one during tests of one of the Douglas airliners. Small dorsal fins appeared earlier than on the Stratoliner, notably on the Douglas DC-3, first produced in 1935, and on the Douglas DC-4, which had its first flight in 1938.

In spite of the small dorsal fin installed on the DC-3, that airplane is still subject to rudder lock in all configurations with power on (Figure 14.14). John A. Harper flew a U.S. Air Force C-47B, the military version of the DC-3, in NACA flying qualities tests in 1950. Harper later speculated that rudder lock might have contributed to some puzzling DC-3 accidents resulting from loss of power on one engine, followed by a stall and spin. In these strange

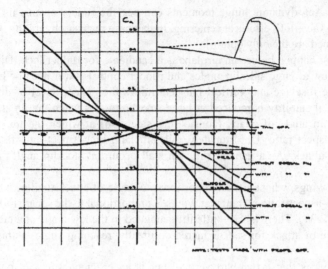

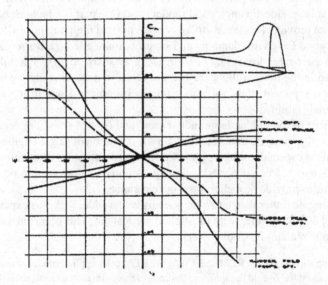

Figure 14.13   The variation of yawing moment coefficient with sideslip angle for the Boeing Stratoliner with original vertical tail (*above*) and revised vertical tail and dorsal fin (*below*). Rudder-free cases are shown by the dashed lines. With the original tail, adverse yawing moment due to the ailerons overcomes the low level of restoring moment at large side-slip angles, and there is rudder lock. (From G. S. Schairer, *Jour. of the Aeo. Sci.*, May 1941)

accidents, the airplane spun into the operating engine, the reverse of what one would expect. Harper argues that rudder lock and high pedal forces for recovery could have occurred if the pilot overcontrolled with the rudder to turn toward the live engine.

Rudder lock was suspected in the early Boeing 707 airplanes, which had manually operated rudders assisted by spring tabs and internal aerodynamic balance. An Air Force test of the XC-135 tanker version reported rudder lock and an American Airlines crash on Long Island may have been due to rudder lock. As a result, the 707 and KC-135 series of airplanes have powered rudders.

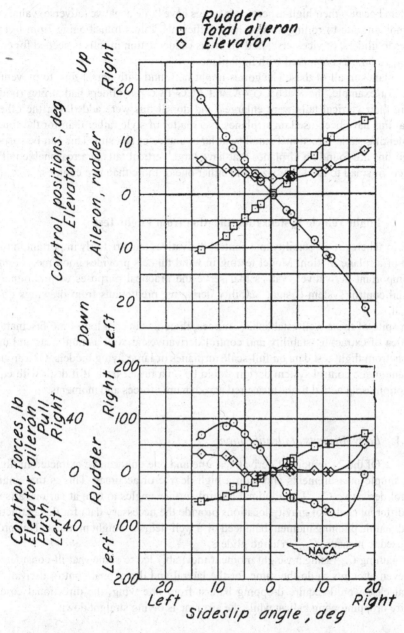

Figure 14.14 Incipient rudder lock on the DC-3 airplane. The rudder force has gone to zero at a right sideslip angle of 18 degrees. The rudder angle is only 21 degrees left, with 9 more degrees of deflection available before reaching the rudder stop of 30 degrees. The rudder locks over at larger rudder and sideslip angles, but these are not reached in this test series because of heavy airplane buffet. (From Assadourian and Harper, NACA TN 3088, 1953)

In addition to the large rudder area requirement for the engine-out condition on multi-engine airplanes, large rudder areas are needed for spin recovery on maneuverable airplanes, to handle heavy crosswinds for airplanes intended to operate out of single-strip airports, and for gliders, to counter adverse aileron yaw. Gliders have a particular adverse yaw problem because their high-aspect-ratio wings have large negative (adverse) values of yaw-ing moment due to rolling at high lift coefficients. Pilots transitioning from light power planes to gliders, or vice versa, find vigorous rudder action in rolls is needed for coordina-tion in gliders, as compared with light planes.

Airplanes in all of these categories might be found with dorsal fins, to prevent rudder lock. For example, the Waco CG-4A and XCG-13 cargo gliders had strong rudder lock before their vertical tails were enlarged and dorsal fins were added. On the other hand, dorsal fins have been used on airplanes as a matter of style rather than for the function of augmenting static directional stability at large angles of sideslip. This can be suspected if dorsal fins are found on airplanes that have large vertical tails at a reasonable tail length, rudders of small to moderate size, and either one or more than two engines.

14.8   Flight Vehicle System Identification from Flight Test

There are 21 stability and control derivatives that are fairly important in the equa-tions of airplane motion. Model testing in wind tunnels provides good measurements of the important derivatives, values that serve the practical purposes of preliminary stud-ies and control system design. Stability derivative predictions from drawings do almost as well.

In spite of these well-established sources, there has been a long-time fascination with the idea of extracting stability and control derivatives as well as nonlinear and unsteady effects from flight test data on full-scale airplanes or large flying models. One argument is that automatic control system design would be on a firmer basis if it dealt with equations of motion using actual flight-measured aerodynamic forces and moments.

14.8.1   *Early Attempts at Identification*

Of the 21 important derivatives, one and one only can be extracted in flight tests with simple measurements and with a high degree of accuracy. This is the longitudinal control derivative C_{m_δ}. Longitudinal control surface angles to trim at various airspeeds at two different center of gravity locations provide the necessary data for this extraction, the aerodynamic pitching moment balanced by a well-defined weight moment. This procedure was used to measure C_{m_δ} on cargo gliders.

Obtaining C_{m_δ} using a weight moment inevitably led to somewhat ill-considered plans and even attempts to do the same for the lateral and directional control derivatives. The lateral case would require dropping ballast from one wing; the directional case would require dropping wing ballast while the airplane is diving straight down.

14.8.2   *Knob Twisting*

Informal and rather elementary stability and control derivative extraction took place starting in the early 1950s, when the first electronic analog computers, such as the Reeves Instrument Company's REAC, were used to get time histories of airplane motions. Numerical values of individual dimensionless stability derivatives, such as C_{n_β}, are represented by potentiomenter settings on analog computers. Computed airplane motions appear on pen-type recorder records. The experimenter can try to match an actual flight record for a given control input by resetting potentiometers and rerunning cases over and over. Since potentiometer settings are controlled by knobs on the face of the analog computer cabinet, this trial-and-error process is known familiarly as knob twisting.

Knob twisting is not altogether a random process, since an experimenter is guided by approximations to the modes of airplane motion. We know, for example, that the period of the Dutch roll oscillation is controlled by the directional stiffness derivative C_{n_β}. The amplitude of the roll oscillation relative to that of sideslip or yawing velocity is controlled by the dihedral effect derivative C_{l_β}, and so on.

14.8.3 *Modern Identification Methods*

The higher powered stability derivative extraction schemes that followed knob twisting engage the interest of many mathematically minded people in the stability and control community. The focus has been broadened beyond the linearized stability derivatives, and the subject is now usually called flight vehicle system identification. The years have seen centers of identification activity at individual laboratories, such as Calspan and the NASA Dryden Flight Research Center, and any number of university graduate students earning doctoral degrees in this area. Kenneth W. Iliff and Richard E. Maine at Dryden are leaders in the identification field in the United States.

The DLR Institute in Braunschweig is particularly active in this area, under the guidance of Dr. Peter Hamel. The state of the art up to 1995 is summarized in a paper having 183 references by Drs. Hamel and Jategaonkar (1996). This summary has been updated (Hamel and Jategaonkar, 1999). A generalized model of the vehicle system identification process is shown in Figure 14.15.

A flow chart for a widely used method known as the maximum likelihood or output error method is given in Figure 14.16. The maximum likelihood method starts with a mathematical model of the airplane, which is nothing more than the linearized equations of airplane motion in state variable form (see Chapter 18). The method produces numerical values for the constants in those equations, the airplane's dimensional stability and control derivatives. A cost function is constructed as the difference between measured and estimated responses, summed over a time history interval. Iliff describes the workings of the maximum likelihood method as follows (see Figure 14.16):

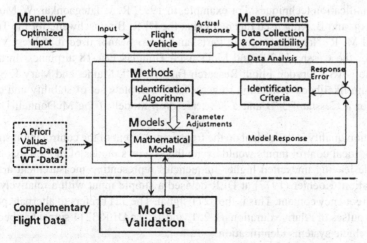

Figure 14.15 The generalized method of vehicle system identification. (From Hamel, RTO MP-11, 1999)

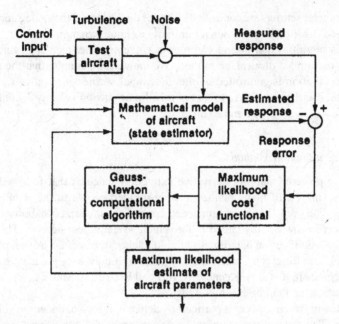

Figure 14.16   Flow chart for the maximum likelihood method of airplane stability and control derivative extraction from flight test data. (From Iliff, *Jour. of Guidance*, Sept.–Oct. 1989)

The measured response is compared with the estimated response, and the difference between these responses is called the response error. The cost functions ... include this response error. The minimization algorithm is used to find the coefficient values that minimize the cost function. Each iteration of this algorithm provides a new estimate of the unknown coefficients on the basis of the response error. These new estimates of the coefficients are then used to update values of the coefficients of the mathematical model, providing a new estimated response and, therefore, a new response error. The updating of the mathematical model continues iteratively until a convergence criterion is satisfied.

At the time of this writing, the maximum likelihood method is the most widely used of the available identification techniques. For example, in 1993, R. V. Jategaonkar, W. Mönnich, D. Fischenberg, and B. Krag used this method at the DLR, Braunschweig, for the Transall airplane; and M. R. Napolitano, A. C. Paris, and B. A. Seanor used it at West Virginia University for the Cessna U-206 and McDonnell Douglas F/A-18 airplanes. In an earlier use at the NASA Dryden Flight Research Facility, Iliff, Maine, and Mary F. Schafer used maximum likelihood estimation to get a fairly complete set of stability and control derivatives for a Cessna T-37B and a 3/8-scale drop model of the McDonnell Douglas F-15.

Identification quality is dependent on the frequency content of the control input signal in Figure 14.16. Ideal control inputs would excite the system's modes of motion that require control, while leaving unexcited higher frequencies representing measurement artifacts, such as vibration. Koehler (1977) at DLR devised a simple input with a relatively wide, but limited, frequency content. This is the 3211 signal. The 3211 refers to alternate positive and negative pulses of relative durations 3, 2, 1, and 1. The DLR 3211 signal has become a standard in vehicle systems identification.

The alternate extended Kalman filter method of stability derivative extraction has the advantage of being suited to real-time operation. That is, it can be used as an element in a

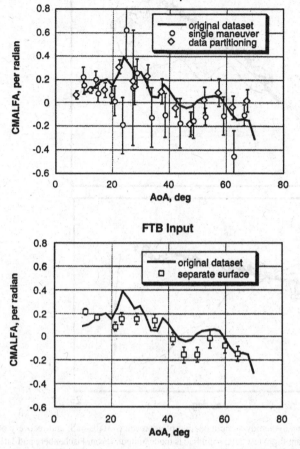

Figure 14.17 Improvement of $C_{m\alpha}$ identification on the X-31A with separate surface inputs. (From Weiss, *Jour. of Aircraft*, 1996)

closed-loop flight control system. Many of the applications of extended Kalman filtering imply or use this feature, as in the 1983 to 1991 work at Princeton University by M. Sri-Jayantha, Dennis J. Linse, and Robert F. Stengel.

A challenging area for identification is that of the aerodynamically unstable airframe, which can be flown only with full-time stability augmentation. Data scatter can be large in these cases, using current methods. Figure 14.17 shows that exciting motion with a separate control surface from that used for closed-loop control reduces identification data scatter.

Rotary derivatives and cross-derivatives, such as the rolling moment due to yawing and the yawing moment due to aileron deflection, are generally the least well-known data in the airplane equations of motion. Identification methods are at their worst for such parameters. Correlations of extracted derivatives with wind-tunnel and theoretical data generally focus on just those derivatives such as C_{l_β} and C_{l_p} whose numerical values are well known from other sources. Flight test data quality must be high for identification algorithms to work well, probably higher than needed for any other application. State noise from atmospheric turbulence and sensor noise are obvious complications.

The power of flight vehicle system identification continues to advance to the point where the derived system models can meet accuracy requirements for high-fidelity flight

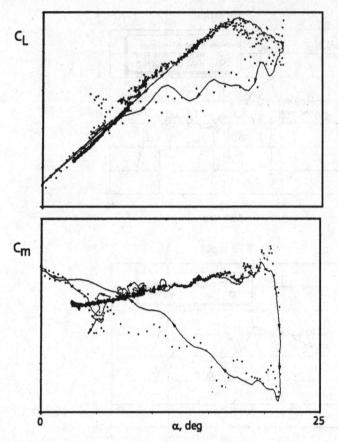

Figure 14.18   Unsteady aerodynamic modeling of approach to stall, stall, and recovery on the Do 328 transport. Dots are flight test data; solid line is model output. (From Fischenberg and Jategaonkar, RTO MP-11, 1999)

simulators, such as those used in research and pilot training. Recent extensions of the theory include using the frequency domain in place of the usual time domain methods, and the application of neural networks.

14.8.4   *Extension to Nonlinearities and Unsteady Flow Regimes*

As mentioned in Chapter 5, Sec. 24, analytical redundancy for fly-by-wire control system safety will become feasible only when vehicle system identification operates well under all flight conditions, and not just where linearized, small perturbation equations of motion apply. For this reason, and to generate practical design information, there has been an effort to extend identification to nonlinear and unsteady aerodynamic regimes. The extended Kalman filter mentioned in the previous section can generate full nonlinear aerodynamic models, such as $C_m(\alpha,\delta,q,\dots)$, and not just aerodynamic stability derivatives at various operating points.

The transfer function model for unsteady flow (Chapter 10, Sec. 6.1) has been used successfully at the DLR to model lift hysteresis at the stall for the Fairchild/Dornier Do 328 transport. The procedure (Fischenberg, 1999) is in several parts, starting with a steady-state approximation for the point of trailing-edge flow separation at high angles of attack, using static wind-tunnel data. Time dependency is introduced by the assumption that the separation point can be modeled as the solution of a first-order differential equation, equivalent to a single-pole transfer function. A final assumption is that the lift coefficient in trailing-edge separated flow is a function of the separation point, using a model proposed by Kirchoff. Four parameters of the model remain to be identified in flight testing, but when these are found, good comparison of flight measurements of lift coeffcient with the modeled values is obtained (Figure 14.18).

The significance of this work is that once the unsteady lift parameters of representative lifting surfaces can be predicted, the basis is in hand for predicting unsteady values of complete airplane stability derivatives. Those derivatives are computed from the forces and moments of the lifting surfaces and the fuselage-type shapes that make up a complete configuration. Future work in this area will presumably deal with leading-edge flow separation models, which are characteristic of thin wings with sharp leading edges, and with predictive models for the unsteady vortex flows that can affect stability derivatives. Stall hysteresis for airfoils with sharp leading edges is discussed by Covert (1993).

14.9   Lifting Body Stability and Control

A lifting body is a wingless vehicle that depends on lift generated from an elongated body or fuselage. Both lifting bodies and ballistic shapes had been studied as space vehicles by NASA before the choice of the Mercury, Gemini, and Apollo ballistic capsule designs. However, lifting body research continued, first at the NASA Ames and Langley Research Centers, then at the NASA Flight Research Center (Reed, 1997). Figure 14.19 is a general arrangement drawing of a typical lifting body design, the NASA/Northrop HL-10 (Heffley and Jewell, 1972). Configurations such as the HL-10 have evolved into the space shuttle Orbiter and follow-up concepts such as the X-33 research vehicle.

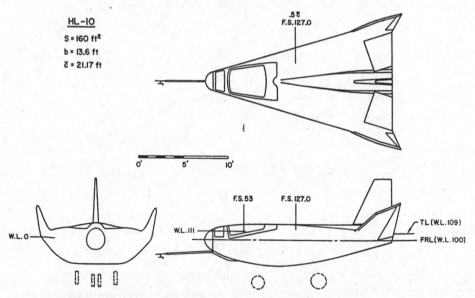

Figure 14.19   Three-view drawing of the NASA/Northrop HL-10 lifting body. (From Heffley and Jewell, NASA CR-2144, 1972)

Stability and control characteristics of a series of lifting bodies were investigated in wind-tunnel and flight tests starting in the mid-1950s in the United States and later in Russia, Japan, and France. All of the problems associated with swept wings and heavy fuselage loadings appeared in the course of these tests, in addition to a number of instances of control oversensitivity and pilot-induced oscillations. Lifting body configurations typically have high dihedral effect, or rolling moment due to sideslip, in proportion to roll damping. The coupled roll-spiral mode of motion (Chapter 18, Sec. 9) may thus exist.

CHAPTER 15

Safe Personal Airplanes

Over the years there have been many innovations that were meant to make airplanes as easy and as safe to fly as cars are to drive. The Guggenheim Safe Airplane Competition (1926–1929) was an early organized attempt to have safe airplanes. Ever since that time, personal airplanes have benefitted from stability and control research that had been directed at larger, heavier airplanes. However, the designers of personal airplanes have not always taken advantage of this body of knowledge.

The prospects for safe personal airplanes is clouded by rapid performance advances, which require the application of advanced stability and control techniques to be safe.

15.1 The Guggenheim Safe Aircraft Competition

In June 1926, the philanthropist and aviation enthusiast Harry Guggenheim funded a Safe Aircraft Competition with the sum of $150,000 to $200,000. The competition was open to airplane manufacturers in any part of the world. Leading aviation personalities of the day helped draft rules for the competition. They included Majors R. W. Schroeder and R. H. Mayo, Professor Alexander Klemin, Lieutenants E. E. Aldrin and James H. Doolittle, airplane designers Anthony H. G. Fokker and G. M. Bellanca, veteran builders and flyers J. D. Hill and Charles Day, and Edward P. Warner, who was then Assistant Secretary of the Navy for Aeronautics.

An extensive set of demonstrations was agreed upon, the main thrust of which was the ability to land in a confined space. Twenty-seven entries came in: five from Great Britain, one from Italy, and the balance from the United States. In the end, 15 airplanes showed up at Mitchell Field, New York, for testing, and 10 were actually demonstrated. The tests and demonstrations took place in 1929.

The competition came down to two biplanes, one built by Handley Page, the other the Curtiss Tanager, designed by a team headed by Dr. Theodore P. Wright. The Tanager had full-span flaps and leading-edge slats, with lateral control provided by isolated or floating ailerons. The leading-edge slats were automatically operated, opening by air loads at a high angle of attack (Figure 15.1). The Handley Page machine, also equipped with flaps and automatic leading-edge slats, was a close second; but the Tanager won with a minimum gliding speed of 37 miles per hour, excellent lateral control, and a total distance from a 35-foot obstacle to a full stop of less than 300 feet.

The Tanager was eventually destroyed in a fire caused by leaking fuel. The thinking and developments initiated by the 1926–1929 Guggenheim Safe Aircraft Competition had results that went far beyond those years. The most obvious benefit was the demonstration of full-span flaps and automatic slats and the longitudinal and lateral control power required for their use.

15.2 Progress after the Guggenheim Competition

Safe personal airplane objectives have been defined as (Upson, 1942):

231

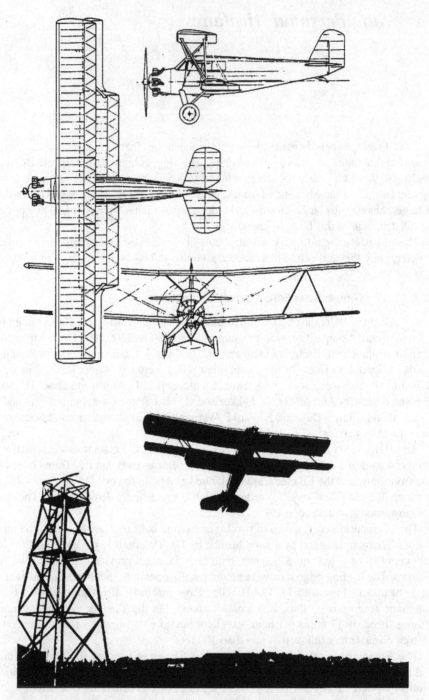

Figure 15.1   The 1929 Curtiss Tanager, with full-span slats and flaps, demonstrating a steep climb over a 35-foot obstacle. This airplane won the Guggenheim Safe Aircraft Competition. (From Pendray, *The Guggenheim Medalists, 1929–1963*)

Outstanding in vision, incapable of spinning, comparable with an automobile in simplicity of control, yet with unquestioned superiority of cross-country performance.

Upson gives as examples no fewer than seven personal airplane designs that by 1942 had tried for these objectives. Which designs were viable, which were not, and what additional tries were made?

15.3   Early Safe Personal Airplane Designs

Aeromarine-Klemm   As imported from Germany, it had unsafe spin characteristics. The wing was modified to have less taper and thicker tip sections. Control movement was restricted, and the center of gravity range was moved forward. All of these modifications, apparently arrived at empirically, were in a direction to improve spin resistance, and this airplane became one of the very first to be called incapable of spinning. Actually, a spin could be forced, but the airplane had to be held into the spin; and with free controls it would recover. Aeromarine-Klemm models were produced with several different engines from the late 1920s to 1932.

Stout Sky Car   Designed in 1931, the Sky Car was one of the first two-control airplanes. It had floating wing tip ailerons that were weight overbalanced, making them float symmetrically with slight negative lift. When deflected for a roll, proverse yaw, or yaw in the direction of the roll, resulted. No rudder control was needed to coordinate the roll. The Sky Car had a tricycle landing gear and limited up-elevator travel. It was a stubby, odd-looking machine, a biplane with a small vertical tail.

Weick W-1A   In 1935 and 1936, this airplane was a test bed for several safety innovations. It had full-span flaps that could be deflected to 80 degrees to make steep descents into small fields. Slot lip spoilers provided lateral control (Figure 15.2). The not-yet-famous Robert T. Jones studied two-control operation and told Weick that the W-1A's spoiler ailerons would be ideal for the purpose, as they turned out to be. As in the Stout Sky Car, elevator control was limited to prevent stall.

Stearman-Hammond Model Y and the Gwinn Aircar   Both of these airplanes were designed with features of the Weick W-1A. The Model Y won a safe airplane competition sponsored by the Department of Commerce. The Aircar had no rudder at all. Its interior looked like an Oldsmobile, with Oldsmobile steering wheel and instruments.

ERCO Model 310 and the Ercoupe   Fred Weick's Ercoupe was the only one of the early safe airplanes to make it into production, which started in 1940 (Figure 15.3). The Ercoupe has the two-control, restricted elevator control and tricycle landing gear features of the W-1A. The U.S. Civil Aeronautics Authority certified the Ercoupe as "characteristically incapable of spinning" and cut the dual time required to solo from 8 to 5 hours and the time for private pilot certification from 35 to 25 hours.

With the yoke hard back, rapid full aileron control deflections from side to side produce nothing more exciting than falling-leaf motions. Cross-wind touchdowns are made with the airplane headed into the relative wind. When the pilot releases the controls the Ercoupe straightens out for its ground roll.

Figure 15.2 The 1935 Weick W-1A airplane, photographed in front of an NACA Langley Field hangar. This innovative airplane had full-span flaps and spoiler ailerons, limited up-elevator travel, and two-control operation. (From Weick, *From the Ground Up*, 1988)

15.4   1948 and 1966 NACA and NASA Test Series

Robert Gilruth's codified requirements for satisfactory flying qualities of 1941 opened the way to apply flying qualities technology to the safe airplane problem. Paul A. Hunter made the first NACA flying qualities measurements specifically on personal-owner airplanes in 1948. This was followed by a second test series on light airplane flying qualities (Barber, Jones, Sisk, and Haise, 1966).

The seven light airplanes tested in 1966 were bigger, heavier, and more complex than the group of five looked at in 1948. Four of the seven were twin-engined; the single-engine ships were the straight and vee-tailed Beech Bonanza and a 285 HP Cessna Skylane RG. In keeping with NASA's practice at that time, data presented are not identified as having come from specific airplanes.

Reported flying qualities problems ranged from rather trivial trim change difficulties to more serious issues. As in the case of the Spitfire and DC-3 (Chapter 3), static longitudinal instability was present for some of these airplanes within their normal loading ranges, especially with flaps down and high power settings. Bobweights and downsprings provided stable force gradients in some cases, without improving stick-fixed stability. Low Dutch roll damping reduced the accuracy of instrument approaches in turbulence (Figure 15.4).

Dangerous stalling characteristics were encountered in the tests. Quoting from the Barber report:

Figure 15.3   The 1940 ERCO Ercoupe, as first produced. (From Weick, *From the Ground Up*, 1988)

Two of the aircraft tested have unacceptable power-on stall characteristics in the landing configuration. The lateral-directional trim changes of one aircraft show that the addition of power introduces a left yawing moment and that the pilot must use full right rudder to maintain heading when near the stall speed. The large yawing moment due to power coupled with the lack of rudder authority causes the aircraft to encounter an uncontrollable left roll/yaw motion at the stall. This motion places the aircraft in a spin that requires 600 to 1200 feet of altitude for recovery. All of the evaluation pilots exceeded the gear and flap placard speeds when recovering from this spin. Another aircraft has a rapid left rolloff in the power-on accelerated stall with landing flaps extended. The rolloff is difficult to stop in less than 60 to 70 degrees of left bank without anticipation and instantaneous recovery control on the part of the pilot. Such a stall may occur when a pilot tightens his final turn in the landing pattern to prevent overshooting the runway. From a left turn, the attendant rolloff, on occasion, proceeded to a nearly inverted attitude that required 200 to 300 feet of altitude to recover.

One is left to wonder how those two airplanes ever got to be certified as airworthy by the Federal Aviation Administration.

15.5   Control Friction and Apparent Spiral Instability

A spirally stable airplane, when disturbed in bank angle from wings-level flight, will return on its own to wings-level flight, although on a different heading than it had before the disturbance. On the other hand, the bank angle of a spirally unstable airplane will increase without limit after an initial disturbance. A spirally unstable airplane's bank angle must be corrected continuously to maintain wings-level flight. However, for the usual case

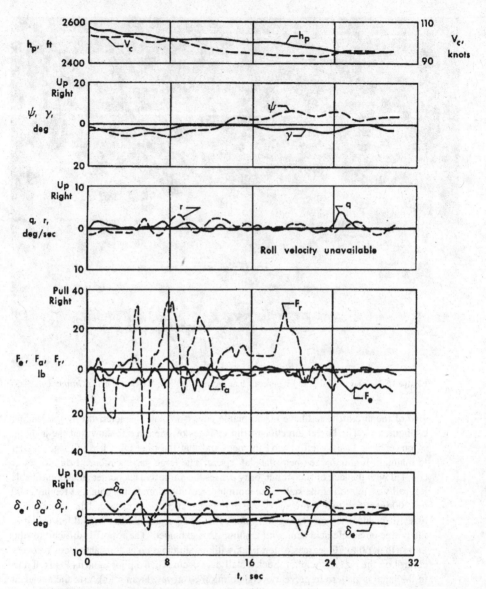

Figure 15.4 Time history of a precision instrument approach in light turbulence, using ILS, for an unspecified general-aviation airplane. This model has low Dutch roll damping and other stability and control deficiencies. The pilot uses large stick and rudder pedal forces, but airspeed, heading and path angle variations are excessive. (From Barber, Jones, Sisk, and Haise, NASA TN D-3726, 1966)

of moderate spiral instability, with times to double amplitude of the order of 20 seconds, corrections are made instinctively. Pilots are generally unaware of the instability.

Therefore, it is not surprising that Ralph Upson's 1942 set of objectives for a safe personal airplane do not include spiral stability. That is, if one were only to make airplanes "as easy to fly as cars are to drive," spiral stability would not necessarily be an objective. Yet, positive spiral stability has over the years been of interest to personal-airplane designers.

One reason for this is that Federal Aviation Regulations for airplanes operating under Visual Flight Rules (91.205) do not require gyroscopic rate of turn indicators. Without one of these instruments the pilot of an airplane that blunders into clouds has no way of maintaining wings-level flight, unless the airplane happens to be spirally stable. In that case, freeing the controls prevents a "graveyard spiral." Another reason for spirally stable personal airplanes is to enable a solo pilot to be able to read a map without finding the airplane in a bank upon looking up.

Unfortunately, an inherently spirally stable airplane can appear to be spirally unstable with rudders and ailerons free, as a result of control friction (Campbell, Hunter, Hewes, and Whitten, 1952). To correct this, NACA researchers designed a rather complicated control-centering device to overcome friction without interfering with normal control activity.

A cylindrical barrel encloses two preloaded compression springs and a shaft passing through the barrel. A shoulder on the shaft and corresponding shoulders on the inside of the barrel are at its midlength. A flat circular pickup ring under the end of each spring is forced against both shoulders with a force equal to the spring preload. The shaft cannot move relative to the barrel without moving one of the pickup rings and consequently compressing one of the springs. Campbell's group installed the preload barrels in both the rudder and the aileron control systems of a Cessna 190. An electric motor provides rudder trim, a jack screw provides aileron trim, and solenoids engage or disengage the preload devices (Figure 15.5).

Without the centering devices, the airplane diverged in the direction of a rudder kick, after a kick and release of all controls. However, the centering devices allowed the Cessna's inherent spiral stability to take effect. After a rudder kick and release of all controls the airplane returned to wings-level flight, and would continue so indefinitely.

15.6   Wing Levelers

Wing levelers are single-axis automatic pilots typically used in general-aviation airplanes to prevent spiral divergence. John Campbell's 1952 NACA aileron centering device flown on a Cessna 190 was improved upon a few years later by another NACA group, which converted it to a wing leveler (Phillips, Kuehnel, and Whitten, 1957).

The Phillips device was an aileron trimmer, added to the earlier aileron centering system. The aileron trim point was moved at a slow rate by the output of a yaw rate gyro (Figure 15.6). The slow 1.5-degree-per-second trim rate left the Dutch roll oscillation unaffected. The Cessna's flight records showed that the trimming device could maintain a safe bank angle indefinitely, and even could provide a wheel-free recovery from banked attitudes, or effective positive spiral stability. Phillips noted (1998) that NACA headquarters failed to file a patent application for the wing leveler, although a patent disclosure had been made. Without coverage by a valid patent, light-plane and autopilot manufacturers would not consider marketing the device.

Modern versions of the Phillips wing leveler device are available. Century Flight Systems, Inc., which started out as Mitchell, became Edo-Aire Mitchell, and finally split off as Century, produces the Century I wing leveler. This device is identical in principle to the Phillips wing leveler flown on the Cessna 190, with yaw rate gyro signals sent to an aileron servo. The ability to command turns and to follow a CDI (Course Direction Indicator) for VOR (VHF Omni-Directional Range), ILS (Instrument Landing System) localizer, or GPS (Global Positioning System) tracking has been added, making the Century I a simple autopilot.

15.7   The Role of Displays

While proper stability and control design, supplemented by artificial means such as control centering devices and wing levelers, are fundamental to safe airplanes, some

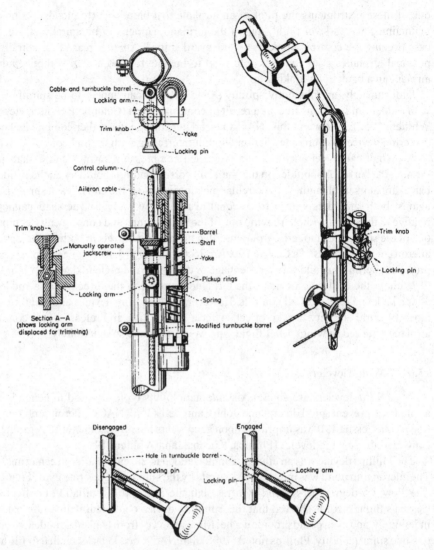

Figure 15.5 Aileron-centering device tested on a Cessna 190. The cylindrical barrel encloses two preloaded compression springs, overcoming control system friction to provide aileron centering. Trim and device engagement are both done manually by the pilot. (From Campbell, Hunter, Hewes, and Whitten, NACA Rept. 1092, 1952)

safety deficiencies can apparently be made up with the right kind of cockpit displays or instruments (Loschke, Barber, Enevoldson, and McMurty, 1974). They reported that on a light twin-engine airplane, a flight director display is of significant benefit during ILS approaches in turbulent air.

Heavy pilot workload during such approaches had been found in an earlier survey of general-aviation airplanes, making precise instrument tracking difficult even for experienced instrument pilots. However, the flight director instrument, which combines inputs from attitude and rate gyros and in effect tells the pilot how to move the controls, reduces somewhat this excessive workload. Even greater improvements in tracking during ILS approaches in turbulence are found when the flight director display is combined with an attitude-command autopilot.

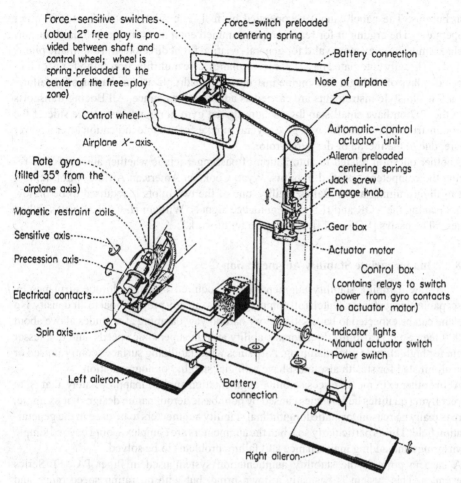

Figure 15.6 A wing-leveler device that works by moving aileron-centering springs at a low fixed rate in response to the measured rate of yaw. Wheel force switches improve maneuvers by precessing the gyro. The device was flown on a Cessna 190. (From Phillips, Kuehnel, and Whitten, NACA Rept. 1304, 1957).

The experiences of an airline pilot in operating heavily automated passenger-jet airplanes is relevant to the improvements provided by flight directors and automatic pilots for general-aviation airplanes. The question is how to use these devices to enhance safety for the average pilot under all operating conditions. William M. Ferree of Mount Vernon, New Hampshire, writes as follows (1994):

> I've been a professional pilot for over 20 years and currently fly the Boeing 757 and 767. A high level of automation gives these excellent airplanes capabilities that would have been remarkable a few years ago. However, the design breaks down when some significant change of plan is introduced, which may happen because of an equipment failure or, more commonly, because of difficulty with the air traffic control system or the weather. The problem is that unless the computer is reprogrammed in these situations, it is useless. And reprogramming must often be done while landing preparation is being completed, which is an extremely busy time.

Ferree goes on to note that the 757 and 767 autopilot/flight director control panel consists of a few knobs for selecting things such as airspeed and altitude and many identical square push buttons. The panel cannot be operated by feel. The pilot must look at it in order to operate it. The argument for readily reprogrammed equipment in commercial-transport airplanes must be even more valid for general-aviation flight directors and automatic pilots.

Today's heavily automated passenger jet airplanes have multifunction cathode ray or flat-screen displays of all flight and engine instruments, quadruplicated as backups for failure. Yet, a few old-style instruments are carried as additional insurance. All Boeing transports from the 707 on have small standby vacuum-driven gyro horizons, just to the side of the central instrument panel. This will give several minutes of reliable indication after a power failure, due to the inertia of the gyro's rotor.

Another question related to automation of passenger jets is whether automation is reducing the competence needed in pilots. Wyatt Cook, an American Airlines pilot, reports that in flight training at the Dallas facility, one of the two pilots is required to be on raw data, meaning the VOR and ILS radio guidance signals. William H. Cook, Wyatt's father, writes, "The basics [VOR and ILS] require a lot of work."

15.8   Inappropriate Stability Augmentation

Yaw damping stability augmentation is required for high-altitude airplanes by inescapable physical facts. Dutch roll damping ratio is directly proportional to air density. No airplane can be expected to have satisfactory natural yaw damping at altitudes above about 35,000 feet. So it is also for directional stability at high supersonic speeds, and to a lesser extent for high-altitude pitch damping. Airplanes with stabilizing surfaces compromised or even eliminated for stealth also must have artificial stability or augmentation.

At the other extreme, one sees stability augmentation applied inappropriately, that is, to correct flying qualities deficiencies caused by poor basic aerodynamic design. For example, there is really no reason for static longitudinal stability augmenters to be used in the general-aviation field. This is particularly so when the augmenters are complex, going beyond simple downsprings and adding maintenance and failure problems to be solved.

A case in point is the stability augmentation system used in Piper PA-31T Series Cheyennes. This system is basically a downspring, but with operating speed range and variable spring tension controlled by an angle of attack vane and a computer. A bobweight completes the installation. All this is needed because the basic Cheyenne airframe was derived from a lower powered Piper model, the Navajo. The Cheyenne's engines are more powerful, but also lighter, leading to aft center of gravity problems. Rebalancing the airplane would have been a better solution than what we consider to be an inappropriate use of stability augmentation.

15.9   Unusual Aerodynamic Arrangements

Unusual aerodynamic arrangements, such as canards, floating main wings (Regis, 1995), fore and aft horizontal tails, tailless configurations, and vee-tails represent particular hazards in the designs of personal airplanes because of the limited opportunity for thorough aerodynamic testing. Designers expect to work out stability and control problems in prototype flight testing, but flight testing can leave many areas unexplored, as compared with systematic wind-tunnel testing.

Possible tail stalling on the vee-tailed Beech Model 35 in sideslips at forward center of gravity positions, flaps down, leading to abrupt noseovers is mentioned in Chapter 14. The special stability and control problems of canard airplanes are mentioned in Chapter 17. Canard applications to military airplanes have a better chance of success than those for personal airplanes. Military canard applications take place under the protection of large wind-tunnel test programs and electronic stability augmentation, protections not available to designers of personal airplanes.

15.10 Blind-Flying Demands on Stability and Control

Blind flying is controlled flight without reference to the outside scene, more specifically, the horizon. Simply put, there is no way that an airplane can be made suitable for blind flight by aerodynamic design alone. A pilot must rely on some form of gyroscopic device to retain control, either as a panel instrument or as part of an automatic pilot.

The need for instrumentation comes from the effects of spiral instability, or aileron control friction on airplanes that are spirally stable, as discussed in Sec. 5. Even if strong spiral stability were to be built into a design, at the expense of other desirable flying qualities, a pilot could lose control over altitude unless trained to damp the phugoid mode (Chapter 18, Sec. 9) by reference to airspeed alone.

Spatial disorientation due to illusions can be prevented by reference to instruments. The FAA's *Aeronautical Information Manual* (1999) lists no fewer than 14 flight illusions that have been identified, such as a Coriolis, graveyard spin, somatogravic, and inversion illusions.

15.10.1 *Needle, Ball, and Airspeed*

U.S. Federal Aviation Regulations forbid flight under Instrument Flight Rules (IFR), or blind flight, without a gyroscopic pitch and bank indicator, or artificial horizon. This instrument was not invented until 1929, yet skillful pilots were flying blind before that. They used the gyroscopic rate-of-turn indicator, invented about 1920 by Elmer Sperry, Jr. That instrument senses yaw rate and displays it to the pilot by a needle that deflects in the turning direction. The instrument, called a turn and slip indicator, incorporates a separate ball in a curved glass tube that acts as a lateral accelerometer or sideslip indicator. When the turn and slip indicator is combined with training to use elevator control and an airspeed meter to damp the phugoid mode, the blind-flying technique is called needle, ball, and airspeed. Charles Lindbergh retained control through clouds on his transatlantic flight by that technique.

Some modern light planes use a tilted-axis form of the turn and slip indicator. When the gyro's gimbal axis, ordinarily aligned with the airplane's longitudinal axis, is tilted nose upward about 30 degrees, the instrument measures rolling velocity as well as yawing velocity. Rolling velocity leads yawing velocity when starting a turn. The tilt provides anticipation of the turn. The tilted-axis turn and slip instrument is called a turn coordinator.

U.S. Federal Aviation Regulations for private-pilot certificates (Part 61, Subpart E) requires pilots to demonstrate the ability to control airplanes solely by reference to the usual blind-flying instruments, which include the artificial horizon and directional gyro. Yet, many of those pilots are still taught needle, ball, and airspeed blind-flying procedures. The limited blind-flying capability required for private pilots is not required for the FAA recreational pilot certificate (Part 61, Subpart D).

15.10.2 *Artificial Horizon, Directional Gyro, and Autopilots*

The history of intentional blind flying began with Jimmy Doolittle's flight in a Consolidated NY-2 in September 1929. The airplane was equipped with two gyroscopic instruments, a gyro horizon and a directional gyro. Both have free gyroscopic wheels, which tend to preserve their position in inertial space. The wheels are mounted in double gimbals so that the airplane is free to rotate about them. The relative rotation is displayed on the instrument faces. Legal IFR flight in an airplane with a standard-category U.S. airworthiness certificate requires airplanes to be equipped with these instruments.

Recent developments in inertial navigation with strapped-down laser gyros and attitude determination by the Global Positioning System (GPS) provide synthetic versions of the artificial horizon and directional gyro. Autopilots use these instruments or their synthetic versions in automatic blind flying.

15.11   Single-Pilot IFR Operations

Making landing approaches completely by instrument is a demanding piloting procedure. Pilots are given the instrument rating only after many hours of ground school, practice in flight, and rigorous ground and flight examinations. In the United States, non-airline instrument-rated pilots must renew their ratings every six months, either by an instrument competency check ride with a flight instructor or by having flown six hours under the hood or in instrument weather conditions, and also having made six instrument approaches in that time period.

Pilots of commercial airliners get a great deal more instrument flying practice than do private instrument-rated pilots. Private instrument-rated pilots must have a minimum of only six hours of blind-flying time and six instrument approaches every six months in order to remain current. Commercial airline pilots get frequent instrument flying practice because, by law, all airplanes operating above an altitude of 18,000 feet must be on an instrument flight plan. Also, jet air carriers have agreed with the FAA to fly under instrument flight plans when below 10,000 feet, although this is not required by law.

This attention to training is not misplaced, when one understands the cockpit environment in a personal airplane during an instrument approach in bad weather. The pilot has to cope with instrument readings of heading, lateral position, velocity, vertical position, and rate of sink, as a minimum. The pilot or pilots must also handle radioed instructions or advisories on headings, intersection crossing altitudes, radio frequency changes, traffic (other aircraft), wind, and airport runway conditions. Instructions, over a frequently busy and static-corrupted radio, must be repeated and copied on a knee pad as a backup to memory, since uplinked and displayed instructions are still in the future. Approach plates and landing checklists must be consulted under often poor lighting. Added to all of these may be concerns over icing conditions and fuel reserves.

By the time of NASA's second flight test series on personal-airplane flying qualities, in 1966, personal airplanes were being used for precision ILS instrument approaches, often with a single pilot. It was and still is reasonable to ask whether the poor Dutch roll and phugoid damping, excessive longitudinal trim changes due to power application, and the badly designed trim systems found in the 1966 NASA tests add significantly to the single-pilot instrument approach problem.

Pilot workload studies in connection with instrument approaches began almost as soon as the procedures themselves. However, the few studies that examine the correlation between flying qualities and single-pilot instrument approach capability have not gone far enough. For example, a variable-stability airplane test at Princeton University used a single, experienced pilot, subjected to artificial distractions (Bar-Gill and Stengel, 1986). Only mild correlations were found between glide slope and airspeed errors and some stability and trim change parameters.

However, one would really like to know whether poor flying qualities can play a role in deteriorating a single-pilot instrument approach under more severe conditions and with less capable but instrument-rated pilots. Single-pilot instrument approaches under stressful conditions have ended in disaster many times, and we usually have no way of knowing what contribution was made by specific flying qualities.

15.12 The Prospects for Safe Personal Airplanes

Commercial production in the United States of newly designed personal airplanes has been neglected for a number of years, the victim of high costs and liability suits. Kit-built airplanes have proliferated instead, with no apparent improvement in stability and control over the last commercial designs. Many kit airplane designs show inadequate tail lengths, spin resistance, and spin recovery features. This prevails in spite of the provisions of FAA regulations Part 21, which call for kit manufacturers to comply with the FAA Part 23 airworthiness requirements for light general-aviation airplanes.

U.S. amateur-built airplanes qualify as experimental under FAA Part 21 and are excluded from the Part 23 airworthiness provisions. A new FAA category of Ultralight Vehicles (Part 103) is also excluded from Part 23 regulations. Part 103 vehicles are VFR (Visual Flight Rules) airplanes weighing less than 254 pounds empty, with stalling speeds not more than 24 knots and maximum airspeeds less than 55 knots.

The hiatus in commercial production of new personal-airplane designs may bring about stability and control improvements. There are two key developments that make this likely: computer-aided design and satellite navigation. First, computer-aided design has been brought down to a level of cost and simplicity that makes it available to the general-aviation industry and conceivably to computer-wise kit builders. There are at least three such systems available.

The Design, Analysis and Research Corporation of Lawrence, Kansas, sells software called "Advanced Aircraft Analysis," which allows users to size tail surfaces for stability, to compute stability and control derivatives, and to generate linearized dynamics. Another available computer-aided design scheme is that developed by Frederick Smetana, as a computerized version of the Perkins and Hage book *Airplane Performance, Stability and Control*. A third group of aerodynamic design and analysis computer programs specifically designed for personal computers is marketed by the Desktop Aeronautics Company of Stanford, California. Desktop's "LinAir Pro" computer program can be used to generate a complete set of stability derivatives for an airplane of arbitrary configuration. Accuracy should be comparable with that for preliminary design handbook methods.

With the availability of personal computer design programs such as those mentioned and others sure to be developed there really is no reason for basic errors in stability and control layouts to be made by kit airplane designers, not to mention the designers of the commercially built personal airplanes of the future. As a last resort, if the machine resembles a well-known existing airplane, and has a center of gravity near 25 percent of the wing chord, it probably will fly and can be made reasonably satisfactory by control surface and other modifications based on flight testing.

GPS satellite navigation is within the reach of general-aviation airplanes, with consequences for stability and control. Many small airports should be able to provide approved, charted instrument approaches. If instrument approaches become more common in general aviation, there will be increased attention needed to handling qualities under those conditions. Simplifying navigation with GPS should allow pilots to pay more attention to airplane control.

CHAPTER 16

Stability and Control Issues with Variable Sweep

Variable wing sweepback is an attempt to combine the best performance, stability, and control characteristics of straight and swept wings. Straight wings have benign low-speed stability and control characteristics, good low-speed maximum lift, and low cruise drag, while sweptback wings have low transonic and supersonic drag and good high-Mach-number stability and control. In a variable-sweep airplane, wings are spread fully, or unswept, at low speeds and are swept back at high Mach numbers.

16.1 The First Variable-Sweep Wings – Rotation and Translation

The designers of the first variable-sweep airplanes, the Messerschmitt P.1101, the Bell X-5, and the Grumman XF10F-1, found it necessary to move the wing inboard ends forward on the fuselage as the wing tips were moved aft. This was to keep the wing's mean chord in about the same fore-and-aft position along the fuselage. This kept the distance from the airplane's cg to the wing aerodynamic center about the same as the wing was swept back.

It can be imagined that the complication of a wing–fuselage attachment that translated as well as rotated was a powerful deterrent to aircraft designers. In fact, while this was the only available variable-sweep method, the concept turned up only in research aircraft.

16.2 The Rotation-Only Breakthrough

The rotation-only concept for variable sweep was pioneered by Dr. Barnes Wallis at Vickers-Armstrongs, Weybridge, around 1954. Starting in 1959, brilliant work by a NASA Langley Laboratory team, including Dr. Wallis, made variably swept wings a practical design option. Team members William J. Alford, Jr., Edward C. Polhamus, and Wallis found a practical way to eliminate the translation, or fore-and-aft motion of the wing inboard ends, drastically simplifying the variable-sweep rotation/translation mechanism to rotation alone.

The clue was to pivot the wings well out from the airplane centerline and to bring the wing trailing edges when fully swept parallel and close to the horizontal tail leading edges. In the Alford-Polhamus-Wallis design, the wing pivots are on the outboard ends of a glove, a diamond-shaped, highly swept inboard fixed-wing section. Wing spanwise loads are carried primarily on the outboard or unswept panels when the wings are in the forward position. The wing's spanwise load shifts relatively inboard, to the glove, when the wings are in the aft position. This relative load shift is exactly what one wants in order to minimize movement of the total wing aerodynamic center when the wings go through its sweeping routine (Loftin, 1985). Alford and Polhamus jointly hold the U.S. patent on this design.

An additional benefit of the Alford-Polhamus-Wallis arrangement arises from downwash changes with wing sweep. Bringing the wing trailing edge close to the horizontal tail drastically increases the downwash rate of change with angle of attack, reducing the tail's stabilizing effect. That is, the tail's increasing up-load with increasing angles of attack is reduced. In effect, the wing acts as a huge turning vane, aligning with itself the airflow into the horizontal tail. Reduced stability from the horizontal tail is just what is needed when the wing is swept back by rotation alone.

Another way of thinking of the Alford-Polhamus-Wallis arrangement is to consider the horizontal tail as an extension of the wing when the latter is fully swept back. Surface area at the rear of a lifting surface carries a smaller airload than does the same amount of area as an independent lifting surface. The lower airloads on the horizontal tail result in reduced static longitudinal stability, again just what is needed.

16.3 The F-111 Aardvark, or TFX

Practical variable sweepback came along just in time for the all-service TFX concept, which later became the Air Force's F-111 (Figure 16.1). The F-111 uses the

Figure 16.1 The F-111 with wings fully swept at 72.5 degrees and fully unswept at 16 degrees. (USAF photos)

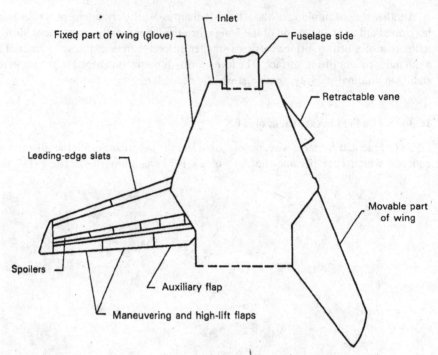

Figure 16.2   The approximate wing planform of the variable-sweep Grumman F-14 Tomcat. The retractable glove vane reduces excessive longitudinal stability with the wings fully swept back. (From Loftin, NASA SP 468, 1985)

Alford-Polhamus-Wallis rotation-only mechanism. The wing sweep range is 16 to 72.5 degrees, with normal subsonic cruise at a sweep of 26 degrees. Triply redundant three-axis stability augmentation is used. There are no particular stability and control problems with this machine.

16.4   The F-14 Tomcat

The next variable-sweep airplane to see service was the Navy's Grumman F-14 Tomcat. Static longitudinal stability is excessive with the wings fully swept back at high Mach numbers, even though the F-14 uses the Alford-Polhamus-Wallis arrangement. Grumman corrected the problem with a small leading-edge extension on the glove, which is called a glove vane (Figure 16.2). The vane starts to extend at a Mach number of 1.0, reaching a full 15-degree extension at a Mach number of 1.5.

Other interesting F-14 stability and control features are the use of horizontal tail differential incidence for lateral control. Wing spoilers also provide lateral control for wing sweep angles below 57 degrees. The spoilers are almost aligned with the wind at sweep angles above 57 degrees, so locking them out loses nothing in roll control power. Like the F-111, the F-14 has triply redundant three-axis stability augmentation. These are analog systems, typical for aircraft of its generation.

16.5   The Rockwell B-1

The Rockwell B-1 strategic bomber is an interesting variable-sweep case in that the aerodynamic center shift when sweeping the wings is large enough to require compensating center of gravity shifts. The wing pivot point is outboard, as in the F-111 and F-14, but the trailing edge of the fully swept wing does not merge with the horizontal tail leading edge. Thus, the B-1 configuration does not take full advantage of destabilizing downwash increases with the wing swept back. The required center of gravity shifting on the B-1 is done by pumping remaining fuel between the forward and aft tanks.

The airplane can be landed if the wing becomes locked in its full aft position of 67.5 degrees (65 degrees for the B-1B), but with full nose-up control minimum airspeed is close to 200 knots, making for a very high-speed landing. The story is that a B-1 based at a U.S. Air Force airfield in Kansas with a wing stuck in its full aft swept position had to fly to Edwards Air Force Base in California to use its extra-long runways.

The B-1 can have a severe stability problem at the other end of the sweep range, the landing position of 15 degrees. This occurs if the wings are swept forward to 15 degrees without waiting for the fuel to be pumped forward as well. This was guarded against originally by a warning light that came on if fuel transfer had not been made before unsweeping. According to Paul H. Anderson, a warning light was used originally instead of a positive interlock that would prevent unsweeping until fuel was transferred because of concern that a failure in the interlock system could lock the wing in its aft position.

However, a tragic accident occurred at Edwards when a pilot apparently ignored the warning light and unswept a B-1's wings without the compensating forward center of gravity shift by fuel pumping. The airplane simply became uncontrollably unstable and was lost. A positive interlock replaced the warning light after that accident.

16.6   The Oblique or Skewed Wing

Another rotation-only variable-sweep concept was invented by the late Robert T. Jones at the NACA Ames Aeronautical Laboratory, around 1945 (Figure 16.3). This is the oblique or skewed wing, in which wing sweepback (and sweepforward) is achieved by rotating the wing at its center, sweeping one side back and the other side forward. With the oblique wing rotated back into symmetry, the configuration avoids the tip stalling and low-speed stability and control problems associated with ordinary wing sweepback. Jones' invention seems to have paralleled other rotating-wing sweep concepts, those of Lachmann of Handley Page and Richard Vogt of Blohm and Voss. Jones expected an additional advantage for the oblique wing as compared with conventionally swept wings, that of higher supersonic lift–drag ratio.

Had the unorthodox oblique-winged configuration been proposed by someone without Jones' immense prestige, it might have been dismissed at once. But, for one thing, Robert Jones was credited with the invention of wing sweepback to alleviate compressibility effects during World War II, independently of the Germans. He also contributed largely to stability and control theory, in all-movable controls, lateral control, two-control airplanes, and in solutions of equations of motion. Like the Wright brothers, Edward Heinemann, and John Northrop, Jones was not university-trained. His considerable mathematics were self-taught.

With the wing rotated, the oblique-wing configuration is that rarest of heavier-than-air machines, one without bilateral or mirror-image left–right symmetry. Birds, dragonflies, and our own flying creations all have bilateral symmetry, as we ourselves do. It seems obvious that the flying qualities of an oblique-winged airplane would be strange, if not dangerous. For one thing, pulling up the airplane's nose to increase angle of attack would create inertial rolling and yawing moments, quite absent in symmetrical airplanes. These moments arise from pitching velocity and acceleration acting on a nonzero product of inertia term I_{xy}.

Figure 16.3 Robert T. Jones, ahead of his time in many areas of aeronautics. He was the inventor in the United States of wing sweepback and of the oblique-winged airplane. Jones contributed to stability and control theory in lateral control, in two-control airplanes, and in all-movable controls. (From Hansen, *Engineer in Charge*, 1987)

The effectiveness of trailing-edge flap-type controls is seriously reduced at large sweep or skew angles. Control deficiencies can be made up if the airplane carries conventional tail surfaces. Control problems are more critical if an oblique wing airplane is always operated in the skewed position, but this would obviate the need for rotating engine pods and vertical tail surfaces.

Wing torsional divergence on the sweptforward panel, discussed in Chapter 19, "The Elastic Airplane," has been raised as an issue for the oblique wing. Jones quite early predicted that rigid-body roll freedom would tend to raise the divergence speed to safe values outside of the flight range. That is, when the leading or sweptforward panel starts to bend upward under high airloads, the lift on that panel would increase, causing a large rolling moment. Airplane roll response to that moment would alleviate the airload and the wing would be safe.

However, the case must be considered in which automatic roll control operates to hold the airplane at zero bank. If the control rolling moment that holds the zero-bank angle comes from a horizontal tail, the wing torsional divergence speed could be close to the body-clamped case. A free-free analysis that includes autopilot loops would seem to be needed. On the other hand, if control rolling moment comes from ailerons on the leading panel, the panel loads would be reduced, as in the case of free-body roll. This would raise torsional divergence speed above the body-clamped value.

Some detailed stability and control data on oblique wings were obtained in NASA Ames Research Center wind-tunnel tests and in a NASA–Navy funded study begun in 1984. The

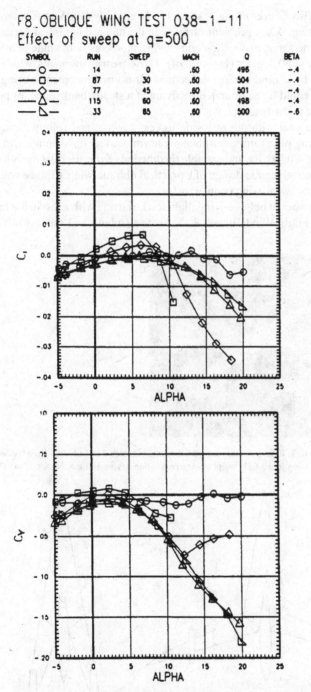

SYMBOL	RUN	SWEEP	MACH	Q	BETA
─○─	14	0	.60	496	-.4
─□─	87	30	.60	504	-.4
─◇─	77	45	.60	501	-.4
─△─	115	60	.60	498	-.4
─▽─	33	85	.60	500	-.6

Figure 16.4 Zero-sideslip variations of rolling moment and side force coefficients for an oblique wing tested on a model of the NASA-Vought F-8 research airplane. Sizable, nonlinear values appear for wing skew angles of 30 degrees and above. (From Kroo, AIAA Short Course Notes, 1992)

study was on the feasibility of converting NASA's F-8 Digital Fly-By-Wire Airplane to an oblique wing configuration. A key problem surfaced in the unusual nonlinear variations at zero sideslip angle in side force, and rolling and yawing moments with angle of attack, at wing skew angles as low as 30 degrees (Figure 16.4). These are trim moments, which would have to be trimmed out by control surface deflections for normal, nonmaneuvering flight. The nonzero side force could be equilibrated by flying at a steady bank angle, or possibly by wing tilt with respect to the fuselage.

Other possibilities to deal with nonzero side forces, yawing, and rolling moments at zero sideslip include wing plan form adjustments, unsymmetrical tip shaping, wing pivot selection, antisymmetric wing twist, and variable tip dihedral (Kroo, 1992). One is left with the impression that the aerodynamic design of a practical oblique-wing airplane will be far more complex than for its swept-wing counterpart.

There have been a number of oblique-wing flight tests, starting with a test in the Langley Research Center's Free Flight Wind Tunnel. R. T. Jones also built and successfully flew a

Figure 16.5   The R. T. Jones invention in flight, the Ames-Dryden AD-1 oblique-wing testbed, flying with its adjustable wing in the fully swept 60-degree position. (From Hallion, NASA SP-4303, 1984)

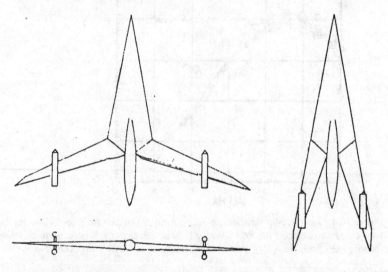

Figure 16.6   The Vickers-Armstrong "Swallow" variable-sweep concept, tested at NASA's Langley Laboratory and found to be longitudinally unstable with wings unswept. (From Polhamus and Toll, NASA TM 83121, 1981)

series of small free-flying oblique-winged model aircraft, culminating in a radio-controlled two-meter span model whose wing and tailplane skew angle could be systematically changed in flight. A considerable number of oblique-wing design studies followed. NASA eventually contracted to have a full-scale oblique-wing test airplane built for low-speed flight tests. This airplane, the AD-1, a single-engine jet, was flown successfully at the NASA Dryden Flight Research Center at Edwards, California (Figure 16.5). Ten degrees of bank angle on the AD-1 are required to cancel the side force produced by a 60-degree wing skew angle (Kroo, 1992).

The Dryden flight tests were followed by a NASA design research contract for an oblique-wing supersonic transport. The contract was awarded to Boeing, McDonnell-Douglas, and Kansas University, around 1992. The study revealed arrangement problems with that particular arrangement. An all-wing version of the oblique-wing eliminates the need for hinging the wing to a fuselage, although engine pods and any vertical tails still require hinging. Another NASA design research contract to Stanford University is for a flying model of a 400-foot-span all-wing supersonic transport, operated as an oblique wing. Stability and control for the all-wing versions of the oblique wing are problematic because of the problem of nonzero side forces, rolling, and yawing moments in oblique cruising flight.

16.7   Other Variable-Sweep Projects

The reason that a British team under Dr. Barnes Wallis at Vickers-Armstrongs was available to work on variable sweep with the NASA Langley group in 1959 was that Wallis could not get British government funding for full-scale variable-sweep tests. Inspired by wartime German research, Wallis had actually started variable-sweep research in Britain in 1945, using models launched from a rocket-powered trolley. After some successful model flights, Vickers-Armstrongs contracted for a small piston-engine variable-sweep test airplane with Heston Aircraft Ltd., but the parts built were never assembled and were eventually broken up.

In 1959, Wallis brought the Vickers-Armstrongs variable-sweep team and data to NASA's Langley Laboratory for further research. At the time, the variable-sweep configuration of interest to the British was a high-aspect-ratio tailless arrangement in which the wing inboard ends required translation. The British called this arrowhead-like configuration "Swallow." The Swallow was to lead to a high-subsonic airliner capable of flying nonstop from London to Australia at 50,000 feet. NASA wind-tunnel tests indicated that the Swallow would be longitudinally unstable with the wings unswept at low subsonic speeds (Figure 16.6). A return to horizontal tails and the successful outboard-hinge rotation-only arrangement followed.

Later practical applications of variable sweep were the Anglo-German-Italian Panavia Tornado and the former USSR's MiG-23 Flogger, MiG-27, Su-17, Su-24, Tu-22M Backfire, and the Tu-160 Blackjack.

CHAPTER 17

Modern Canard Configurations

The 1903 Wright brothers' Flyer was of course a canard airplane, with its horizontal tail in front. However, in the years that followed, up to quite recent times, canard configurations were definitely a curiosity, out of the mainstream. Conservative stability and control engineers think this is just as well.

17.1 Burt Rutan and the Modern Canard Airplane

Elbert L. (Burt) Rutan is an original thinker, a classic inventor, who left jobs at the Edwards Air Force Base and with Jim Bede in Kansas to build experimental personal airplanes at Mojave Airport, California. His Rutan Aircraft Factory, in a barracks-style building, became the home of the VariEze and Long EZ fiberglass canard airplanes. These designs have been built in large numbers from Rutan Aircraft plans by home builders. The excellent speed and climb performance of these little machines, compared with mass-produced, all-metal general-aviation airplanes, led to several major canard projects by Scaled Composites, Inc. One of these was the Beech Model 2000 Starship 1, an 8- to 11-seat business aircraft.

Rutan's successes with the VariEze and Long EZ, with the canard round-the-world Voyager, and with the Beech projects have inspired many new canard home-built sport aircraft projects in the United States. Among them are the American Aircraft Falcon series, the Beard Two Easy, the Co Z Development Cozy, the Diehl XTC Hydrolight, the Ganzer Model 75 Gemini, and so on. What can we say about this trend? Some corrective notes on the supposed advantages of the modern canards, and on canard stability and control pitfalls, seem in order, in the hope that future designers will be fully informed.

17.2 Canard Configuration Stall Characteristics

Canard aircraft are characteristically difficult to stall at all. The canard surface is generally designed to stall before the main, or aft, wing does, when the angle of attack is increased at a normal, gradual rate. When the canard does stall, with the main wing still unstalled, the airplane tends to pitch down, recovering normal flight. However, William H. Phillips comments that in the airplane pitch-down following canard stall, the canard surface's angle of attack is increased again by the airplane's angular velocity. This could delay recovery from an unstalled condition until the airplane has reached a steep nose-down attitude. It can be argued that an aft-tailed airplane also tends to recover automatically from a stall. On aft-tailed airplanes the horizontal tail, operating in the wing's downwash, experiences a relative upload when the wing stalls. This is because the wing downwash drops off when the wing stalls.

The main concern in canard airplane stalls is the dynamic stall, entered at a high rate of angle of attack increase. Pitching momentum could carry the angle of attack up to the point where the main wing stalls, as well as the canard. In combination with unstable pitching moments from the fuselage, this could produce a total nose-up pitching moment that cannot be overcome by available canard loads. Wing trailing-edge surfaces that augment canard pitching moment control would be ineffective with the main wing stalled. Thus, a canard airplane's main wing stall could produce deep stall conditions, in which a recovery to unstalled flight cannot be made by any forward controller motion (see Chapter 14). Deep stall at aft center-of-gravity positions and high power settings was identified in NASA tests of a tractor propeller canard configuration (Chambers, 1948).

The possibility of dynamic stalling on canard airplanes is minimized if the configuration is actually a three-surface case: main wing, canard, and aft horizontal tail. Examples of three-surface configurations are the Piaggio P.180, the Sukhoi Su-27K, the DARPA/Grumman X-29A forward-swept research airplane, and the many three-surface airplanes designed by G. Lozino-Lozinsky, of MiG-25 fame. Even at extreme angles of attack that stall the main wing, the aft horizontal tail may be in a strong enough downwash field to remain unstalled, or it may be unstalled by nose-down incidence. With an unstalled aft horizontal tail, longitudinal control can be maintained.

Another way to minimize the possibility of dynamic stalling of canard airplanes is to operate them at centers of gravity far forward enough so that elevator power cannot produce high nose-up rotation rates. This amounts to restriction of the available center-of-gravity range and a reduction in the airplane's utility.

17.3   Directional Stability and Control of Canard Airplanes

Vertical tail length, or the distance from the center of gravity to the vertical tail aerodynamic center, is typically short for canards as compared with tail-aft configurations. Plan-view drawings for two Beech aircraft illustrate this (Figure 17.1). The Super King Air B200 and Starship 1 have similar gross weights (12,500 to 14,000 pounds) and have fuselage lengths of about 44 feet. The canard Starship's vertical tail length is 18 feet; that for the tail-last King Air is 25 feet, or about 40 percent greater. The tail-last configuration would have better directional stability and control, assuming equal vertical lifting surface effectiveness for both aircraft.

Note however that the canard Starship's vertical tails are at the wing tips. In that location, vertical tail effectiveness is not increased by fuselage end plating. The fuselage-mounted vertical tail for the King Air tail-last configuration benefits from fuselage end plating to the extent of about a 50-percent increase in lift curve slope and in effectiveness.

Lower directional stability and control levels in canard configurations can be corrected by large vertical surfaces, at the expense of higher weight and cost. The original tip-mounted vertical tails of the Rutan Vari Eze were found to be too small in NASA wind-tunnel tests (Yip, 1985). Low directional stability levels are associated with adverse yaw in rolling and poor lateral control. Low directional control power leads to control problems in takeoff and in landings in crosswinds or with asymmetric power.

According to Professor Jan Roskam of Kansas University, directional stability on the Piaggio P.180 Avanti business airplane, which has a canard surface and a relatively short vertical tail length, was improved greatly at high angles of attack with strakes located at the fuselage rear.

17.4   The Penalty of Wing Sweepback on Low Subsonic Airplanes

Extra vertical tail length is obtained in canard configurations with wing-tip-mounted vertical tails by using wing sweepback. While we have learned how to provide good stall characteristics and a stable pitching moment stall break on sweptback wings,

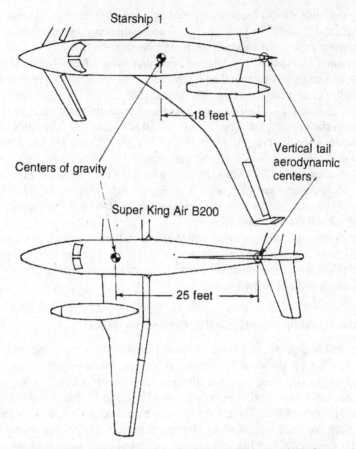

Figure 17.1 Drawings of the tail-last Beech Super King Air B200 (*above*) and canard Starship 1 (*below*). The two airplanes are of similar size and gross weight, but the B200's vertical tail length is 40 percent greater than the Starship's. (From Jane's *All the World's Aircraft*, 1987–1988)

these come at a cost in wing twist, special airfoil sections, or stall control devices such as slats, fences, and slots. Thus, wing sweepback used on a canard configuration to improve directional stability and control brings cost and weight penalties relative to tail-last configurations.

17.5   Canard Airplane Spin Recovery

Aerodynamic and mass criteria for good spin recovery of tail-last configurations are well known, as a result of years of experience and testing. A builder of a conventional tail-last configuration can rely on NASA spin recovery design charts with a fair degree of confidence. The NASA design charts specify minimum rudder and fuselage areas in certain locations, depending on calculated airplane moment of inertia parameters. The point is that airplane designers who cannot afford the expense of testing their designs in specialized spin-tunnel facilities or in model drop tests can still be reasonably assured of safe spin recoveries by using the NASA design charts and other guidelines.

The NASA spin recovery design charts do not specifically apply to canard configurations and can only offer the most general guide in those cases. A canard airplane designer should count on spin-tunnel or drop model tests, in order to ensure safe spin recoveries. Canard

Figure 17.2 The canard six-place Jetcruzer, the first airplane to be granted a spin-resistant certification under FAA Part 23. Numerous attempts to spin the airplane were unsuccessful. (From *AOPA Pilot*, Aug. 1994)

surface tests by Neihouse in 1960 showed prospinning or propelling yawing moments for some canard sizes and locations on the fuselage nose.

Possible spin recovery problems are of course avoided if the airplane's longitudinal control power is limited sufficiently so that the airplane cannot be stalled. An airplane's main wing must be stalled and autorotation (Jones, 1934; McCormick, 1979) must be initiated before an airplane can spin. Even if the airplane stalls, spinning might still be avoided if rudder power is limited or coordinated with the ailerons, as in two-control airplanes such as the Ercoupe. Control limiting without penalty to the airplane's utility is altogether feasible for modern computer-controlled fly-by-wire machines, such as the Northrop B-2 and Grumman X-29A.

For ordinary fly-by-cable airplanes, limiting longitudinal control power to that just needed to attain maximum lift coefficient can be defeated by loading the airplane to have a more aft center of gravity. Otherwise stated, limiting longitudinal control power to avoid stalling and spinning inevitably cuts down an airplane's usable center-of-gravity range, reducing its utility. Some form of control limiting through center-of-gravity range has apparently been used in a recently certificated canard airplane. This is the six-place turboprop Jetcruzer, a product of Advanced Aerodynamics and Structures, Inc., of Burbank, California (Figure 17.2). Numerous attempts by test pilots to stall and spin the airplane were unsuccessful. The FAA granted the airplane a "spin-resistant"-type certificate, under the Federal Airworthiness Standards, Part 23.

17.6 Other Canard Drawbacks

Canard surfaces usually reduce forward and downward visibility for the flight crew, a drawback in landing approaches and landings. A canard surface added to a conventional tail-last configuration results in three lifting surfaces in tandem. This is objectionable from an analysis and testing standpoint. That is, there are two downwash and sidewash fields to account for, adding to design and testing complexity.

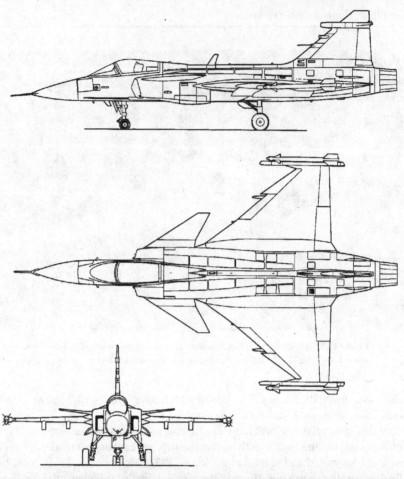

Figure 17.3 A modern canard airplane, the SAAB JAS 39 Gripen. (From Jane's *All the World's Aircraft*, 1987–1988)

17.7 Pusher Propeller Problems

Although canard airplanes can have propellers in front, in the so-called tractor position, canard propeller-driven aircraft generally wind up with pusher propellers. Thus, in the context of discussing design, stability, and control problems of canards, it is appropriate to bring up some design problems of pusher propellers as well.

A tail-down landing touchdown attitude is often desired, for a minimum energy landing. In fact, many nose landing gears are not expected to take landing impact loads and are noticeably lighter and weaker than main landing gear assemblies. Pusher propellers tend to have relatively small diameters, just to provide clearance for tail-down landings. This is a constraint on propeller design, leading to lower propulsive efficiency. Alternatively, airplanes with pusher propellers tend to have relatively long, heavy main landing gear legs.

Pusher propellers generally act in the wakes of either wings or horizontal tails. While there may be no appreciable propulsive efficiency loss for such arrangements, a distinctive propeller noise generally results, which could be a problem for people on the ground. Pusher propellers have vibration problems, and their engines can have cooling problems.

17.8 The Special Case of the Voyager

The 1986 nonstop round-the-world flight of Burt Rutan's Voyager brought deserved high praise for its designer and courageous flight crew. However, the account of that historic flight shows that the pilots were handicapped by the instability of the airplane at high gross weights. The Voyager is a canard configuration whose tips were joined to the main wing by parallel fuselages (Yeager, Rutan, and Patton, 1987). The statement is made, "Hand-flying Voyager required almost all our concentration, and flying it on autopilot still required most of our concentration."

A note from Brent W. Silver, a consulting member of the Voyager design team, points to a likely cause of this problem. Apparently, bending of the Voyager's main wing in turbulence coupled into the canard tips through the parallel fuselages. This caused canard twisting in phase with main wing bending and considerable pitch changes. The same main wing flexibility in a conventional tail-last arrangement should have not caused such a pitch reaction.

17.9 Modern Canard Tactical Airplanes

The canard disadvantages enumerated above either do not apply or are overwhelmed by other considerations in the case of tactical airplanes designed for supermaneuverability, or for controllable flight beyond the stall. The stability and control of tactical airplanes in the supermaneuverability regime are covered in Chapter 10, "Tactical Airplane Maneuverability."

Control of the vortex system shed from the fighter nose is known to be critical for controllable flight beyond the stall. Forebody strakes have been found valuable for this purpose. Canards offer another means for shaping the forebody vortex system. They are used in some modern fighter designs, such as the Sukhoi Su-35, the Saab JAS 39 Gripen (Figure 17.3), the IAI Lavi, the Rockwell/MBB X-31A Enhanced Fighter Maneuverability (EFM), and the Eurofighter 2000.

CHAPTER 18

Evolution of the Equations of Motion

There is a reproduction in Chapter 1 of George H. Bryan's equations of airplane motion on moving axes, equations developed from the classical works of Newton, Euler, and Lagrange. This astonishingly modern set of differential equations dates from 1911. Yet, Bryan's equations were of no particular use to the airplane designers of his day, assuming they even knew about them.

This chapter traces the evolution of Bryan's equations from academic curiosities to their present status as indispensable tools for the stability and control engineer. Airplane equations of motion (Figure 18.1) are used in dynamic stability analysis, in the design of stability augmenters (and automatic pilots), and as the heart of flight simulators.

18.1   Euler and Hamilton

One of the problems faced by Bryan in developing equations of airplane motion was the choice of coordinates to represent airplane angular attitude. Bryan chose the system of successive finite rotations developed by the eighteenth-century Swiss mathematician Leonhard Euler, with a minor difference. In Bryan's words:

> In the [Eulerian] system as specified in Routh's *Rigid Dynamics* and elsewhere, the axes are first rotated about the axis of z, then about the axis of y, then again about the axis of z. The objection to this specification is that if the system receives a small rotation about the axis of x, this cannot be represented by *small* values of the angular coordinates.

Bryan chose instead to rotate by a yaw angle Ψ about the vertical axis, a pitch angle Θ about the lateral axis, followed by a roll angle Φ about the pitch axis – a sequence that has been followed in the field ever since. However, Bryan's orthogonal body axes fixed in the airplane are rotated by 90 degrees about the X-axis as compared with modern practice. That is, the Y-axis is in the place of the modern Z-axis, while the Z-axis is the negative of the modern Y-axis (Figure 18.2).

Bryan's Eulerian angles have served the stability and control community well in almost all cases. However, there were other choices that Bryan could have made that would have avoided a singularity inherent in Euler angles. The singularity shows up at pitch angles of plus or minus 90 degrees, the airplane pointing straight up or straight down. Then the equation for yaw angle rate becomes indeterminate.

The Euler angle singularity at 90 degrees is avoided by the use of either quaternions, invented by Sir W. R. Hamilton, or by direction cosines. The main disadvantage of quaternions and direction cosines as airplane attitude coordinates is their utter lack of intuitive feel. Flight dynamics time histories calculated with quaternions or direction cosines need to be translated into Euler angles for intelligent use. Except for simulation of airplane or space-vehicle vertical launch or of fighter airplanes that might dwell at these attitudes, the Euler angle singularity at 90 degrees is not a problem.

As the term implies, there are four quaternion coordinates; there are nine direction cosine coordinates. Since, as Euler pointed out, only three angular coordinates are required

Notes: 1. Body axes, with origin at the cg.
2. s and c stand for sine and cosine
3. X-Z plane of symmetry.

Displacements

$$\dot{X}_e = Uc\Theta c\Psi + V(c\Psi s\Phi s\Theta - s\Psi c\Phi) + W(c\Psi c\Phi s\Theta + s\Psi s\Phi)$$

$$\dot{Y}_e = Uc\Theta s\Psi + V(s\Psi s\Phi s\Theta + c\Psi c\Phi) + W(s\Psi c\Phi s\Theta - c\Psi s\Phi)$$

$$\dot{Z}_e = -Us\Theta + Vs\Phi c\Theta + Wc\Phi c\Theta$$

Linear Velocities

$$X - mgs\Theta + THR \cdot AT = m(\dot{U} - VR + WQ)$$

$$Y + mgs\Phi c\Theta + THR \cdot BT = m(\dot{V} - WP + UR)$$

$$Z + mgc\Phi c\Theta + THR \cdot CT = m(\dot{W} - UQ + VP)$$

Euler Angles

$$\dot{\Psi} = (Qs\Phi + Rc\Phi)/c\Theta$$

$$\dot{\Phi} = P + \dot{\Psi}s\theta$$

$$\dot{\theta} = Qc\Phi - Rs\Phi$$

Angular Velocities

$$L + THR \cdot DT = \dot{P}I_x - \dot{R}I_{xz} + (I_z - I_y)QR - PQI_{xz}$$

$$M + THR \cdot ET = \dot{Q}I_y + (I_x - I_z)RP + (P^2 - R^2)I_{xz}$$

$$N + THR \cdot FT = \dot{R}I_z - \dot{P}I_{xz} + (I_y - I_x)PQ + QRI_{xz}$$

Figure 18.1 The 12 equations of airplane rigid-body motion, used extensively in flight simulation. All but three are in classical state-variable form, suitable for sequential application of computer integrating subroutines. Substituting the seventh into the eighth equation puts that one into state-variable form. A matrix inversion of the tenth and twelfth equations puts them into state-variable form.

to specify rigid-body attitudes, quaternion and direction cosine coordinates have some degree of redundancy. This redundancy is put to good use in modern digital computations to minimize roundoff errors in an orthogonality check. Another advantage to quaternion as compared with Euler angle coordinates is the simple form of the quaternion rate equations, which are integrated during flight simulation. Euler angle rate equations differ from each other, are nonlinear, and contain trigonometric functions. On the other hand, quaternion rate equations are all alike and are linear in the quaternion coordinates.

The nine direction cosine airplane attitude coordinates are identical to the elements of the 3-by-3 orthogonal matrix of transformation for the components of a vector between two

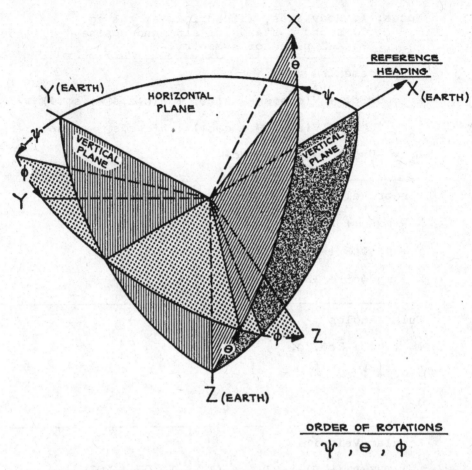

ORDER OF ROTATIONS

ψ , θ , ϕ

Figure 18.2 The Euler angle sequence in most common use as airplane attitude coordinates in flight dynamics studies. This sequence was defined by B. Melvill Jones in Durand's *Aerodynamic Theory*, in 1934. (From Abzug, Douglas Rept. ES 17935, 1955)

coordinate systems. As in the quaternion case, all nine direction cosine rate equations have the advantage of being alike in form, and all are also linear. The direction cosine rate equations are sometimes called *Poisson's equations*. Airplane equations of motion using quaternions are common; those using direction cosine attitude coordinates are rare.

The Euler parameter form of quaternions uses direction cosines to define an axis of rotation with respect to axes fixed in inertial space. A rotation of airplane body axes about that axis brings body axes to their proper attitude at any instant (Figure 18.3). This goes back to one of Euler's theorems, which states that a body can be brought to an arbitrary attitude by a single rotation about *some* axis. There is no intuitive feel for the actual attitude corresponding to a set of Euler parameters because the four parameters are themselves trigonometric functions of the direction cosines and the rotation angle about the axis.

The first published report bringing quaternions to the attention of airplane flight simulation engineers was by A. C. Robinson (1957). Robinson's contribution was followed in 1960 by D. T. Greenwood, who showed the advantages of quaternions in error checking numerical computations during a simulation. A detailed historical survey of all three attitude coordinate systems is given by Phillips, Hailey, and Gebert (2001). The flight simulation

DEFINITIONS:

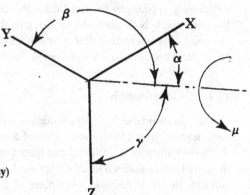

$$e_1 = \cos\mu/2$$

$$e_2 = \cos\gamma \, \sin\mu/2$$

$$e_3 = \cos\beta \, \sin\mu/2$$

$$e_4 = \cos\alpha \, \sin\mu/2$$

$$e_1^2 + e_2^2 + e_3^2 + e_4^2 = 1.0 \text{ (orthogonality)}$$

DIRECTION COSINE MATRIX

If $\{x\}_{body} = [L_{EB}]\{x\}_{earth} = [l_{ij}]\{x\}_{earth}; \ i,j = 1,3$
then

$$[L_{EB}] = \begin{bmatrix} e_1^2 - e_2^2 - e_3^2 + e_4^2 & 2(e_1e_2 + e_3e_4) & 2(e_2e_4 - e_1e_3) \\ 2(e_3e_4 - e_1e_2) & e_1^2 - e_2^2 + e_3^2 - e_4^2 & 2(e_2e_3 + e_1e_4) \\ 2(e_4e_2 + e_3e_1) & 2(e_2e_3 - e_1e_4) & e_1^2 + e_2^2 - e_3^2 - e_{42} \end{bmatrix}$$

QUATERNION RATES

$$2\dot{e}_1 = -e_4P - e_3Q - e_2R$$
$$2\dot{e}_2 = -e_3P + e_4Q + e_1R$$
$$2\dot{e}_3 = e_2P + e_1Q - e_4R$$
$$2\dot{e}_4 = e_1P - e_2Q + e_3R$$

QUATERNIONS TO EULER ANGLES:

$$\Theta = \sin^{-1}[2(e_1e_3 - e_2e_4)]$$

$$\Phi = \tan^{-1}[2(e_3e_2 + e_4e_1)/(-e_4^2 - e_3^2 + e_2^2 + e_1^2)]$$

$$\Psi = \tan^{-1}[2(e_4e_3 + e_2e_1)/(e_4^2 - e_3^2 - e_2^2 + e_1^2)]$$

EULER ANGLES TO QUATERNIONS

$$e_1 = \cos\Psi/2 \, \cos\Theta/2 \, \cos\Phi/2 + \sin\Psi/2 \, \sin\Theta/2 \, \sin\Phi/2$$
$$e_2 = -\cos\Psi/2 \, \sin\Theta/2 \, \sin\Phi/2 + \sin\Psi/2 \, \cos\Theta/2 \, \cos\Phi/2$$
$$e_3 = \cos\Psi/2 \, \sin\Theta/2 \, \cos\Phi/2 + \sin\Psi/2 \, \cos\Theta/2 \, \sin\Phi/2$$
$$e_4 = \cos\Psi/2 \, \cos\Theta/2 \, \sin\Phi/2 - \sin\Psi/2 \, \sin\Theta/2 \, \cos\Phi/2$$

Figure 18.3 The Euler parameter form of quaternions used in some flight simulations to calculate airplane attitude. The upper group of equations defines the Euler parameters in terms of an axis of rotation of XYZ to a new attitude. $\{x\}_{body}$ are vector components on the rotated axes; $\{x\}_{earth}$ are the same components on the original axes. Transformations between Euler parameters and Euler angles are given in the lower two sets of equations.

community appears to be divided on the choice between Euler angles and quaternions. In some cases, both are used in different flight simulators within a single organization. However, it is interesting that so many modern digital computations of airplane stability and control continue to use Euler angle coordinates in the 1911 Bryan manner.

18.2   Linearization

In their basic form, the equations of airplane motion are a set of nine simultaneous nonlinear differential equations. One of the most far-reaching steps taken by Bryan was the development of a perturbation, linearized, form of these equations. The perturbation motion of a simple mechanical object, such as a pendulum, about a state of rest is a familiar concept. In his *Mécanique Analytique* of 1788, J. L. Lagrange developed the theory of small perturbation motions of systems having many degrees of freedom about a position of stable equilibrium. Bryan extended Lagrange's work by replacing the position of stable equilibrium by a steady equilibrium motion.

The utility of Bryan's linearization arises from the nature of airplane perturbed motions. Under normal operating conditions, such as personal-airplane and airliner climbs, cruises, and landing approaches, airplanes are among the most linear dynamic systems known. Aerodynamic force and moment are quite closely proportional to airplane perturbed motions, without any equivalent to coulomb friction. Small-perturbation or linearized equations are perfectly suitable to describe the motions experienced by the crew and passengers, and for the design of stability augmenters and automatic pilots.

Bryan analyzed small perturbations about steady, symmetric, rectilinear flight, either level, climbing, or diving. Most of the subsequent literature on airplane dynamics is based on the same model. Equations of perturbed airplane motion about steady turning and steady sideslipping flight came soon after Bryan, in an important 1914 report by Leonard Bairstow. A further extension to general curvilinear flight was made using earth-referred coordinates (Frazer, Duncan, and Collar, 1938). Still later investigators (Abzug, 1954; Billion, 1956) used the more useful body-fixed coordinates. Then, in a series of NASA papers dating from 1981 to 1983, Robert T. N. Chen applied linearization to the case of perturbations from uncoordinated turns. Chen's immediate goal was to represent perturbation motions of single-rotor helicopters in low-airspeed, steep turns, in which appreciable amounts of sideslip are quite normal.

The 1914 linearization work by Bairstow suffered the fate of theory that was too far ahead of its time. The later investigators mentioned above seemed to have been unaware that Bairstow had already extended the original Bryan equations.

Bryan's linearization of the equations of airplane motion reduces them to two sets of three simultaneous linear differential equations, each set of fourth order. The linearized equations shown in Figure 18.4 illustrate three typical features of these equations. Differentiation is indicated by the Laplace variable s, operating on the small-perturbation quantities such as u, w, θ, and β. Aerodynamic variations with small-perturbation quantities, called stability derivatives, are in the "dimensional" form, suitable for closed-loop system studies and for simulation.

Finally, the derivatives are the primed form such as Lp' rather than Lp for the rolling moment due to rolling velocity. Primed derivatives combine inertial terms with aerodynamic terms, simplifying the lateral set and putting these equations into state-variable form (Sec. 11).

The fact that the linearized equations of motion separate into two independent sets is of enormous significance. Engineers can treat airplane dynamics as two individual problems:

$$\begin{bmatrix} s - X_u^* & -X_w & W_0 s + g \cos \theta_0 \\ -Z_u^* & (1 - Z_{\dot{w}})s - Z_w & -U_0 s + g \sin \theta_0 \\ -M_u^* & -(M_{\dot{w}} s + M_w) & s^2 - M_q s \end{bmatrix} \begin{bmatrix} u \\ w \\ \theta \end{bmatrix} = \begin{bmatrix} X_{\delta e} \\ Z_{\delta e} \\ M_{\delta e} \end{bmatrix} \begin{bmatrix} \delta_e \end{bmatrix}$$

$$q = s\theta$$

$$\hbar = -w \cos \theta_0 + u \sin \theta_0 + (U_0 \cos \theta_0 + W_0 \sin \theta_0)\theta$$

$$a_z = sw - U_0 q + (g \sin \theta_0)\theta$$

$$a_z' = a_z - l_x s^2 \theta$$

$$\begin{bmatrix} s - Y_v & -\dfrac{W_0 s + g \cos \theta_0}{V_{T_0}} & \dfrac{U_0 s - g \sin \theta_0}{V_{T_0} s} \\ -L_\beta' & s(s - L_p') & -L_r' \\ -N_\beta' & -N_p' s & s - N_r' \end{bmatrix} \begin{bmatrix} \beta \\ \dfrac{p}{s} \\ r \end{bmatrix} = \begin{bmatrix} Y_{\delta a}^* & Y_{\delta r}^* \\ L_{\delta a}' & L_{\delta r}' \\ N_{\delta a}' & N_{\delta r}' \end{bmatrix} \begin{bmatrix} \delta_a \\ \delta_r \end{bmatrix}$$

$$v = V_{T_0} \beta \qquad\qquad a_y = sv + U_0 r - W_0 p - g(\cos \theta_0)\varphi$$

$$\varphi = \frac{p}{s} + \frac{r}{s} \tan \theta_0 \qquad\qquad a_y' = a_y + l_{x_{lat}} sr - l_z sp$$

$$\psi = \frac{1}{\cos \theta_0} \frac{r}{s}$$

Figure 18.4 Dimensional forms of the small-perturbation equations of airplane motion, suitable for closed-loop system studies. The longitudinal set is above, the lateral set below. Output equations, for calculation of some sensor readings, are listed below the matrix sets. (From Teper, Systems Technology, Inc. Rept. 176–1, 1969)

longitudinal stability and control, arising from the symmetric equation set, and lateral stability and control, arising from the asymmetric set. However, separation into independent longitudinal and lateral sets fails for perturbations from curvilinear or sideslipping flight. Coupled lateral-longitudinal equations of up to eighth order result. Bairstow (1920) treated perturbations from circling flight.

18.3 Early Numerical Work

Useful solutions to Bryan's equations of airplane motion for scientific or engineering uses are either roots or eigenvalues or actual time histories, which give airplane responses to specific control or disturbance inputs. Either type of solution was essentially out of the question with the means available in 1911. However, by 1920 Bairstow had found useful approximations that served as starting points for developing eigenvalues from the Bryan equations.

When, later on, research engineers in both the United States and in Britain generated time history solutions to the linearized Bryan equations, it was only with great labor. Early step-by-step numerical solutions were published for the S.E.-5 airplane of World War I fame by F. Workman in 1924. A year later, B. Melvill Jones and A. Trevelyan (1925) published step-by-step solutions for the lateral or asymmetrical motions.

As an advance over step-by-step methods, B. Melvill Jones (1934) applied the formal mathematical theory of differential equations to the linearized Bryan equations, producing a marvelously complete set of time histories for the B.F.2b Bristol Fighter at an altitude of 6,000 feet (Figure 18.5). A generation of pre–electronic-computer engineers struggled through those formal solutions. The complementary function is found first. In addition to using a considerable amount of algebra, one has to find the real and complex roots of a fourth-degree polynomial. The complementary function gives the time histories of the variables of motion under no applied forces and moments, but with arbitrary initial conditions.

The last step in the formal solution is finding a particular integral of the equations. This adds to the complementary function the effects of constant applied moments, such as are produced by deflections of the airplane's control surfaces. In Jones' own words, "The numerical computations involved...are heavy, they involve amongst other things, the solution of four simultaneous equations with four variables." It is little wonder that numerical time history calculations languished for years, until electronic analog computers were commercially available, about the year 1950.

18.4   Glauert's and Later Nondimensional Forms

Hermann Glauert's contribution to the evolution of the equations of airplane motion was to introduce a dimensionless system based on the time unit $\tau = \mu l / V$. In the expression for τ, μ is the airplane relative density $m / \rho Sl$, and ρ is the air density. l and S are the airplane's characteristic length and area, respectively. Typically l is the wing span and S the wing area. V is the airspeed. The relative density μ is the ratio of airplane mass to the mass of air contained in a volume $S \times l$, determined by airplane size. Under Glauert's system, time solutions come out in units of τ seconds.

When the Glauert process is carried out, the numerical values of all symbols that appear in the equations (except for μ) depend only on the airplane's shape, mass distribution, attitude, and angles of attack and sideslip. Airplane size, velocity, mass, and the air density, or altitude of flight, are all represented by the single parameter μ.

Glauert defined new boldfaced dimensionless symbols such as $\mathbf{t}$, $\mathbf{w}$, $\mathbf{q}$ for time, vertical velocity, and pitching velocity, and $\mathbf{k}$ with an appropriate subscript for moment of inertia divided by l^2 times m. The stability derivatives are likewise nondimensionalized. For example, $\mathbf{x}_w$ stands for $(dX/dw)/\rho VS$. As B. Melvill Jones (1934) says:

> When it is desired to convert the solutions so as to apply to a specified flight of a specified aeroplane in terms of specified units, it is merely necessary to multiply $\mathbf{u}$, $\mathbf{v}$, $\mathbf{w}$, by V/μ; $\mathbf{p}$, $\mathbf{q}$, $\mathbf{r}$, by $V/\mu l$, and $\mathbf{t}$ by τ (or $m/\rho VS$); where μ, $\mathbf{V}$, l, S relate to the specified flight and are expressed in terms of the specified units.

If this is confusing to the reader, it was also confusing to the generation of stability and control engineers who practiced their art before electronic analog and then digital computers transformed the picture. Airplane time history calculations are now easy to make, so that there is no longer a premium on allowing a single dimensionless computation to represent many altitude, velocity (but not Mach number), size, and mass cases.

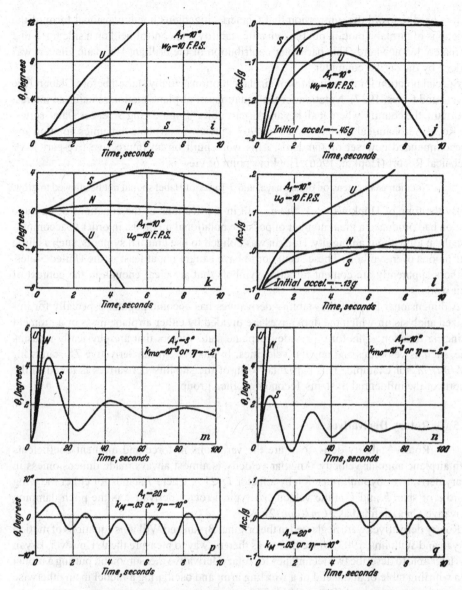

Figure 18.5 Examples of equation of motion solutions for the Bristol F.2b (Bristol Fighter) produced in the 1930s for B. Melvill Jones' section in Durand's *Aerodynamic Theory*. The pitch attitude and normal acceleration solutions are for initial airspeed and vertical velocity perturbations and step elevator angles, at different equilibrium angles.

Aerodynamic data are generated in dimensionless form as the computed output of wind-tunnel tests and are so presented to the engineers that use equations of airplane motion. However, the special time unit τ has all but disappeared from the scene. Airplane motions are calculated in terms of actual, rather than dimensionless, velocity units, except for the angles of attack and sideslip. In Glauert's day dimensionless aerodynamic coefficients in Britain were based on ρV^2, not $(\rho/2)V^2$. Thus, Glauert's dimensionless stability derivatives were half as big as NACA dimensionless stability derivatives, except for the pitching moment rate derivatives m_w and m_q, based on the time to fly a chord, not a half-chord.

It is hard to avoid the impression that Glauert's ingenious nondimensional form of the equations of airplane motion put the dynamic stability and control field on a side track that ultimately led nowhere. This particular contribution of the brilliant Hermann Glauert was undone by the digital computer.

Special notation for the equations of airplane motion actually started before Glauert (see Bryant and Gates, 1937). Notation for the equations of airplane motion remains an interest in Britain, the country where it all began. As part of its "Engineering Sciences Data" series, the Royal Aeronautical Society issued in 1967 a review of airplane dynamics notation and a recommended new set of standards. This work built on an impressive five-part RAE Technical Report (Hopkin, 1966). Hopkin's point of view is

> Notation is an extension of language, and a Tower of Babel should not be allowed to grow.

By the time of Hopkin's work, the growth in applications of the equations of airplane motion had produced a great number of possible notational methods. In order to accommodate them all without ambiguity, Hopkin was obliged to use unusual symbols, such as little half-moons over symbols. These seem not to have caught on, at least in the United States. Authors apparently are content to define symbols that are clear enough in the context of their work.

A dimensional form of the stability derivatives has become popular especially for linearized analysis, in which the derivatives are divided by either airplane mass or a moment of inertia function. This form provides airplane state vectors that are physically measurable, such as velocities and angular velocities. In this system the derivative Z_u stands for $(\partial Z/\partial u)/m$, for example. This particular form of the stability derivatives is found in the reports of the influential Systems Technology, Inc., group.

18.5   Rotary Derivatives

Rotary stability derivatives are the variations in force and moment coefficients with airplane angular velocity. Angular velocity is almost always made dimensionless in rotary derivatives by multiplication by a factor $1/(2V)$, where 1 stands for either the wing chord c or span b and V is the velocity. A typical rotary derivative is the pitch damping derivative Cm_q, defined as $\partial Cm/(\partial qc/2V)$.

Rotary derivatives were neglected in the original Bryan and Williams equations of motion (Bryan and Williams, 1903), since there was then no way to measure them. However, Bryan was later able to describe two techniques for rotary derivative measurement: putting a model on a whirling table or at the end of a whirling arm, and oscillating a model in an otherwise conventional wind tunnel (Bryan, 1911).

The oscillation technique survived right up to modern times. It is used in supersonic as well as in low-speed wind tunnels. An ingenious forced oscillation technique for measuring rotary derivatives uses feedback control to stabilize the amplitude and frequency of a forced oscillation regardless of the model's level of stable or unstable derivatives (Beam, 1956).

An additional feature of Beam's forced oscillation method is the separation of pitch and yaw damping derivatives from the cross-rotary derivatives, such as the rolling moment due to yawing, by oscillating the model around different axes. In- and out-of-phase torque measurements are solved simultaneously for the answers. The drawback in Beam's work is that the damping derivatives such as Cm_q and Cn_r are inseparable from angle of attack and side-slip rate derivatives, such as $Cm_{\dot{\alpha}}$ and $Cn_{\dot{\beta}}$. This separation is possible in specialized forced-motion wind-tunnel tests.

One of the few wind tunnels that produced pure damping derivatives was the NACA Langley Stability Wind Tunnel. The past tense is used because the Stability Wind Tunnel was dismantled some years ago and shipped to the Virginia Polytechnic Institute. The Stability Wind Tunnel had curved test sections in which the forces and moments on an ordinary model were the result of rotary flows. This yielded the rotary derivatives uncombined with attitude rate derivatives. The same effect was produced with curved airship models tested in ordinary wind tunnels back in the 1920s.

The Stability Wind Tunnel also used radial turning vanes ahead of the test section to produce rolling flow. Flow angularity with respect to the wind-tunnel centerline was a linear function of distance from the centerline to the tunnel walls. The aerodynamic forces on a model held rigidly at the tunnel center would be identical to those of a rolling model in an ordinary wind tunnel, except for some transverse boundary layer motion caused by radial pressure gradient. The DVL in Germany experimented with rolling flow in wind tunnels in the 1930s.

The whirling arm as a device for measuring rotary derivatives had a rebirth of sorts at the Cranfield College of Aeronautics in the early 1960s (Mulkens and Ormerod, 1993). The motivation is support of a Royal Aircraft Establishment flight research program called HIRM, for High-Incidence Research Model. Carbon-fiber–reinforced plastic, foam, and fiberglass models are whirled on an 8.3-meter arm inside a toroidal test channel. Moving the models at constant angle of attack along circular paths provides pure rotary derivative data, equivalent to that gotten from curved flow wind tunnels.

18.6   Stability Boundaries

Until the advent of electronic analog and digital computers, numerical solutions of the equations of airplane motion were essentially limited to finding stability boundaries, the combinations of airplane stability derivatives and other parameters that divide stability from instability. Stability boundaries are found by Routh's Criterion, developed by the Briton E. J. Routh in the early 1900s.

Airplane stability boundaries were first calculated in Britain (Bryant, Jones, and Pawsey, 1932). This was in a study of dynamic stability beyond the stall. Bryant and his co-authors found stability derivatives for a number of airplanes up to an angle of attack of 40 degrees. With these data, they produced stability boundaries as functions of static directional and lateral stability derivatives, both nondimensionalized by Glauert's airplane relative density parameter μ.

There was an earlier British paper by S. B. Gates that presented contours of constant damping ratio and natural frequency for the longitudinal phugoid, as functions of tail volume and center-of-gravity position (Gates, 1927). While not strictly a stability boundary analysis, the Gates work certainly laid the groundwork for Bryant's boundaries.

Two NACA reports by Charles H. Zimmerman (1935 and 1937) carried on Gates' and Bryant's pioneering stability boundary work. Zimmerman's ambitious goal was to produce charts for the rapid estimation of the dynamics of *any* airplane. The Zimmerman reports have charts for both longitudinal and lateral motions, 40 of the former and 22 of the latter (Figure 18.6). As in Bryant's work, the results are normalized using Glauert's airplane density parameter μ. The Zimmerman charts include period and damping estimates for the phugoid and Dutch roll motions.

18.7   Wind, Body, Stability, and Principal Axes

One of the most distressing experiences for beginning stability and control engineers is to be faced with at least four alternate sets of reference axes for the equations of airplane motion. The original Bryan set, called body axes, is perhaps the most easily

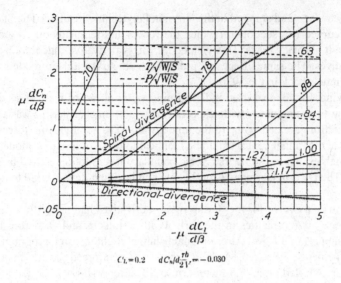

$$C_L = 0.2 \quad dC_n/d\frac{rb}{2V} = -0.030$$

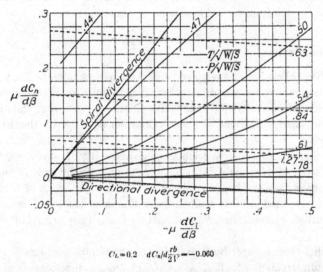

$$C_L = 0.2 \quad dC_n/d\frac{rb}{2V} = -0.060$$

Figure 18.6 Representative lateral-directional stability boundaries. Spiral and directional divergence boundaries are given, along with approximations for Dutch Roll period and damping. The airplane relative density μ is used in the chart coordinates. (From Zimmerman, NACA Rept. 589, 1937)

grasped. Orthogonal reference axes are fixed in the airframe as if they were painted on, remaining in place through all subsequent motions. To be fair, even body axes can migrate with respect to the airframe, since the most common form has its origin at the airplane's center of gravity, which shifts about with different loadings.

Body axes have the practical virtue that the variables of motion that are calculated, such as the linear and angular velocities, are easily related to the readings of flight instruments, which are, after all, also fixed to the body. However, in the early days of stability and control analysis, there were advantages to wind axes (Zimmerman, 1935).

In wind axes, the forward or X-axis points into the wind during the entire motion, rotating about the center of gravity with respect to the airframe. The independence of translatory and rotational motions allows this to happen without affecting the calculation of pitching motions. An advantage of wind axes is that the X and Z forces are the exact negatives of the familiar drag and lift forces presented in wind-tunnel test reports and used in airplane performance calculations.

Stability axes came into the picture in the 1940s, as a device to simplify calculation of small-perturbation airplane motions. Stability axes are a special set of body axes. The X stability axis points into the relative wind in the equilibrium flight that precedes the disturbed motion, but remains fixed in the body during the calculated motions around equilibrium. All that is accomplished by stability axes is the elimination of a few terms in the equations that include initial angle of attack. With the advent of powerful new digital computers stability axes have become mostly a curiosity, except for the fact that the primed derivatives mentioned in Sec. 2 have their basis in stability axes. Duane McRuer notes that

> Primed derivatives based on stability axes often have a remarkably simple connection with the basic motions of the aircraft. . . . [For example] the square of the Dutch roll undamped natural frequency is usually given to a high degree of accuracy by N_β'. . . . stability axes are appropriate for determining the characteristic modes [of motion] and their predominant constituents.

To complicate things, the term *stability axes* sometimes has quite another meaning than that of a special set of body axes for flight dynamics studies. Wind-tunnel data are quite often produced in what are called *stability axes*, but for clarity should be named *wind-tunnel stability axes*. The Z-axis is in the plane of symmetry and normal to the relative wind; the X-axis is in the plane of symmetry and is normal to the Z-axis; the Y-axis is normal to both X- and Z-axes.

Principal axes are another curiosity in present-day practice, since they are used only to eliminate the product of inertia terms in the equations of motion. As with stability axes, principal axes have been obsoleted by powerful digital computers. A few added terms in the equations seem to add nothing to computing time.

The hybrid case in which wind axes are used for the three force equations and body axes for the three moment equations can be found in some simulations. The first hybrid application the authors are aware of was made by Robert W. Bratt at the Douglas Aircraft Company's El Segundo Division, about 1955, in connection with inertial coupling studies. A more recent example of hybrid axes is NASA's SIM2, which actually uses three sets of axes, wind, wind-tunnel stability, and body (Figure 18.7). SIM2 was first put to use at the NASA Dryden Flight Research Center for real-time digital simulation of the McDonnell Douglas F-15. The aerodynamic data base was filled in to an angle of attack of 90 degrees, to allow simulation of stalls and spins. Later SIM2 applications were to the space shuttle Orbiter and to the Northrop B-2 stealth bomber.

With three axes systems carried along simultaneously in the solution, the angular relationships among the SIM2 axes sets must also be continuously computed. The fundamental force vector equation on moving axes used in SIM2 uses the vector cross-product of angular velocity of wind axes and the velocity vector. A key vector equation solves for the angular velocity of wind axes as the angular velocity of body axes minus two terms, the angular velocity of wind-tunnel stability axes with respect to wind axes and the angular velocity of body axes with respect to wind-tunnel stability axes.

Wind axes differ from wind-tunnel stability axes only by a positive sideslip angle rotation about the Z stability axis, so that the second of the three terms in the vector equation for wind axes angular velocity has only one nonzero element, the sideslip angle rate. Likewise, wind-tunnel stability axes are derived from body axes by a single angle of attack rotation

$$\dot{V} = X_{\mathrm{w}}/m$$

$$\dot{\beta} = (Y_{\mathrm{w}}/mV) - R_{\mathrm{s}}$$

$$\dot{\alpha} = Q - P_{\mathrm{s}}\tan\beta + Z_{\mathrm{w}}/(mV\cos\beta)$$

where

$$\begin{bmatrix} X_{\mathrm{w}} \\ Y_{\mathrm{w}} \\ Z_{\mathrm{w}} \end{bmatrix} = \begin{bmatrix} \cos\beta & \sin\beta & 0 \\ -\sin\beta & \cos\beta & 0 \\ 0 & 0 & 1 \end{bmatrix} \begin{bmatrix} \cos\alpha & 0 & \sin\alpha \\ 0 & 1 & 0 \\ -\sin\alpha & 0 & \cos\alpha \end{bmatrix} \begin{bmatrix} X \\ Y \\ Z \end{bmatrix}$$

and

$$\begin{bmatrix} P_{\mathrm{s}} \\ Q_{\mathrm{s}} \\ R_{\mathrm{s}} \end{bmatrix} = \begin{bmatrix} \cos\alpha & 0 & \sin\alpha \\ 0 & 1 & 0 \\ -\sin\alpha & 0 & \cos\alpha \end{bmatrix} \begin{bmatrix} P \\ Q \\ R \end{bmatrix}$$

Note on symbols: Subscript w refers to wind axes, subscript
                   s refers to (wind tunnel) stability axes.
                   Unsubscripted symbols refer to body axes.

Figure 18.7   One form of the force equations of motion for hybrid axes, in which wind axes are used for the force equations and body axes are used for the moment equations. This particularly compact form is used in the NASA-Northrop SIM2 digital simulation for the space shuttle Orbiter and the B-2 bomber.

along the negative Y-body axis. The required vector transformations are made in component form, always taking care to add components in the same axis systems.

The sideslip and angle of attack variables that define the difference among the three axis sets in SIM2 have one of the two possible definitions. The SIM2 convention happens to agree with the most common definition, in which wind axes are derived from body axes by an initial negative angle of attack $-\alpha$ rotation followed by a positive sideslip angle rotation β (Figure 18.8). The reverse convention is rare but not unknown.

Extended airplane axes sets that allow for flight at extreme speeds and altitudes, taking into account the earth's actual shape, are treated in Sec. 15.

18.8   Laplace Transforms, Frequency Response, and Root Locus

One of the minor mysteries in the evolution of the equations of airplane motion is why it was not until 1950 that the Laplace transformation appeared in the open literature as a solution method for the airplane equations of motion. This was in an NACA Technical Note by Dr. G. A. Mokrzycki (1950), who later anglicized his name to G. A. Andrew. Laplace transforms were common among servomechanism engineers and in a few aeronautical offices for at least ten years before that. Laplace transforms provide a much simpler, more organized method for finding time history solutions than the classical operator methods described by B. Melvill Jones (1934) and Robert T. Jones (1936). Laplace transforms also provide a formal basis for airplane transfer functions, frequency responses, time vector analysis, and root loci, all used in the synthesis of stability augmentation systems, as described in Chapter 20, "Stability Augmentation."

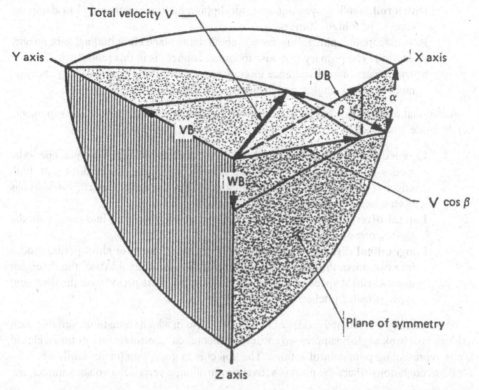

Figure 18.8   Usual angle-of-attack α and sideslip angle β convention, used in the NASA SIM2 flight simulation. X, Y, and Z are body axes. (From Abzug, Northrop paper, 1983)

18.9   The Modes of Airplane Motion

Small-perturbation airplane motions are characterized by modes, just as the disturbed motions of two spring-coupled masses are a composite of a high-frequency mode of motion in which the masses move toward and away from each other, and a low-frequency mode in which the masses move in the same direction. The five classical modes of airplane motion are found as factors of the airplane's longitudinal and lateral characteristic equations (Jones, 1934).

The characteristic equations are of fourth or higher degree, so that factors must be found by successive approximations, rather than in closed form. The factors are either real or they occur in pairs, in conjugate complex form. The real factors are characterized by times to double or halve amplitude following a disturbance or by the inverse of the factor, the time constant. The complex factors are usually characterized by their periods or frequencies (damped or undamped) and by their dimensionless damping ratios. The five classical modes are

> **Phugoid**, a low-frequency motion involving large pitch attitude and height changes at essentially constant angle of attack. Damping is low, especially for aerodynamically clean airplanes.
>
> **Longitudinal short period**, a rapid, normally heavily damped motion at essentially constant airspeed. Damping is provided by wing lift in plunging, as well as horizontal tail lift in rotation. Rapid pitch maneuvers occur in this mode.

Dutch roll, a rolling, yawing, and sideslipping motion of generally low damping, especially at high altitudes.

Roll, essentially a pure rolling motion about the airplane's longitudinal axis, heavily damped. The primary response to lateral controls is in this mode.

Spiral, a very slow divergence or convergence involving large heading changes, moderate bank angles, and near-zero sideslip.

Additional or combined modes appear in special circumstances, such as the supersonic height mode, discussed in Chapter 11. Notable combined modes are

Coupled roll-spiral or lateral phugoid, the conversion of two simple, aperiodic modes into one oscillatory mode. This mode occurs on airplanes with high effective dihedral and low roll damping (Ashkenas, 1958; Newell, 1965). It has been observed on some V/STOL and high-speed airplanes.

Lateral divergence, a degeneration of the Dutch roll mode into two aperiodic modes, one divergent.

Longitudinal divergence, a degeneration of the phugoid or short-period modes into two aperiodic modes, one divergent. For the phugoid case, the divergent mode is called speed instability or tuck; for the short-period case the divergent mode is called pitchup.

Kinematically constrained modes of motion are those in which some flight variable such as altitude or bank angle is suppressed entirely by theoretical control surface or thrust closed loops, representing pilot control actions. The object is to get approximate stability criteria for flight conditions where the pilot is actively controlling a variable. Two such modes are

Constrained airspeed mode, in which altitude is maintained by some control moment, such as would be produced by the elevator. This produces a mathematical demonstration of speed stability (Neumark, 1957). The constraint results in a first-order differential equation in perturbation airspeed. There is an unstable real root for flight on the back side of the lift–drag polar, corresponding to lift coefficients above that for minimum drag. Section 2 of Chapter 12 discusses the implications of speed stability for naval aircraft.

Constrained yaw mode, in which zero bank angle is maintained by the ailerons (Pinsker, 1967). This constraint results in a first-order differential equation in perturbation yawing velocity. Pinsker demonstrated an aperiodic divergence at angles of attack greater than 18 degrees for an airplane with a low-aspect-ratio wing. This is similar to the nose slice experienced by some modern fighters. Stability of this aperiodic mode is governed by the LCDP parameter (Chapter 9, Sec. 15) $N_v - (N_{\delta a}/L_{\delta a}) L_v$, where N_v and L_v are the yawing and rolling moments due to sideslip and $N_{\delta a}$ and $L_{\delta a}$ are the yawing and rolling moments due to aileron deflection.

The useful concept of airplane modes of motion has been extended to rotary-wing aircraft. In forward flight, their modes of motion are similar to those of fixed-wing aircraft. However, many of the usual stability derivatives disappear in hovering flight, giving quite different results for the modes of motion in hover.

By adding apparent mass effects to the stability derivatives, one can obtain modes of motion for lighter-than-air vehicles. Cook (2000) used earlier models by Lipscombe, Gomes, and Crawford and recent wind-tunnel data to derive modes of motion for a modern nonrigid airship.

18.9.1 Literal Approximations to the Modes

A literal approximation to a mode of airplane motion is defined as an approximate factor that is a combination of stability derivatives and flight parameters such as velocity or air density. This approximation is quite distinct from the factors that are obtained from the airplane's fourth- or higher degree characteristic equations, factors that are necessarily in numerical form. Literal approximations to the modes have a long history, starting with Lanchester in 1908. A feedback systems analysis approach to developing and validating approximate modes was developed by Ashkenas and McRuer (1958).

A well-known and usually quite accurate literal approximation to the roll mode is for the roll mode time constant T_R. The roll mode time constant is the time required for rolling velocity to rise to 63 percent of its steady value following an abrupt aileron displacement. The approximation is $T_R = -1/L_p$. The symbol $L_p = C_{lp}q\,Sb^2/(2\,VI_x)$, where

C_{lp} = dimensionless roll damping derivative, a function of wing planform para-
      meters such as aspect ratio and sweep angle;
q = flight dynamic pressure, $(\rho/2)V^2$;
S = wing area;
b = wing span;
V = flight velocity;
ρ = air density;
I_x = roll moment of inertia.

Note that all of the individual parameters in the roll mode approximation would normally be known to an airplane designer. A large literature has been produced on literal approximations to the modes. McRuer (1973) lists four reasons for this interest, as follows:

1. Developing the insight required for the determination of airframe/automatic-control combinations that offer possible improvements on overall system complexity.
2. Assessing the effects of configuration changes on aircraft response and on airframe/autopilot/pilot system characteristics.
3. Showing the detailed effects of particular stability derivatives (and their estimated accuracies) on the poles and zeros and hence on aircraft and airframe/autopilot/pilot characteristics.
4. Obtaining stability derivatives from flight test data.

To this list one might add that mode approximations provide a reasonableness check on complete solutions generated within massive digital-computer programs, assuring that no input errors have been made. Literal approximations to the modes are obtainable only if the equations of motion themselves are simplified in some way, or if the factorization itself is approximated.

Mode approximations are useful in the ways McRuer lists as long as the approximations are simple ones, easy to grasp. One can improve the approximations, bringing the numerical values closer to the actual factors of the characteristic equation. This can provide additional insight into aircraft flight mechanics. However, if the literal expressions are lengthy, their utility suffers. The improvement to the classical Lanchester result for the phugoid mode period made by Regan (1993) and others (see Chapter 11, Sec. 13), which adds only one simple term but greatly improves accuracy at high airspeeds, is an example of a useful improved approximation, in the context of McRuer's comments.

On the other hand, the improved modal approximations of Kamesh (1999) and Phillips (2000), while demonstrating considerable mathematical skills and adding to our understanding of flight dynamics, are probably too complex for the applications mentioned by McRuer.

18.10   Time Vector Analysis

The time vector analysis method provides an excellent insight into the modes of airplane motion. The method came about as an incidental result of debugging one of the world's first electronic analog computers, built to represent generalized airplane longitudinal dynamics. This computer's inventor was Dr. Robert K. Mueller; his device, now in the MIT Museum, was built to support his 1936 MIT ScD thesis.

The fundamental concept of time vector analysis is that for any oscillatory transient generated by a linear system having a certain undamped natural frequency and damping ratio:

1. the amplitude of the transient derivative is the transient amplitude multiplied by the undamped natural frequency, and
2. the phase of the transient derivative is the phase of the transient advanced by 90 degrees plus the angle whose sine is the damping ratio.

With this concept, one can construct time vector polygons representing each term in any system equation corresponding to a particular modal solution of the characteristic equation. The time vector polygons show which terms are dominant and how the amplitude and phase relations among the variables arise (Figure 18.9). In Mueller's thesis example, the time vector polygons give insight into the wind axis equations of longitudinal motion and suggest correction of the phugoid mode instability with pitch attitude feedback. At the urging of his then-supervisor at the Glenn L. Martin Company, James S. McDonnell, he presented a paper on the topic at a meeting of the Institute of Aeronautical Sciences (Mueller, 1937).

In Germany, Dr. Karl-H. Doetsch used the time vector method to study lightly damped airplane–autopilot combinations. Working at the Royal Aircraft Establishment (RAE) after World War II, K-H. Doetsch and W. J. G. Pinsker applied time vector analysis methods to the Dutch roll problems of jet airplanes.

There was an early application of the time vector analysis method by Leonard Sternfield of the NACA Langley Laboratory to the Dutch roll oscillation. Around 1951 he built two bridge-table–size mechanical analogs of the roll and yaw time vector polygons to predict the Dutch roll characteristics of new airplanes. Around the same time E. E. Larrabee made what he thought was the first use of time vector analysis to extract stability derivatives from flight time history measurements, although Doetsch had done much the same in England.

18.11   Vector, Dyadic, Matrix, and Tensor Forms

Bryan used quite conventional cartesian coordinates in the derivation of the equations of airplane motion on moving axes, in 1911. Cartesian coordinates were used as well by subsequent investigators, such as B. Melvill Jones (1934), Charles Zimmerman (1937), and Courtland Perkins (1949). The first author who applied vector methods to the derivation of these equations appears to have been Louis M. Milne-Thomson, in his book *Theoretical Aerodynamics* (1958).

The most notable thing about the Milne-Thomson vector derivation is the way in which a fundamental moving axis rate equation is developed. This is the relationship between vector

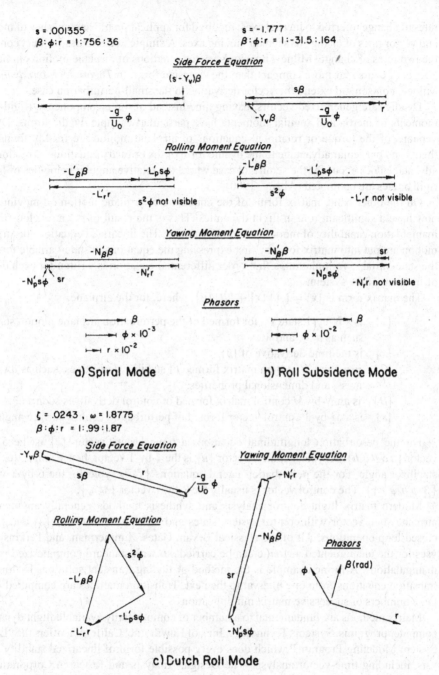

a) Spiral Mode

b) Roll Subsidence Mode

c) Dutch Roll Mode

Figure 18.9 Time vector diagrams for a conventional airplane, from *Aircraft Dynamics and Automatic Control* by McRuer, Ashkenas, and Graham (1973). Sideslip angle β is almost nonexistent in the spiral mode. Bank angle Φ dominates in the roll subsidence mode. All motions are of the same order of magnitude in the Dutch roll mode.

rates of change referred to inertial axes, required for application of Newton's law of motion, and vector rates of change as seen on moving axes. A simple vector cross-product connects the two rates of change. Milne-Thomson's vector equations of airplane motion on moving axes is of course far more compact than the cartesian form. In *Theoretical Aerodynamics*, Milne-Thomson did extend the vector derivation to the small-perturbation case.

Dyadics are generalized vectors, having nine instead of three components. Rigid-body moments of inertia and angular momenta have particularly simple dyadic forms. Dyadic versions of the torque or rotational equations of airplane motion are readily found, but there is no particular advantage to the dyadic form of the ordinary equations of motion. An advantage does occur for the semirigid case where the relative angular velocities of linked rigid bodies are computed (Abzug, 1980).

On the other hand, matrix forms of the equations of airplane motion on moving axes now have a significant role in flight dynamics. This is the result of the marvelous matrix manipulation capability of modern digital computers. The linearized equations of airplane motion are put into matrix form by first expressing the equations in state-variable form. In the state-variable formulation, a first-order differential equation is written for each degree of freedom of the system.

The matrix form is $\{\dot{x}\} = [A]\{x\} + [B][u\}$, where, for the airplane,

$\{x\}$　is a N-by-1 state vector formed of the perturbation airplane motion states, such as u, v, and w.

$\{\dot{x}\}$　is the time derivative of $\{x\}$.

$[A]$　is a N-by-N system matrix formed of stability derivatives, such as $\partial X/\partial u$, mass, and dimensional properties.

$[B]$　is an N-by-M control matrix formed of control derivatives such as $\partial X/\partial \delta$.

$\{u\}$　is a M-by-1 control vector formed of perturbation control surface angles.

For the perturbation longitudinal equations a typical state vector $\{x\}$ is the 5-by-1 vector $\{u\,\alpha\,\theta\,q\,h\}$. A typical control vector $\{u\}$ is the 1-by-1 vector, hence scalar, $\{\delta_h\}$, the stabilizer angle. For the perturbation lateral equations $\{x\}$ is typically the 6-by-1 vector $\{\beta\,\phi\,p\,\psi\,r\,y\}$. The control vector is usually the 2-by-1 vector $\{\delta_a\,\delta_r\}$.

Modern matrix flight control analysis and synthesis methods generally augment the airplane state vector with control system states and manipulate matrices [A] and [B] in closed-loop operations. All of the classical Bryan, Gates, Zimmerman, and Perkins analyses for the unaugmented airframe can be carried out with standard computerized matrix manipulations. A prime example is the method of finding transient solutions by forming transition equations from one interval to the next. Transition matrices are computed using large numbers of successive matrix multiplications.

Matrix methods are fundamental to a number of commercially available flight dynamics computer programs. Systems Technology, Inc., of Hawthorne, California, offers the "Linear System Modeling Program," which does every possible form of linearized stability analysis, including time-vector analysis. The Design, Analysis and Research Corporation of Lawrence, Kansas, produces the "Advanced Aircraft Analysis" program, which does stability and control preliminary design, trim, and flight dynamics. Large general-purpose matrix manipulation computer programs such as "MATLAB" from MathWorks and "Mathcad" from MatSoft are also commercially available to the stability and control engineer.

The remarks about the dyadic form of the equations of airplane motion apply as well to tensor forms. That is, there is no special advantage in expressing the ordinary equations of rigid-body airplane motion in tensors, as compared with cartesians or vectors. Zipfel (2000) uses a tensor form of the rigid-body equations of airplane motion on moving axes.

18.12   Atmospheric Models

A mathematical model of the earth's atmosphere is needed for stability and control flight simulation and other computer programs. These programs typically use dimensionless stability derivatives in setting up equations of motion for flight dynamics studies, and stability augmenter and autopilot analyses.

Standard atmospheric mathematical models were published by NACA starting in 1932. A 1955 model covered altitudes up to 65,800 feet (ICAO, 1955). NACA, the U.S. Air Force, and the U.S. Weather Bureau extended that model to an altitude of 400,000 feet (ICAO, 1962). The AIAA publishes a guide to standard atmosphere models (1996). For all its utility, the standard atmospheric model is based on quite simple assumptions: The air is dry, it obeys the perfect gas law, and it is in hydrostatic equilibrium.

Standard atmosphere computer codes for stability and control computer programs normally accept as inputs the airplane's altitude and true speed at each computing time. A minimum set of outputs at each computing time would include atmospheric density, Mach number, dynamic pressure, and equivalent airspeed. Additional outputs that could be generated are static pressure and calibrated airspeed.

The standard atmosphere FORTRAN computer code shown in Figure 18.10 represents one of the two methods used in stability and control programs. In this example, air density (RHO) is curve-fitted with exponential functions. Four function fits give satisfactory accuracy over the entire range of −4,000 to 400,000 feet. Speed of sound (ASPE), from which the Mach number is calculated, requires eight curve-fitted linear equations. The alternate standard atmosphere coding is ordinary interpolation from stored tables of density and speed of sound.

It has become increasingly important to represent wind gusts, shears, downbursts, and vortex encounters in stability and control flight simulation. Early flight simulations relied on a very simplified approach in which an additional gust angle of attack or sideslip is simply added to the values calculated at each instant from the airplane's motions in an inertial space reference.

The sounder approach, now in common use, is an inertially fixed model for the wind environment, including gusts, shears, downbursts, and vortices. A NASA downburst model uses the conservation of mass principle to calculate wind velocity at all points within a downburst (Bray, 1984). A central core is surrounded by an annular mixing region and a region of outflow parallel to the ground (Figure 18.11). In the Bray inertially fixed wind environment, an airplane penetrates the wind model as it moves along its path, just as in reality.

Earlier ad hoc wind shear models were proposed by NASA for flight simulation. A boundary layer shear model represents a low-level temperature inversion overlaid by strong winds. Two additional shear models, with the colorful names of the Logan and Kennedy shears, represent meteorologists' best estimates of conditions existing at those airports during specific airplane wind shear encounters.

In Bernard Etkin's terminology, the Bray downburst model and the NASA shear models are usually used as point atmospheric models, in which variations of local wind velocity over the airplane's dimensions are neglected. Otherwise stated, the airplane is assumed to be vanishingly small with respect to the wavelengths of all spectral components in the turbulent atmosphere. This assumption obviously fails for gust alleviation systems that depend on sensing devices that sample air turbulence ahead of the main structure.

Etkin (1972) provides a thorough study of the finite airplane case, in which local wind velocities vary over the airplane's dimensions. The required mathematics are surprisingly complex because atmospheric turbulence is a random process, and only a statistical, probabilistic

```
        SUBROUTINE ATMOS(ALT,VEL,RHO,AMACH,DYN,VEKT)
      ; NASA/USAF/USWB STANDARD ATMOSPHERE, -4,000 TO 400,000 FT.
        DIMENSION X(3)
        X(3)=ALT
        IF(-X(3)-35.E3) 300,300,310
  300 Z1=342.5E2+X(3)*4.3/35.
        GO TO 500
  310 IF(-X(3)-45.E3)320,320,400
  320 Z1=299.5E2-.285*(-X(3)-35.E3)
        GO TO 500
  400 IF(-X(3)-60.E3) 405,405,410
  405 Z1=271.E2-.12*(-X(3)-45.E3)
  500 RHO=.002377*EXP(X(3)/Z1)
        GO TO 490
  410 IF(-X(3)-140.E3) 415,415,420
  415 RHO=EXP(-4.67263E-5*(-X(3)-6.E4)-8.41364)
        GO TO 490
  420 IF(-X(3)-240.E3) 425,425,430
  425 RHO=EXP(-3.8712E-5*(-X(3)-14.E4)-12.151584)
        GO TO 490
  430 RHO=EXP(-5.18378E-5*(-X(3)-24.E4)-16.022785)
  490 CONTINUE
        IF(-X(3)-362.E2)600,600,700
  600 ASPE=1117.-149.*(-X(3)/362.E2)
        GO TO 800
  700 IF(-X(3)-66.E3) 710,710,720
  710 ASPE=968.
        GO TO 800
  720 IF(-X(3)-105.E3) 730,730,740
  730 ASPE=6.71282E-4*(-X(3)-66.E3)+968.
        GO TO 800
  740 IF(-X(3)-1555.E2) 750,750,760
  750 ASPE=1.73941E-3*(-X(3)-105.E3)+994.18
        GO TO 800
  760 IF(-X(3)-172.E3) 770,770,775
  770 ASPE=1082.02
        GO TO 800
  775 IF(-X(3)-200.E3) 780,780,785
  780 ASPE=-1.215714E-3*(-X(3)-172.E3)+1082.02
        GO TO 800
  785 IF(-X(3)-2625.E2) 790,790,795
  790 ASPE=-2.62368E-3*(-X(3)-200.E3)+1047.98
        GO TO 800
  795 ASPE=884.
  800 AMACH=VEL/ASPE
        DYN=(RHO/2.)*VEL**2
        VEKT=17.1861216*SQRT(DYN)
        RETURN
        END
```

Figure 18.10 FORTRAN digital computer subroutine for the NASA/USAF/USWB standard atmosphere. Air density (RHO) and speed of sound (ASPE) are curve-fitted in altitude bands from -4,000 to 400,000 feet. The subroutine requires inputs of altitude (ALT) and true speed (VEL). The subroutine outputs density, Mach number (AMACH), dynamic pressure (DYN), and equivalent airspeed (VEKT). (From ACA Systems, Inc. FLIGHT program)

treatment can be made (Ribner, 1956). Local wind velocity is a random function of both space and time. It simplifies things to assume stationarity, homogeneity, isotropy, and Gaussian distributions. Experimental data exist that provide adequate turbulence models for both high altitudes and near the ground, where isotropy does not hold.

More exotic atmospheric disturbances are significant for operation at very high altitudes and at hypersonic speeds. Flight disturbances due to temperature shears experienced by the

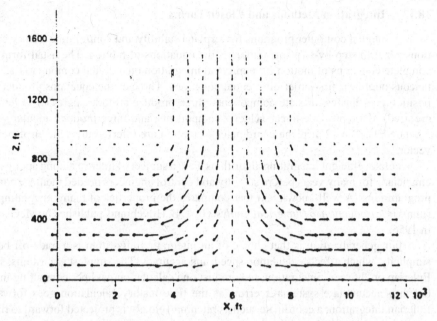

Figure 18.11 Vertical cross-section through the Bray model of a down-burst. Arrow lengths are proportional to air flow velocity. (From Bray, NASA TM 85969, 1984)

Lockheed SR-71A and the North American XB-70 are discussed in Chapter 11, "High Mach Number Difficulties." In anticipation of a National Aerospace Plane (NASP) that would fly hypersonically, NASA laboratories at Dryden, Marshall, and Langley and the McDonnell Douglas Houston operation collaborated on a sophisticated FORTRAN atmospheric model called the NASP Integrated Atmospheric Model (Schilling, Pickett, and Aubertin, 1993).

The model is suitable for real-time simulations as well as batch programs. It provides global coverage, from the ground to orbital altitudes. The NASP model features of particular stability and control interest are the small-scale perturbations that include continuous turbulence and the "thermodynamic" perturbations of density, pressure, and temperature. Gusts and thermodynamic perturbations can be selected either in patches or as discrete upsets.

Isolated mountain ridges that lie at right angles to the direction of strong prevailing winds can generate a so-called mountain wave. Air over the top of such a ridge cascades in huge volumes onto the lower terrain to the leeward. It then seems to bounce off, rising, then falling, and then rising again in a series of diminishing waves, all parallel to the ridge line. A huge rotor, or horizontal vortex, forms to the leeward of the first bounce. The characteristic mountain wave structure is well known to glider pilots, since glider altitude records are set by maneuvering into the rising air of the first bounce. Glider pilots also know to avoid the rotor, whose lower edge generally just brushes ground level. The National Transportation Safety Board (NTSB) concluded that a rotor was a possible cause for at least one airline accident. This was United Airlines Flight 585, a Boeing 737 lost at Colorado Springs in 1991. Various jet aircraft encounters with a rotor are modeled by Spilman and Stengel (1995).

Vortex wakes left in the atmosphere by airplanes flying ahead can be a severe hazard, although the principles for avoiding vortex wakes are known. Vortex wake fields can be modeled for flight simulation (Johnson, Teper, and Rediess, 1974).

18.13   Integration Methods and Closed Forms

Digital computer programs for airplane stability and control time history calculations perform step-by-step integration of the equations of motion. The usual form of the complete equations of motion for numerical integration on a digital computer is 12 simultaneous nonlinear first-order differential equations. Three of the equations produce linear position coordinates, or state components, three produce attitude angles (if Euler angles are used), three produce linear velocity components, and three produce angular velocity components. The 12 airplane coordinates of motion are referred to as the airplane's state vector.

Accurate, efficient integrating algorithms were a subject of interest among applied mathematicians for many years before stability and control engineers needed them for computer programming. A well-known text that compares the properties of many integrating algorithms is *Introduction to Numerical Analysis* by F. B. Hildebrand, published by McGraw-Hill in 1956.

A fair generalization is that choice of an integrating algorithm is a trade-off between simplicity, which affects calculation speed, and accuracy. The simplest algorithms, such as Eulerian or "boxcar" integration, require just one calculation pass per computing interval, but they accumulate systematic errors as the time history calculation goes forward. In Eulerian integration, a coordinate such as pitching velocity is projected forward to the next time interval simply by adding to the present value the product of the present value of pitching acceleration and the time interval length, which is usually of the order of 0.05 second. In general terms, the state vector at the next time interval is the current state vector plus the product of the state derivative vector and the time interval.

More accurate integration requires the calculation of intermediate values in order to take the same time step, adding to the computing time but improving accuracy. The best known accurate integration method, in effect a standard for stability and control time history calculations, is the fourth-order Runge-Kutta method. This method can be adapted in FORTRAN to the integration of multiple states, such as the 12 airplane coordinates or state vectors (Melsa and Jones, 1973, and Figure 18.12).

While digital computers became available in engineering offices for stability and control time history calculations around the time of the inertial coupling crisis, or 1950, it was not until many years later that computing speed had increased to the point that the calculations could be made in real time and could thus support flight simulation. One of the earliest such applications was at the Ling-Temco-Vought plant in Arlington, Texas, in the late 1960s. An all-digital flight simulator that went on-line a little later was Northrop Aircraft's Large Amplitude Simulator, which progressed from analog to hybrid to all-digital in late 1975. With the introduction of real-time digital flight simulation accurate, but slow, integration methods such as the fourth-order Runge-Kutta routine have become something of an obstacle. There is a premium on the development of fast integration methods that still have a fair degree of accuracy. Fast but accurate integration methods have been developed all over the United States to meet that need: methods generally starting with a classical scheme and modified by mathematical tinkerers.

The Adam-Bashford method was the starting point for the algorithm used for the projector gimbals in the Northrop Large Amplitude Simulator. A different set integrates the airplane equations of motion. Another integration method developed specifically for flight simulation modifies the second-order Runge-Kutta method, replacing the second state derivative vector calculation with a prediction based on a weighted average of previous mid- and endframe values. The modified second-order Runge-Kutta method seems to be almost as accurate as

```
      SUBROUTINE INTG1
 C  FOURTH-ORDER RUNGE-KUTTA INTEGRATION
      DIMENSION Y1(40),E1(40),E2(40),E3(40)
      TSTEP=DT/2.
      DO 2 I=1,N
    2 Y1(I)=X(I)
      CALL DERV1
      DO 4 I=1,N
      E1(I)=TSTEP*F(I)
    4 X(I)=Y1(I)+E1(I)
      T=T+TSTEP
      CALL DERV1
      DO 5 I=1,N
      E2(I)=DT*F(I)
    5 X(I)=Y1(I)+.5*E2(I)
      CALL DERV1
      DO 7 I=1,N
      E3(I)=DT*F(I)
    7 X(I)=Y1(I)+E3(I)
      T=T+TSTEP
      CALL DERV1
      DO 8 I=1,N
    8 X(I)=Y1(I)+(E1(I)+E2(I)+E3(I)+TSTEP*F(I))/3.
      RETURN
      END
```

Figure 18.12   FORTRAN digital computer subroutine for the integration of a state derivative vector *x*. This is the widely used fourth-order Runge-Kutta method. COMMON input–output statements have been removed for generality. (From ACA Systems, Inc. FLIGHT program)

the fourth-order Runge Kutta, while requiring only one calculation of the state derivative vector per interval.

Aerodynamic forces and moments are involved in the calculation of the state derivative vector. This requires huge amounts of table lookup on modern flight simulations that cover large Mach number, altitude, and control surface position ranges and uses computer time more than any other part of the computation. Thus, a single calculation of the state derivative vector, as in the modified second-order Runge-Kutta method, is very efficient for real-time digital flight simulation. A modified second-order Runge-Kutta method was developed in 1972 by Albert F. Myers of NASA; it was then improved by him in 1978 for the HIMAT vehicle flight simulation (Figure 18.13).

Another important advance in digital flight simulation is the use of closed-form solutions for the first- and second-order linear differential equations that typically represent analog flight control elements, such as control surface valves and actuators. Closed-form solutions for these elements remove them from the state vector that has to be integrated, reducing the order of that vector to perhaps no more than is required by the airplane equations of motion themselves, or 12. The nonlinearities of control position and velocity limiting are easily represented. This technique is attributed to Juri Kalviste, although there may be other claimants to priority.

18.14   Steady-State Solutions

Steady-state solutions to the equations of airplane motion are defined as motions with zero values of body axis linear and angular accelerations. Steady straight flight includes climbing, level flight, and diving and allows the airplane to have a nonzero sideslip angle. Steady turning flight allows constant values of the three body axis angular velocities, yawing, pitching, and rolling.

```
      SUBROUTINE INTG2
C  MODIFIED SECOND-ORDER RUNGE-KUTTA INTEGRATION
      DIMENSION Q(40),DQ(40)
C  STORE STARTING VALUES
      DO 1 I=1,N
      Q(I) = X(I)
      DQ(I) = F(I)
C ESTIMATE STATES AT MIDPOINT
      X(I) = X(I) + (DT/2.)*F(I)
    1 CONTINUE
C CALCULATE STATE DERIVATIVES AT MIDPOINT
      CALL DERV1
C UPDATE STATES, START TO ENDPT, WITH MIDPT DERIV'S.
      DO 2 I=1,N
      X(I) = Q(I) + DT*F(I)
C  PREDICTOR FOR STATES AT MIDPOINT
      F(I) = 1.5*F(I)-.5*DQ(I)
    2 CONTINUE
C  UPDATE THE TIME
      T = T + DT
      RETURN
      END
```

Figure 18.13   A modified second-order Runge-Kutta integration subroutine developed to run quickly, for use in real-time digital flight simulation. This FORTRAN subroutine was developed by A. F. Myers for NASA's SIM2 simulation. X is the state derivative vector. COMMON input–output statements have been removed for generality. (From ACA Systems, Inc. FLIGHT simulation)

Steady flight conditions are used as reference values for the perturbations of linearized analysis (Sec. 18.2). Applications are to root locus, frequency response, covariance propagation, and optimization. Steady flight conditions also establish initial state variables for nonlinear transient analysis, such as landing approaches, gust response, and pilot-initiated maneuvers. Finally, basic stability conditions can be deduced from the control surface angles required for steady flight. For example, spiral instability is implied when opposite aileron angle is required in a steady turn, such as left aileron to hold trim in a steady right turn.

Steady flight solutions are usually obtained for the nonlinear equations of motion by driving to zero selected body axis linear and angular accelerations. Stevens and Lewis (1992) apply a minimization algorithm called the simplex method for trim in steady, straight, symmetric (unsideslipped) flight. A cost function is formed from the sums of squares of the forward, vertical, and pitching accelerations. A multivariable optimization adjusts thrust, elevator angle, and angle of attack to minimize the cost function.

A closed-loop trimming method (Abzug, 1998) is an alternative to the simplex method. The nonlinear state equations are solved in sequence, together with control equations that adjust thrust, angles of attack and sideslip, and control surface angles to minimize accelerations. In the control equations, thrust is adjusted in small steps to minimize longitudinal acceleration, angle of attack is adjusted in small steps to minimize vertical acceleration, elevator angle is adjusted in small steps to minimize pitching acceleration, and so on.

18.15   Equations of Motion Extension to Suborbital Flight

Suborbital flight is flight within the atmosphere but at extremely high altitudes. In this regime, flight speeds are very high, and the curving of constant-altitude flight trajectories around the earth's surface adds appreciable centrifugal force to wing lift. Bryan's equations of rigid-body motion are for flight over a flat earth. Flat-earth equations of motion generally are inadequate for airplanes that operate in a suborbital mode.

A derivation of nonlinear airplane equations of motion for the spherical-earth case can be found in Etkin (1972). The main distinction between the spherical- or oblate-earth cases and the classical Bryan flat-earth equations lies in additional kinematic (nondifferential) equations. As in the ordinary flat-earth equations, 12 state equations must be integrated. In the Etkin approach, linear accelerations are integrated in airplane body axes, producing the usual inertial velocity and angle of attack and sideslip variables. However, this is only one of several possible choices for the linear accelerations. The angular acceleration equations of motion are integrated in airplane body axes, as for the flat-earth case. This is the only practical choice, since airplane moments and products of inertia are constant only in body axes.

Full nonlinear equations of airplane motion about a spherical or oblate rotating earth were produced somewhat later at Rockwell International in connection with the Space Shuttle Orbiter and still later for studies of the National Aerospace Plane (NASP). The earliest set is found in Rockwell Report SD78-SH-0070, whose authors we have been unable to identify. Six distinct reference axes systems are used. The Rockwell set integrates linear accelerations and velocities in an earth-centered inertial axis system, making transformations to the other axes, such as the body and airport reference sets.

Still another approach was followed at the NASA Dryden Flight Research Center (Powers and Schilling, 1980, 1985) for the Space Shuttle Orbiter, in order to build on an earlier flat-earth 6-DOF computer model. A heading coordinate frame is centered at the orbiter's center of gravity, with the Z-axis pointed to the earth's center and the X-axis aligned with the direction of motion. X and Z define the orbit plane through the geocenter. Linear accelerations and velocities are integrated in heading coordinate and earth axes frames, respectively. Vehicle vertical and horizontal velocities in the orbit plane and body axis heading relative to the orbit plane replace the ordinary body axis velocity coordinates in the airplane's state vector. Altitude above a reference sphere of equatorial radius, latitude, and longitude replace the ordinary altitude, downrange and cross-range position coordinates in the airplane's state vector. High precision data, such as FORTRAN double precision with 15 significant figures, are needed.

Attitude deviations from the Rockwell/Dryden heading coordinate frame produce Euler angles in the classical sense: yaw, then pitch, then roll. Use of this particular heading coordinate system also for space or re-entry vehicles would produce a consistent set of aerospace flight mechanics axes, which would seem to be an advantage.

The oblate earth version of the equations of airplane motion is sometimes used even when there is no question of hypersonic or suborbital flight operations. This is in flight simulators when one wishes to have only one set of airplane equations for both flying qualities and long-range navigation studies. A single, unified airplane mathematical model for both purposes avoids duplication of costly manned flight simulators and the problem of keeping current two different data bases during airplane development. For simulated flights lasting on the order of hours, correct latitude and longitude coordinates can be calculated as inputs to flight data computers.

The almost incredible capacity of modern digital computers makes it feasible to expend computing capacity by including high-frequency airplane dynamics terms in the flight simulation of an hours-long navigational mission, as compared with spending engineering time to develop a special simulation without the high-frequency terms. This was the route chosen for the Northrop B-2, according to our best information.

18.15.1 *Heading Angular Velocity Correction and Initialization*

The inertial reference for body axis heading angular velocity relative to the orbit plane in the Rockwell/Dryden formulation can be thought of as true north, defined by the local meridian. However, a local meridian cannot be used as an inertial reference unless its motion as the earth turns is accounted for. Powers and Schilling (1980, 1985) derive this correction.

The earth's atmosphere is carried around with earth rotation, causing side winds relative to the orbit plane. This requires special initialization for starting transient response calculations at zero sideslip. Closed-form initialization formulas are available using initial angle of attack, velocity, and flight path angle.

18.16 Suborbital Flight Mechanics

The effects of the earth's curvature are quite negligible on the airplane modes of interest to the stability and control engineer under ordinary flight conditions. However, some significant effects are expected for the suborbital case. A number of investigators have extended the flat-earth equations to spherical or oblate models in order to examine these effects.

Linearized airplane motions have been examined in perturbations from great-circle and minor-circle trajectories about a spherical earth (Myers, Klyde, McRuer, and Larson, 1993). In principle, this is the same procedure followed by Bairstow (1914) in his extension of the Bryan equations of motion to perturbations from steady turning flight. An extra longitudinal mode of motion is found, in addition to the usual short-period and phugoid modes. This is a first-order density mode, also referred to as an *altitude mode*. Aside from this extra complexity, with a typical hypersonic configuration at Mach numbers from 3 to 20 the density mode occasionally couples with real phugoid poles.

There is also an extra lateral-directional real mode, in addition to the usual Dutch roll, spiral, and roll modes. This is called a *kinematic mode*, generally of very long time constant. At some high Mach numbers, the kinematic mode couples into the spiral mode, producing a very low-frequency stable oscillation.

18.17 Additional Special Forms of the Equations of Motion

Trajectory or point-mass equations of airplane motion, lacking the torque or moment equations, have been found useful for flight performance studies. In these applications, angles of attack and sideslip are assumed functions of time or are found in simple closed loops, instead of being the result of attitude adjustments influenced by control surface angles. Trajectory equations of motion have only 6 nonlinear state equations, as compared with 12 for the complete rigid-body equations. The savings in computer time are unimportant with modern digital computers, but there is a conceptual advantage for performance studies in needing to specify only lift, drag, and thrust parameters.

Another special form of the equations of airplane motion puts the origin of body axes at an arbitrary location, not necessarily the center of gravity. The first use of such equations seems to have been for fully submerged marine vehicles, such as torpedoes and submarines. With the center of body axes at the center of buoyancy, there are no buoyancy moment changes due to changes in attitude (Strumpf, 1979). An equivalent set for airplanes came later (Abzug and Rodden, 1993).

Apparent mass and buoyancy terms in the equations of airplane motion are discussed in Chapter 13, "Ultralight and Human-Powered Airplanes." The various special forms of the equations of airplane motion for representing aeroelastic effects are discussed in the next chapter, "The Elastic Airplane."

Equations of motion for an airplane with an internal moving load that is then dropped were developed by Bernstein (1998). The motivation is the parachute extraction and dropping of loads from military transport airplanes. A control strategy using feedback from disturbance variables to the elevator was able to minimize perturbations in airplane path and airspeed during the extraction and dropping process.

CHAPTER 19

The Elastic Airplane

Aeroelasticity deals with the interactions of aerodynamic and inertial forces and aircraft structural stiffness. Additional significant interactions with aerodynamic heating and automatic control systems give rise to the Germanic-length terms *aerothermoelasticity* and *aeroservoelasticity*. Aeroelasticity concerns stability and control, dealt with here, but also flutter and structural loads arising from maneuvers and atmospheric turbulence. Aeroelasticity affects airplane stability and control in a number of areas. Prediction of aerodynamic data at the design stage (Chapter 6), tactical airplane maneuverability (Chapter 10), the equations of motion (Chapter 18), and stability augmentation (Chapter 20) are all affected.

Aeroelastic effects are considered as distractions by many stability and control engineers, obscure problems that get in the way of the real work at hand. Aeroelastic methods are certainly abstract, involving such arcana as normal modes. How does one fix body axes in a flexible structure? What is its angle of attack? We trace this difficult but important branch of stability and control from the early days of Samuel Langley, the Wright brothers, and Anthony Fokker to the present.

The early days were dominated by isolated occurrences of aeroelastic problems and ad hoc solutions. The advent of large-scale digital computers and finite-element or panel methods for the first time provides, if not a general theory, at least an organized approach to prediction and solution of stability and control aeroelastic problems.

19.1 Aeroelasticity and Stability and Control

Bernard Etkin (1972) gives a succinct description of the way stability and control engineers handle the effects of airframe distortion or elasticity. There are two basic categories into which all treatments fall. Etkin calls these categories "The method of quasistatic deflections" and "The method of normal modes." Here are his words:

> **Method of Quasistatic Deflections**   Many of the important effects of distortion can be accounted for by simply altering the aerodynamic derivatives. The assumption is made that changes in aerodynamic loading take place so slowly that the structure is at all times in static equilibrium. (This is equivalent to assuming that the natural frequencies of vibration of the structure are much higher than the frequencies of the rigid-body motions.) Thus a change in load produces a proportional change in the shape of the vehicle, which in turn influences the load.
>
> **Method of Normal Modes**   When the separation in frequency between the elastic degrees of freedom and the rigid-body motions is not large, then significant inertial coupling can occur between the two. In that case, a dynamic analysis is required, which takes account of the time dependence of the elastic motions.

In the latter case Etkin goes on to describe the application of normal mode analysis to the stability and control problem. The important distinction between the quasistatic and normal mode treatments holds as well for the approximate normal modes generated by quasi-rigid models.

19.2   Wing Torsional Divergence

Wing torsional divergence, in which the wing structure itself becomes unstable, tip incidence increasing without limit, is a structural rather than stability and control problem. Torsional divergence occurs with increasing airspeed if the wing's aerodynamic center is ahead of its shear center, or elastic axis. Although wing torsional divergence is not strictly a stability and control problem, it is the first known phenomenon that can be analyzed with methods used in aeroelastic stability and control.

According to Bisplinghoff, Ashley, and Halfman (1955), the wing failure that wrecked Samuel P. Langley's Aerodrome on the Potomac River in 1903 was a wing torsional divergence. While there has been some controversy on this point, a torsional divergence occurrence that appears quite certain was on the Fokker D-8 monoplane in 1917. When the first D-8 was sandbag-loaded the wing was proved to be sufficiently strong, but the German government's engineering division called for rear spar strength equal to that of the front spar. The change was made and three D-8 airplanes, one after the other, were lost when their wings failed in flight. The story is picked up in Anthony H. G. Fokker's book *Flying Dutchman*:

> I took a new wing out of production and treated it to a sandload test in our own factory. As it was progressively loaded, the deflections of the wing were carefully measured from tip to tip. I discovered that with the increasing load, the angle of incidence at the wing tips increased perceptibly. I did not remember having observed this action in the case of the original wings, as first designed by me. It suddenly dawned on me that this increasing angle of incidence was the cause of the wing's collapse, as logically the load resulting from pressure in a steep dive would increase faster at the wing tips than in the middle, owing to the increased angle of incidence.

This is a classic wing torsional divergence, since the increasing wing tip incidence increased tip aerodynamic load, which further increased the incidence, and so on. The problem was solved when the government permitted the front spar to be reinforced to bring back the original ratio of stiffness between the front and rear spars. This moved the shear center forward.

The Wright brothers and a few other aviation pioneers used the wing's elastic properties in a positive sense, warping them for lateral control. The Wrights had no problems with aeroelasticity, aside from an unimportant loss in propeller thrust due to blade twisting.

19.3   The Semirigid Approach to Wing Torsional Divergence

In the semirigid approach to wing torsional divergence and related problems a reference section of the wing is selected to represent the entire three-dimensional wing. This simplification works quite well for slender wings, that is, wings of high-aspect ratio.

Semirigid analyses of wing torsional divergence are given in a number of textbooks (for example, Duncan, 1943; Fung, 1955). Fung shows a wing section that rotates about a pivot and is acted upon by a lift load. The pivot represents the chordwise location in the section of the wing's elastic axis, or location where lift loads will not produce twist. The lift load can be taken as acting through the section's aerodynamic center. The aerodynamic center, near the section's quarter-chord point, is the point about which section pitching moments are invariant with angle of attack (Figure 19.1).

The wing section will come to a static equilibrium angle at some angle of attack under the combined action of the lift load and a spring restraint about the pivot. The spring restraint represents the wing's elastic stiffness. If the pivot, representing the elastic axis, is behind

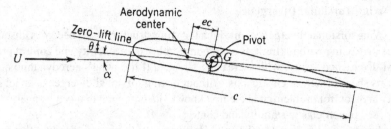

Figure 19.1 Semirigid model for wing torsional divergence. The wing is replaced by a typical section, pivoted about a point that represents the wing's elastic axis. The spring represents elastic stiffness. In this illustration, the wing's aerodynamic center, where the lift acts, is forward of the pivot point. Increasing airspeed eventually leads to a torsional divergence. The angle of attack α increases without limit. (From Fung, *The Theory of Aeroelasticity*, Dover, 1969)

the wing's aerodynamic center, the equilibrium angle of attack increases with increasing airspeed, which gives higher wing lift loads.

For any eccentricity, or distance of the aerodynamic center ahead of the pivot, and for given spring constants and wing lift curve slopes, or variations of wing lift with angle of attack, there is an airspeed at which the semirigid model diverges. That is, the equilibrium solution fails. Twist angle and angle of attack increase without limit. This is the calculated wing torsional divergence speed.

Wing torsional divergence problems were encountered on the Republic F-84 and Northrop F-89 airplanes, both equipped with large tip tanks (Phillips, 1998). Fixed fins on the outside rear of the F-84's tanks moved the wing's aerodynamic center aft, eliminating the problem.

19.4   The Effect of Wing Sweep on Torsional Divergence

One of the rare benign effects of wing sweepback, aside from its function in reducing transonic drag and instability, is the virtual elimination of the possibility of wing torsional divergence. A wing that is swept back bends under lift loads in a direction that reduces or washes out incidence at the wing tips. This provides automatic load relief.

By the same token, a wing that is swept forward bends under lift load in the opposite sense, increasing the wing tip incidence and load. This adds to the wing-bending deflection and the load in the classic feed-forward sense, producing torsional divergence at sufficiently high airspeeds. Thus, although swept forward wings are free of the premature tip flow separation problems mentioned in Chapter 11, "High Mach Number Difficulties," they had for many years been dismissed from consideration for new high-speed airplanes.

A classic paper gave the first published account of the effects of sweepback and sweep-forward on wing torsional divergence (Pai and Sears, 1949). A striking aspect of this early paper is the statement of the fundamental equation of aeroelasticity in matrix form. This is an integral equation for the local, or section lift, coefficient. The choice of aerodynamic theory is left free. In 1949, the choices were strip theory, which neglects aerodynamic induction; Prandtl theory; and Weissinger's approximation.

The advent of composite materials as aircraft structural elements has reopened the door to the sweptforward wing. In 1972, Professor Terrence A. Weisshaar at Purdue University studied the divergence and aeroelastic optimization of forward-swept wings, under a NASA grant. His student, a returning Vietnam veteran fighter pilot named Norris Krone, proposed a PhD thesis on fighter sweptforward wings.

With additional help from Professor Harry Schaeffer, Krone proposed building swept-forward wings in which layers of fiber–plastic composites are oriented to increase greatly torsional stiffness. Such wings could have torsional divergence speeds well out of the flight range. Later, as an official in the Defense Advanced Research Projects Agency, Krone had the unusual opportunity to help turn his research into a practical airplane, the successful Grumman X-29A research airplane.

19.5 Aileron-Reversal Theories

Aileron reversal is closely related to wing torsional divergence, involving also quasi-static wing twist. Aileron reversal limits roll maneuverability at high airspeeds and low altitude. At airspeeds where wings are still within structural limits, torques exerted on wings by deflected ailerons can twist the wing in the opposite direction enough to cancel much of the aileron's lift or rolling moment, and even to reverse the aileron's effect. The airspeed at which complete rolling moment cancellation occurs is called the aileron-reversal speed (Figure 19.2). Like a wing's flutter or torsional divergence speeds, aileron-reversal speed should be only a theoretical number, quite outside of the airplane's operating envelope and with an adequate safety margin.

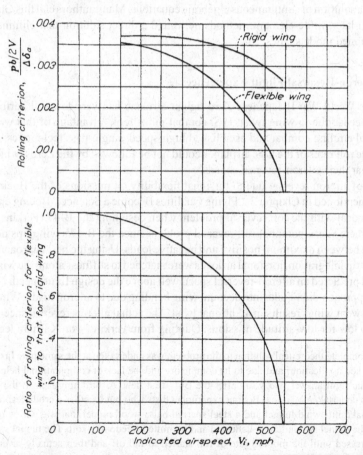

Figure 19.2 An example of the effect of wing flexibility on available rolling velocity. Data for the Republic P-47C-1-RE at an altitude of 4,000 feet. Aileron reversal occurs at 545 miles per hour. (From Toll, NACA Rept. 868, 1947)

Theoretical work on aileron reversal fits into Etkin's quasi-static category. The earliest published work on aileron reversal was necessarily simplified, with the computing resources available at that time (Cox and Pugsley, 1932). The wing is represented in the same semi-rigid manner described for the wing torsional divergence problem. That is, a reference section along the wing is selected and the elastic restoring moment is related to the angular deflection at that station. A modern example of the semi-rigid approach to aileron reversal is given by Bisplinghoff and Ashley (1962).

An extension of the semi-rigid approach yields one of the most useful concepts in static aeroelasticity, the ratio of elastic to rigid control surface effectiveness. For the simple two-dimensional case, this ratio depends only on the flexibility influence coefficient, or twist angle per unit applied torque, the changes in section lift coefficient per unit section angle of attack and aileron deflection, the change in section moment coefficient about the aerodynamic center with aileron deflection, and the ratio of the distance between the elastic axis and aerodynamic center to the section chord length.

In a sense, everything that came after the two-dimensional, or semi-rigid, approach was refinement, to deal adequately with problems of real airplanes. For example, in 1945, Dr. Alexander H. Flax expressed the wing twist spanwise distribution as a superposition of modes with undetermined magnitudes. Mode magnitudes are found by the virtual work principle, in a solution of simultaneous algebraic equations. Many authors call this Galerkin's method. Still later, finite-element methods permitted a direct solution that eliminated the need for assumed mode shapes.

19.6   Aileron-Reversal Flight Experiences

The World War II Japanese Zero fighter airplane had very low roll performance at high airspeeds, due to wing twist. U.S. combat pilots took advantage of this weakness. They avoided circling combat and established high-speed, single-pass techniques. At high airspeeds, the roll rates of the U.S. airplanes could not be followed by the Zeros, which were operating near their aileron-reversal speeds.

The role of aileron reversal due to torsional flexibility on missions of the Boeing B-47 Stratojet is mentioned in Chapter 3, "Flying Qualities Become a Science." Boeing engineers attempted to deal with the roll reversal problem when designing the B-47 (Perkins, 1970). They knew there was a potential roll reversal problem since the B-47's wing tips deflected some 35 feet between maximum positive and negative loads. Using the best approach known at the time, strip integration, torsional airloads were matched to stiffness along the wing span. This method predicted an aileron-reversal speed well above the design limit speed. Unfortunately, this approach didn't take into account wing bending due to aileron loads. Wing bending on long swept wings results in additional twist. The actual aileron-reversal speed turned out to be too low for low-altitude missions. Quoting from Perkins' von Kármán lecture:

> A complete theoretical solution to the problem was undertaken at the same time [as the strip method application] and due to its complexity and the lack of computational help, arrived at the right answer two years after the B-47 first flew. A third approach to the problem was undertaken by a few Boeing experimentalists who put together a crude test involving a makeshift wind tunnel and a steel sparred balsa wood model that was set on a spindle in the tunnel with ailerons deflected and permitted freedom in roll. The tunnel speed was increased until the model's rate of roll started to fall off and then actually reverse. This was the model's aileron reversal speed and came quite close to predicting the full-scale experience. The test was too crude to be taken seriously and again results came too late to influence the design of the B-47.

According to William H. Cook, the B-47 not only had excessive wing torsional deflection due to aileron forces, but also slippage in the torsion box bolted joints. The wings would take a small permanent shape change after every turn. These problems led to a test of spoiler ailerons on a B-47, although the production airplane was built with normal flap-type ailerons.

19.7 Spoiler Ailerons Reduce Wing Twisting in Rolls

Spoiler ailerons as a fix for wing aeroelastic twisting in rolls apparently had their first trial on a Boeing B-47. Spoiler ailerons have lower section pitching moments for a given lift change than flap-type ailerons, which means lower wing twisting moments. The proposal to use spoilers on the B-47 came from Guy Townsend, who had experience with spoilers on a Martin airplane. From an unpublished Boeing document by Cook:

> In order to test them quickly, we first tried a pop-up scheme, where the several segments would fully extend in sequence as the pilot's wheel was rotated. Electric solenoid valves at each hydraulic cylinder were programmed to open in sequence. However, this was too jerky, and proportional control was found to be required. This was later done on the B-52 mechanically.
>
> The next step in complication was on the −80 [prototype Boeing 707], when it was decided to use the spoiler not only for roll control, but also for drag brakes in the air, and on the ground to unload the wing for better braking. This required a "mixer box." While this system has proved reliable on 707 and subsequent models, the programming of spoilers electrically saves space and weight, and probably would provide roll control with safety by using the redundancy provided by the multiple segmented spoilers. The mechanically controlled aileron still provides a good backup for emergency [Figure 19.3].

Upper surface spoilers for lateral control, sometimes augmented by flap-type ailerons for low-speed control, are a standard feature on modern high-aspect-ratio swept-wing jets. They can be seen on a great variety of airplanes, such as the Douglas A3D-1; the Lockheed L-1011 and C-5A; the Convair 880M; the McDonnell-Douglas DC-8, DC-9, DC-10, and MD-11; the Airbus A310 and A320; and the Boeing B-52, 727, 737, 747, 757, 767, and 777 (Figure 19.4). When installed just ahead of slotted wing flaps, spoilers become slot-width control devices when the flaps are down, providing an additional bonus of powerful low-airspeed lateral control.

Aileron reversal is still a potential problem even in this modern age of supersonic airplanes and digital computers, on airplanes with straight as well as swept wings. This is indicated by the chart comparing various aileron designs for Boeing's 2707 SST proposal. Spoiler ailerons would be needed to avoid major losses in aileron control power due to wing twist at high airspeeds, even for the 2707's low-aspect-ratio wing (Figure 19.5).

19.8 Aeroelastic Effects on Static Longitudinal Stability

There had been several published studies of the effects of aeroelasticity on static longitudinal stability, going back to 1942. But the subject really came to wide attention with the appearance of the very advanced, and flexible, Boeing B-47 Stratojet, first flown in 1947. Richard B. Skoog of NACA reported on the details of the stability and control static aeroelastic effects on this airplane (Skoog, 1957) based on classified work done six years earlier.

Strangely, while some individual effects are large, Skoog found that the overall aeroelastic modification to longitudinal stability and control is small (Figure 19.6). Wing symmetric bending causes the wing tips to wash out at increasing angles of attack. This moves the airload

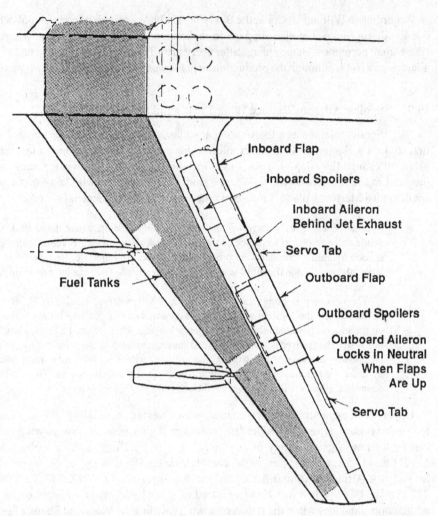

Figure 19.3 Arrangement of ailerons, spoilers, and flaps on the Boeing 707 airplane. The outboard and inboard flap-type ailerons are manually controlled, with the help of internal aerodynamic balance and balancing or geared tabs (here called servo tabs). The spoilers are of the slot-lip variety, located just ahead of the flaps. (From Cook, *The Road to the 707*, 1991)

relatively inboard, resulting in a forward, or destabilizing, shift of the wing aerodynamic center. However, there is a net loss in lift at positive angles of attack, a reduction in the lift curve slope. This is stabilizing, increasing the relative effect of the tail lift.

Fuselage bending under tail aerodynamic loads is destabilizing. That is, for upward tail loads, the fuselage bends upward at the rear, decreasing the tail angle of attack and the restoring moment of the tail. However, this effect is largely canceled by the downward bending of the aft fuselage under its own weight and the weight of the tail assembly, at the lower airspeeds associated with higher angles of attack. Just as wing torsion leads to aileron reversal at a sufficiently high airspeed, so does vertical bending of the aft fuselage lead to elevator or longitudinal control reversal. In the case of elevator controls, stabilizer twist adds to the problem (Collar and Grinsted, 1942).

The basic static aeroelastic analysis methods used up to the time when finite-element methods were introduced was the method of influence coefficients. Early expositions of the

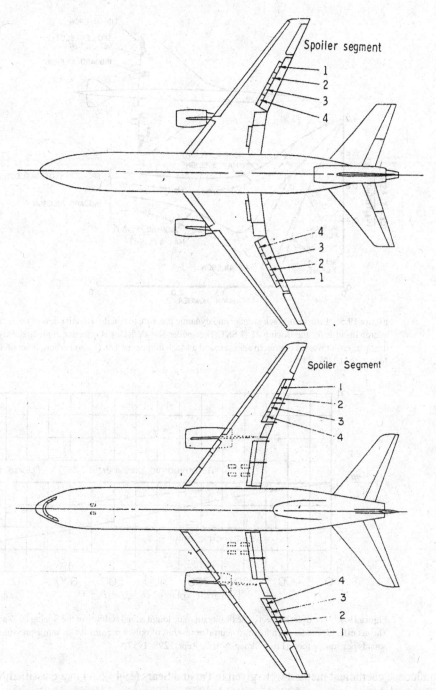

Figure 19.4 Two airplanes with slot-lip spoiler lateral controls to minimize loss in control power at high airspeeds due to wing twist: the McDonnell-Douglas DC-10 (*above*) and the Lockheed 1011 (*below*). In each case small outboard flap-type ailerons are used only at low airspeeds. (From NASA TN D 8373 and TN D 8360, 1977)

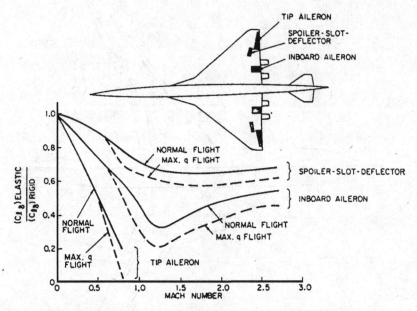

Figure 19.5 Effects of Mach number and dynamic pressure (q) on the effectiveness of three alternate aileron designs for the Boeing 2707 SST. The spoiler-slot-deflector is effective at all airspeeds, while the tip aileron reverses in effectiveness around a Mach number of 1.0. (From Perkins, *Jour. of Aircraft*, July–Aug. 1970)

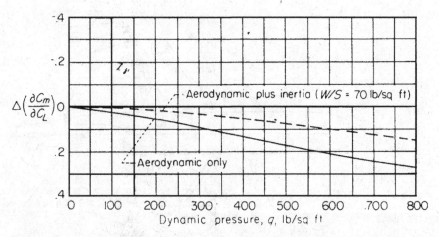

Figure 19.6 The overall effect of flexibility on static longitudinal stability of the Boeing B-47 airplane. The net effect is moderate, a forward neutral point shift of only 7 percent at a dynamic pressure of 500 pounds per square foot. (From Skoog, NACA Rept. 1298, 1957)

influence coefficient method were given in Pai and Sears (1949) and in a classified NACA Research Memorandum of 1950, written by Richard Skoog and Harvey H. Brown.

As early as 1954 an important relationship was stated between frequency response and static aeroelastic characteristics. If aeroelasticity were a branch of pure mathematics, this relationship would be stated as a theorem, in these terms:

> Airplane frequency response at frequencies below the lowest structural bending or torsional modes should agree with calculated rigid-body transfer functions when quasi-static aeroelastic effects are included.

This relationship, proved experimentally with the Boeing B-47 (Cole, Brown, and Holleman, 1957), provides an important check on static aeroelastic methods. In the 1980s, this relationship provided the basis for a comparison of alternate quasi-static aeroelastic methods for the Northrop B-2 Stealth Bomber.

19.9   Stabilizer Twist and Speed Stability

Collar and Grinsted (1942) showed that stabilizer setting can have an effect at high airspeeds on the variation with airspeed of the elevator stick forces required for trim. This variation is called *speed stability*. Push stick forces should be needed for trim at increasing airspeeds, so that if the stick is released, it will come aft, nosing the airplane up and reducing airspeed.

Airplanes with leading-edge-up rigged stabilizers will require trailing-edge-up elevator angles for trim in cruising flight. The up-elevator angles will put a down load on the stabilizer rear spar, tending to twist the stabilizer further in the leading-edge-up direction. Increasing airspeeds will increase the down load and twist, requiring increasing up-elevator angles and pull stick forces. This amounts to speed instability, pull forces needed for trim at increasing airspeeds rather than push forces. The Douglas A2D-1 Sky Shark, with an adjustable stabilizer, had this problem until its elevator tab was rigged trailing-edge up. This caused the elevator to float trailing-edge down in cruising flight and the stabilizer to be carried more leading-edge down, correcting the problem.

A reverse problem can occur on airplanes whose stabilizers are rigged leading-edge down. A leading-edge-down rig is used on some airplanes to improve nose-wheel liftoff for takeoffs at a forward center of gravity. A leading-edge-down rig can lead to excessive speed stability, requiring large push forces to trim in dives. If an airplane with a leading-edge-down stabilizer rig gets into an inadvertent spiral dive and push forces are not supplied, normal acceleration can exceed the structural limit. This effect is thought to be responsible for some in-flight structural failures, unfairly attributed to pilot inexperience in high-performance airplanes. W. H. Phillips considers this may be the cause for some failures of the Beech Bonanza (Phillips, 1998).

An aeroelastic problem related to stabilizer twist is spurious control inputs that result from airframe distortion under maneuvering loads. Fuselage deflection under positive load factor caused control inputs that increased load factor on the Vought F8U-1 airplane (Phillips, 1998). This was destabilizing in maneuvers. Reversing the position of a link in the elevator control system reversed the effect, providing stability instead.

19.10   Dihedral Effect of a Flexible Wing

Dihedral effect C_{l_β}, the rolling moment due to sideslip, is determined chiefly by wing dihedral. However, the dihedral angle of airplanes with flexible wings varies noticeably according to the wing's lift loads. Wing tips that droop when an airplane is parked can rise above the wing roots in flight, creating a positive dihedral angle. High-performance fiberglass sailplane wings bend upward alarmingly in flight, especially when the sailplane is in high-load-factor turns or pullups (Figure 19.7).

The first published analysis of the additional dihedral effect due to load factor showed a near doubling of dihedral effect for a wing of aspect ratio 10, at a hypothetical airplane's limit load factor of 4.0. (Lovell, 1948). In a later published analysis of this effect, W. P. Rodden showed that measured aeroelastic effects on a flexible model of the Douglas XA3D-1 could be correlated with a simple theory that assumes a parabolic distribution of wing bending

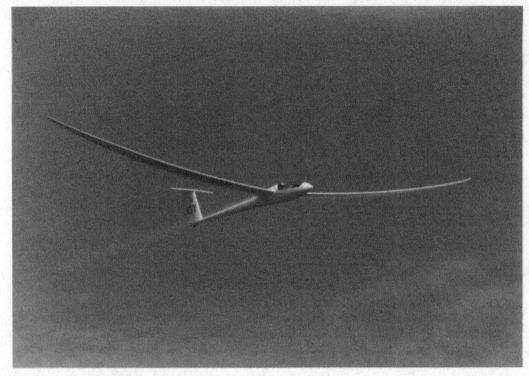

Figure 19.7 Upward bending under airload of the flexible wings of a high-performance fiberglass sailplane. The bending adds to its effective dihedral angle. (Photo by Steve Hines)

(Rodden, 1955). This leads to a linear asymmetric spanwise distribution in additional angle of attack, as for the pure rolling case. An approximation to the flexibility increment to C_{l_β} is thus proportional to well-known roll damping derivative C_{l_p} values.

Rodden's 1955 simplified analysis considers only the effects of symmetric air loads on wing bending. The analysis was extended to include the effects of asymmetric air loads (Rodden, 1965). Asymmetric loads cause an amplification of the dihedral angle. Matrix methods are used in this case, with aerodynamic and structural influence coefficients (Figure 19.8).

19.11   Finite-Element or Panel Methods in Quasi-Static Aeroelasticity

Analyzing quasi-static aeroelastic effects requires balancing air loads against structural stiffness and mass distributions. Because of the complexity of the problem, only approximate methods were available for many years. The advent of finite-element or panel methods both in structural analysis and in aerodynamics made accurate quasi-static aeroelastic analysis really possible for the first time.

In the aerodynamic finite-element approach, the airplane's surface is divided into many generally trapezoidal panels, or finite elements. Under aerodynamic and inertial loadings, the structure finds an equilibrium when boundary conditions are satisfied at control points such as the center of the 3/4-chord line of an aerodynamic panel or at the edges of a structural panel. Finite-element methods in structural analysis preceded those for aerodynamic analysis by many years.

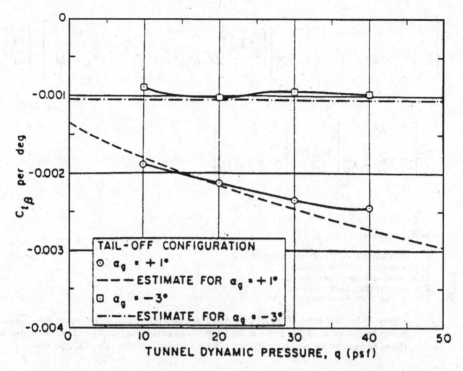

Figure 19.8   Effect of dynamic pressure on the dihedral effect of the Douglas XA3D-1 airplane, at two angles of attack. The wing lift is close to zero at a (fuselage) angle of attack of -3 degrees, and there is little wing bending and change in dihedral effect. (From Rodden, AGARD Report 725, 1989)

The earliest aerodynamic finite-element method, called vortex lattice analysis, appears to have been developed independently by two people. Vortex lattice analysis is documented in internal Boeing Company and Swedish Aeronautical Research Institute reports by P. E. Rubbert in 1962 and Sven G. Hedman in 1965, respectively, and in a few other reports of the same period. Dr. Arthur R. Dusto and his associates at the Boeing Company combined these structural and aerodynamic finite-element methods into an aeroelastic finite-element system they call FLEXSTAB (Dusto, 1974) in the period 1968 to 1974.

Finite-element methods in quasi-static aeroelasticity require generation of mass, structural influence, and aerodynamic influence matrices. The mass matrix is the airframe mass assigned to each element. The structural influence coefficient matrix transforms deflections at control points in an element to elastic forces and moments at the other elements. The aerodynamic influence coefficient matrix transforms angle of attack at one element to aerodynamic forces and moments acting on the other elements.

It is interesting that the advent of finite-element quasi-static aeroelastic methods coincided with the need for methods that account for significant chordwise structural distortions. Quasi-static aeroelastic methods based on lifting line theory were appropriate for flexible airplanes of the Boeing B-47 and Douglas DC-8 generation, subsonic airplanes with long, narrow wings. Proper quasi-static aeroelastic analysis of the lower aspect ratio, complex wing planforms of the Northrop B-2 stealth bomber and supersonic-cruise transport airplanes, requires panel methods.

NASTRAN is a widely used finite-element structural analysis computer program. The MacNeal-Schwendler Corporation's proprietary version, called MSC/NASTRAN, adds

$$
\begin{bmatrix} K^a_{\ell\ell} & K^a_{\ell r} \\ D^T K^a_{\ell\ell} + K^a_{r\ell} & D^T K^a_{\ell r} + K^a_{rr} \end{bmatrix} \begin{Bmatrix} u_\ell \\ u_r \end{Bmatrix} + \begin{bmatrix} M_{\ell\ell} & M_{\ell r} \\ D^T M_{\ell\ell} + M_{r\ell} & D^T M_{\ell r} + M_{rr} \end{bmatrix} \begin{Bmatrix} \ddot{u}_\ell \\ \ddot{u}_r \end{Bmatrix} =
$$

$$
- \begin{bmatrix} K^a_{\ell x} \\ D^T K^a_{\ell x} + K^a_{rx} \end{bmatrix} \{u_x\} + \begin{Bmatrix} P_\ell \\ D^T P_\ell + P_r \end{Bmatrix}
$$

In this equation:

u = displacement vectors or column matrices
K = structural stiffness matrices
M = structural mass matrices
P = aerodynamic force matrices
D = a rigid body mode matrix

Figure 19.9 One form of the NASTRAN quasistatic aeroelastic matrix equations. Additional manipulations are needed to arrive at the unrestrained aeroelastic stability and control derivatives. (From Rodden and Johnson, eds., *MSC/NASTRAN Aeroelastic Analysis User's Guide*, 1994)

aerodynamic finite-element models to the existing structural models with splining or interpolation techniques to connect the two. This version can perform quasi-static aeroelastic analysis (Figure 19.9). This accomplishment is credited to a number of people, including Drs. Richard H. MacNeal and William P. Rodden, and E. Dean Bellinger, Robert L. Harder, and Donald M. McLean.

19.12   Aeroelastically Corrected Stability Derivatives

An important by-product of both the early and modern quasi-static aeroelastic methods is a set of aeroelastically corrected stability and control derivatives, such as C_{m_α} and C_{m_δ}, which can be used in the ordinary equations of rigid-body motion. For example, Etkin (1972) derives the quasi-static aeroelastic contributions of symmetrical first-mode wing bending to tail and wing lift, which become ingredients in stability derivatives.

The wind tunnel provides a complete set of rigid-body aerodynamic stability and control data for most new airplane projects. These data are usually corrected for quasi-static aeroelastic effects using the concept of elastic-to-rigid ratios (Collar and Grinsted, 1942). Elastic-to-rigid ratios preserve in the aeroelastically corrected data all of the nonlinearities and other specific detail of the rigid data. Finite-element methods provide a modern source of elastic-to-rigid ratios for this purpose.

Wind-tunnel tests of elastic models have also been used to obtain aeroelastically corrected stability derivatives. Still another approach is the wind-tunnel test of a rigid model that has been distorted to represent a particular set of airloads, such as those caused by a high load factor. A distorted model of the Tornado was tested in a wind tunnel, to determine aeroelastic effect on stability derivatives.

19.13 Mean and Structural Axes

Even after the advent of panel methods, there remain controversial aspects of the quasi-static aeroelastic problem, related to the choice of axes. Structural distortions must be referred to some set of reference axes. There are essentially two sets of reference axes that will serve. One choice, called *structural axes*, corresponds to a natural reference for laboratory structural deflection tests or their analytical equivalent. Structural axes are aligned with a central hard section of the airplane, such as the wing interspar structure at the airplane's centerline.

The second choice, which is the only one that is consistent with the ordinary pitch–plunge equations of airplane motion, are mean axes. Mean axes are a familiar concept in normal mode analysis. They correspond to the midpoint of normal mode oscillations, the point at which all transverse deflections are momentarily zero. While structural influence coefficients may well be measured or calculated in an arbitrarily chosen structural axis system, pitch and plunge motions of the aeroelastic airplane must be calculated in mean axes, to avoid systematic error (Milne, 1964, 1968). A refinement of mean axes is the use of principal axes in which distributed moments of inertia are accounted for in addition to longitudinal mass distributions.

John H. Wykes and R. E. Lawrence used both mean and structural axes in a 1965 study of aerothermoelastic effects on stability and control, but they noted the difficulties involved in relating airplane angle of attack in the two systems. The angle of attack difficulty found by Wykes and Lawrence is resolved in an offline transformation of the results, such as pitch attitude and angle of attack time histories, from mean to structural axes (Rodden and Love, 1984). The transformation is feasible at the end of the dynamics calculations. The Rodden and Love paper, corrected in Dykman and Rodden (2000), also presents transformation equations from mean to structural axes.

The Rodden papers have an interesting proof of the fallacy of using the more convenient structural axes for dynamics studies in place of mean axes, as has been done by investigators unwilling to face angle of attack difficulties. In a simple swept-forward airplane example using structural axes, load factor and pitching acceleration time histories depend on the fixity choice of the axes, an evidently incorrect result. This error is avoided with mean axes. Mean axes are used in the FLEXSTAB program.

19.14 Normal Mode Analysis

Normal mode analysis, as applied to aeroelastic stability and control problems, is actually a form of the small oscillation theory about given states of motion. This goes back to the British teacher of applied mechanics E. J. Routh, in the nineteenth century. A body is supposed to be released from a set of initial restraints and allowed to vibrate freely. It will do so in a set of free vibrations about mean axes, whose linear and angular positions remain unchanged. The free vibrations occur at discrete frequencies (eigenfrequencies), in particular mode shapes (eigenvectors).

Of course, the airplane does not vibrate freely, but under the influence of aerodynamic forces and moments. These forces and moments are added to the vibration equations through a calculation of the work done during vibratory displacements. Likewise, the changes in aerodynamic forces and moments due to distortions must have an effect on the motion of mean axes, or what we would call the rigid-body motions.

According to Etkin's criterion, if the separations in frequency are not large between the vibratory eigenfrequencies and the rigid-body motions such as the short-period longitudinal or Dutch roll oscillations, then normal mode equations should be added to the usual rigid-body equations. Each normal mode would add two states to the usual airframe state matrix (Figure 19.10). A useful example of adding flexible modes to a rigid-body simulation is provided by Schmidt and Raney (2001). Milne's mean axes are used.

Normal mode aeroelastic controls-coupled analyses were made in recent times for the longitudinal motions of both the Northrop B-2 stealth bomber and the Grumman X-29A research airplane. In both cases, the system state matrix that combines rigid-body, normal mode, low-order unsteady aerodynamic and pitch control system (including actuator dynamic) states was of order about 100 (Britt, 2000).

19.15   Quasi-Rigid Equations

While the normal modes are by far the most common way to account for aeroelastic effects on airplane stability and control, the less abstract, approximate normal mode method called quasi-rigid analysis deserves mention. In quasi-rigid analysis, the flexible airplane structure is represented by a chain of linked rigid bodies, held in position by springs. An approximate normal mode is introduced for each link.

The earliest dynamic stability and control analysis in which an airplane bending mode was represented appears to have been the quasi-rigid analysis of the Boeing XB-47 (White, 1947). A single yaw pivot was assumed behind the wing trailing edge. This effectively represents the airplane's first asymmetric bending mode.

Ground vibration tests had established the B-47's first bending mode frequency at 2.3 cycles per second. Using the known weight of the airplane's aft section, the effective spring was sized to produce the measured bending frequency. A closed-loop servo analysis agreed with flight results that showed some vibration at the first bending mode frequency when the yaw damper rate gyro was located near the vertical tail. Moving the gyro near the airplane's center of gravity corrected the problem.

Just as quasi-rigid analysis has provided a simple, approximate alternative to normal mode analysis, it has done the same for the quasi-static aeroelastic problem. The longitudinal neutral and maneuver points of the Northrop YF-17 were found in this way (Abzug, 1974).

19.16 · Control System Coupling with Elastic Modes

Coupling of the B-47's yaw damper system with the airplane's fuselage side-bending mode was resolved simply when the yaw damper's rate gyro was relocated. That is, the rudder's yaw damping action cut off at a low enough bandwidth that the side-bending mode itself was not reinforced.

The coupling of stability augmentation systems and airplane elastic modes takes on a new dimension for high-bandwidth control systems. If the flight control system is capable of interacting with the airplane's structural modes, the stability of the combination must be assured. A conventional approach is gain stabilization, in which control system response at structural mode frequencies is attenuated by notch filters. The notch filters reduce rate gyro and accelerometer outputs in a narrow band around modal frequencies.

While effective, notch filtering invariably introduces lag at lower frequencies, which can adversely affect flying qualities. Phase stabilization (Ashkenas, Magdaleno, and McRuer, 1983) attempts to replace or supplement notch filtering by creating dipoles out of the

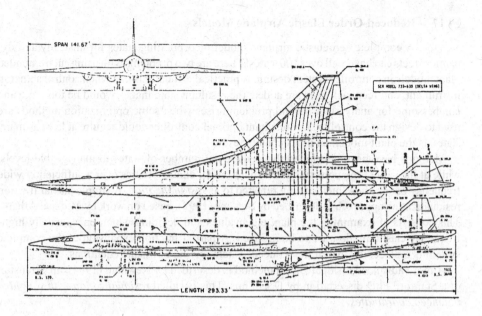

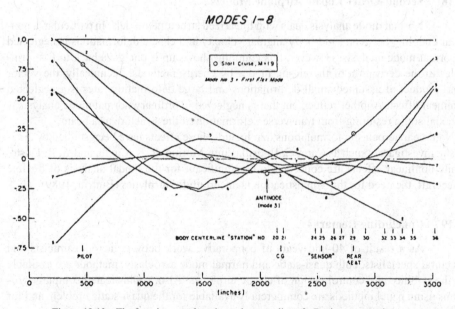

Figure 19.10 The first six normal modes at the centerline of a Boeing supersonic transport proposal, typical of the data used in the normal-mode method for the effect of aeroelasticity on stability and control. The modes are normalized in amplitude. Modes 1 and 2 are rigid-body plunge and pitch. (From Ashkenas, Magdaleno, and McRuer, NASA CR-172201, Aug. 1983)

structural bending poles. The dipoles are the stable type referred to in the "Transfer Function Dipoles" section of Chapter 20, with the zero below the pole in the s-plane. Zero location for particular modes can be controlled by sensor location, but locations that produce stable dipoles for some modes will be wrong for others.

19.17   Reduced-Order Elastic Airplane Models

A complete aeroelastic airplane model incorporating states for unsteady aerodynamic effects can have well over 100 states. There are two problems with using such extended plant models in controller system design. A practical difficulty is that when control actuator, instrument, and feedback laws are added, the resultant state matrix would be too large and cumbersome for analysis. A second problem arises when some optimization methods are used to design the controller. The resultant optimal controller could require at least as many states as the plant model, an unacceptable result.

Methods have been developed that reduce the number of states to manageable levels, while at the same time preserving dominant modal characteristics over a sufficiently wide frequency band. These methods are referred to as *residualization*. A simple partial fraction residualization approach is described by Stevens (1992), based on work by Michael Athans. This consists of examining the plant eigenvalues and deciding which ones, usually high-frequency elastic modes, are to be dropped. A partial fraction expansion of the complete system is replaced by that of the reduced system, giving new state equations.

A more advanced residualization method is described by Newman (1994), an extension of a Stanford PhD dissertation by D. F. Enns. This is called *residualization with weighted balanced coordinates*.

19.18   Second-Order Elastic Airplane Models

Normal mode analysis deals with linearized structural models. In particular, a nonlinear coupling between rigid-body angular velocity and elastic deformations is neglected in normal mode analysis. However, this coupling shows up if one goes back to first principles in the derivation of the elastic modes by energy methods. Technically, the vector cross-product of assumed small deformations and large deformation rates is a neglected nonlinear effect. Another refinement that is neglected in ordinary normal mode analysis is the axial strain resulting from transverse deformation of the structure as a beam.

Nonlinear aeroelastic formulations that include these effects have been studied by aeroelastic investigators such as Carey S. Buttrill, Luigi Morino, R. K. Cavin, and A. R. Dusto. While nonlinear effects are conceded to be significant for the modal analysis of flexible spacecraft, the need for this sophistication is uncertain for airplanes (Buttrill, 1989).

19.19   Concluding Remarks

As a result of 40-odd years of cooperative work between aerodynamicists and structural specialists, both quasi-static and normal mode aeroelastic methods are available for the stability and control design of elastic airplanes. Also, sophisticated computer programs using panel methods are commercially available for the quasi-static problem and for normal mode determination.

As is usual in any technical discipline, there remains room for improvement. The average airplane stability and control engineer is more at home with rigid models of the airframe. Thus, there seems to be a need for increased acceptance and applications of, and publications in the new panel methods, to make this attitude change. Wider use of composites in airplane structures and more sophisticated stress analysis methods will probably result in more flexible airframes in the future. This will make it important for the average stability and control engineer to be a reasonably proficient aeroelastics engineer.

CHAPTER 20

Stability Augmentation

Stability augmentation is the artificial improvement, generally by electromechanical feedback systems, of airplane stability and control *while the airplane remains under the control of the human pilot*. Stability augmentation generally changes the airplane's stability derivatives and modes of motion.

We make the important distinction between stability augmentation, artificial feel systems, and airplane automatic pilots. While artificial feel systems, discussed in Chapter 5, may alter stick-free stability for the better, their main function is providing manageable control forces. Automatic pilots replace the human pilot when they are in use.

20.1   The Essence of Stability Augmentation

To be a true stability augmenter, the device must change the airplane's flight characteristics without the pilot's perception. This means that augmenter outputs must add to those of the pilot in a series fashion. Augmenter outputs put into the primary control circuit between the cockpit and the control surfaces must move only the control surfaces, and not the cockpit controls. The requirement to not move the pilot's controls is sidestepped if the augmenter is not inserted into the primary control circuit but moves a separate, or dedicated, control surface. Still another way around the need for augmenters not to move the pilot's controls is the integrated control surface actuator (Chapter 5), used in fly-by-wire control systems. Integrated servo actuators accept and add electrical signals from both cockpit controls and stability augmenters.

In fly-by-cable control systems, isolation of primary-control-circuit stability augmenter outputs from the cockpit controls is a surprisingly difficult mechanical design problem. Control valve friction in control surface actuators acts to hold the surfaces fixed for small stability augmenter signals. When this happens, the augmenter in effect backs up and moves the cockpit controls instead. The result is an unaugmented airplane for small disturbances and limit cycle oscillations, such as yaw snaking. One cure for excessive valve friction can be as bad as the small signal backup problem. This is to center the cockpit controls with husky spring detents, which have to be overcome by the pilot in normal control use.

The degree of authority of stability augmentation systems is another important design consideration. Since augmenters operate ideally without moving the pilot's controls, the pilot will be unaware of abrupt failures to the limit of augmenter authority until the airplane reacts. Then, there should be enough pilot control authority left to add to and cancel the failed augmenter inputs, with something to spare. This was the design philosophy until the advent of redundant, self-correcting augmentation systems, which make feasible augmentation at full authority or control surface travel.

Automatic pilots, which replace the human pilot when they are in use, are expected to move the cockpit controls. Abrupt full autopilot failures are instantly apparent to an attentive flight crew. Larger control authority than for stability augmenters is feasible, even for systems without the redundant, self-correcting feature.

20.2   Automatic Pilots in History

Stability augmentation goes back only to about 1945, while the history of airplane and missile automatic pilots, or autopilots (that word happens to be a trademark of a particular manufacturer), actually begins before the Wright brothers, with Sir Hiram Maxim's 1891 designs. That history has been told by several authors, including Bollay (1951) and the scholarly but very readable account of automatic pilot development in the first chapter of *Aircraft Dynamics and Automatic Control* by McRuer, Ashkenas, and Graham, dated 1973.

An additional historical account of airplane automatic pilots is that of W. Hewitt Phillips, in his Dryden Lecture in Research (1989). All of these authors refer to the remarkable 1913–1914 demonstration of the Sperry "stabilizer," which provided full automatic control of a Curtiss Flying Boat. However, the present chapter deals only with stability augmentation.

Gust-alleviation systems are a specialized form of airplane automatic pilots, designed to reduce structural loads and to improve ride quality in rough air. These systems are of less interest now than formerly because modern airplanes can fly above turbulence or use weather radar to avoid storms. A complete historical review of gust-alleviation systems is available in a NASA Monograph (Phillips, 1998).

20.3   The Systems Concept

The concept of the airplane's airframe as only one object in a complete dynamical system is part of the thinking of today's stability and control engineer, when faced with the need for stability augmentation. Yet, early researchers in airplane stability augmentation did not approach the problem that way (Imlay, 1940). Imlay enlarged on the classical Routh criterion for stability by the use of equivalent airplane stability derivatives. The equivalent derivatives are the basic control-fixed stability derivatives plus the products of control derivatives such as the yawing moment coefficient due to rudder deflection and a gearing ratio. The gearing ratio is an assumed control deflection per unit airplane motion variable. For example, Imlay studied gearings of 0.356 and 1.116 degrees of rudder angle per degree of bank and yaw, respectively.

The point is that the Imlay stability augmentation analysis method deals only with a modified airplane. No other dynamical elements are represented, although the lag effects of the servomechanism that would drive the control surfaces are suggested by a somewhat awkward representation of a simple time lag as the first three terms of the power series for the exponential.

The key mathematical concept that leads to modern augmentation analysis methods is the control element, which is represented graphically by a box having an input and an output. Control element boxes are linked one to another, with the output of one serving as the input to another. Control elements include sensors such as gyros; pneumatic, electric, or hydraulic control actuators; and, of course, the dynamics of the airframe, or control surface angle as an input and motion such as pitch or yaw rate as an output. Summing and differencing junctions act on inputs and outputs as needed, most notably to create an error signal. This is the difference between the commanded and actual system outputs.

20.4   Frequency Methods of Analysis

Frequency methods of analysis ushered in the age of modern airplane stability augmentation and autopilot analysis. In his 1950 Wright Brothers Lecture, Dr. William Bollay reminded us that the application of frequency-response methods to the airplane case came a full ten years after their use in the development of anti-aircraft gun directors. The footprints of the electrical engineering community in this field are still evident in the use of terms such as decibels and octaves in some airplane frequency-response studies.

Frequency response is the steady-state sinusoidal airplane motion perturbation in response to steady-state sinusoidal control surface input perturbation. Only the amplitude ratio and phase relationship of the two sinusoids are of interest. Frequency response of mechanical and electrical devices is readily found from the parameters of the linearized differential equations that describe the device's motion or electrical properties. The formal mathematics that do this rest on the Laplace transformation, as explained in Chapters 3 and 4 of the classic 1948 text *Principles of Servomechanisms*, by Gordon S. Brown and Donald P. Campbell.

Frequency-response analysis led stability and control engineers to an entirely new way to describe airplane dynamics, the transfer function. The transfer function is the mathematical operator by which any input function is multiplied to obtain the output function for that element. Transfer functions are numerical or literal expressions in the Laplace variable s. Transfer-function denominators are nothing but the characteristic equation, with Routh's and Bryan's operator λ replaced by the complex Laplace variable s. Transfer-function numerators are governed by the input. Thus, the classical transfer function that converts elevator angle disturbances to pitch attitude disturbances is a second-degree polynomial in s divided by a fourth-degree polynomial in the variable s.

One of the first known applications of frequency response in airplane stability augmenter design was made by Roland J. White for the XB-47 yaw damper. White used the inverse frequency-response diagrams described by H. T. Marcy in 1946, in an electrical engineering context. Over the years, frequency-response analysis has never gone out of fashion. For example, the demanding X-29A flight control system was designed using Bode frequency plot techniques by Grumman engineers led by Arnold Whitaker, James Chin, Howard L. Berman, and Robert Klein.

Frequency-response methods are used in some of the latest airplane flight control design methods, giving frequency response another lease on life. As detailed in a later section, singular value methods, associated with robust control theory, use frequency response.

20.5   Early Experiments in Stability Augmentation

The first stability augmenters appeared during World War II. Little detailed information is available about them. A German Blohm and Voss Bv 222 flying boat was thought to have had a pitch damper acting through a small, separate elevator surface. In a paper delivered in 1947, M. B. Morgan described an experimental yaw damper installed on a Gloster Meteor jet airplane. Other notable early designs were the Boeing B-47 and the Northrop YB-49 yaw dampers and the Northrop F-89 sideslip stability augmenter, which are discussed below.

20.5.1   *The Boeing B-47 Yaw Damper*

The B-47 Stratojet was a radical airplane in its time, a six-jet bomber with very flexible sweptback wings. Early flight tests disclosed that damping in yaw at low airspeeds was much less than pilots could deal with in landing approaches. The main pilot objection was to the rolling portion of the motion, caused by the dihedral effect of the swept wings at high angles of attack. After discarding other alternatives, Boeing engineers decided to attack the rolling motion indirectly, by artificial yaw damping using a rate gyro and the airplane's rudder. That is, by suppressing side-slipping motions, the airplane's rolling moment due to sideslip would not cause the objectionable wallowing in landing approaches.

The engineers who were chiefly responsible for the XB-47 yaw damper design were William H. Cook and Edward Pfafman. Roland J. White, who made a frequency-response analysis of the XB-47 yaw damper design, provides a complete account of the development (White, 1950). In White's account one can find all of the elements that go into modern stability augmenter designs, even though in unfamiliar form in some cases. These are

> the application of servomechanism analysis, using the equations of airplane motion;
> airframe mathematical model includes aeroelastic bending effects;
> irreversible power controls;
> stability augmentation series servo, isolating the pilot from the servo action;
> artificial feel system.

Roland White's XB-47 yaw damper servomechanism analysis, using inverse frequency response, was advanced for its time. However, the all-important matter of loop gain, or commanded rudder angle per unit yaw rate, was apparently settled in flight test. William Cook remembers that Robert Robbins, the XB-47 test pilot, had a rheostat that varied yaw damper gain, and that Robbins chose the value that seemed to work best.

With no fund of stability augmenter design information to draw upon, Cook and Pfafman improvised the yaw damper both in terms of design requirements and hardware. A short phone call from Cook at the Moses Lake flight test site to Pfafman laid out the key design requirements of rudder damping authority (one-fourth of full travel) and series actuation. The yaw damper servo was an electric motor and amplifier that had been used for B-29 turbo-supercharger waste gate control (Figure 20.1).

White's paper was delivered at the Design Session of the Institute of the Aeronautical Sciences 1949 Annual Summer Meeting in Los Angeles. The concept of stability augmentation as a normal design feature for swept-wing airplanes had not yet been established, and White's paper irritated at least one purist. According to Duane McRuer, this person, a respected professor of design at Cal Tech, got off the following comment during the paper's discussion period:

> If the B-47 had been designed properly, it would not have needed electronic stability augmentation.

William H. Cook (1991) reports a similar reaction from an MIT professor, unhappy that an "artificial" solution had been used on the B-47 to solve an aerodynamic stability problem. Of course, there is a perfectly sound aerodynamic reason why yaw stability augmentation is needed on jet airplanes and is not an evidence of poor design. Approximately, Dutch roll damping ratio is directly proportional to atmospheric density. An airplane with a satisfactory damping ratio of 0.3 at sea level will have a damping ratio of only 0.06 at an altitude of 45,000 feet.

20.5.2   *The Northrop YB-49 Yaw Damper*

The Northrop YB-49 shares with the Boeing B-47 the distinction of being one of the first stability-augmented airplanes in the modern sense (Figure 20.2). Duane T. McRuer (1950) described the YB-49's yaw damper as follows:

Figure 20.1 The series-type actuator (a surplus turbo waste gate servo) used in the Boeing XB-47 Stratojet's rudder push rod, to provide yaw damping. (From White, *Jour. of the Aeronautical Sciences*, 1950)

For the sensing part of the system, a Honeywell Autopilot rate gyro was chosen.... An electrical signal is then produced which is proportional to this speed or yaw rate. This signal is fed back through an electrical amplifier and reversible motor. Here the signal is transferred mechanically to a linkage that actuates the rudder cable system. The heavy work, that of opening the clamshell rudder to drag the wing back in line, then falls to the fully-powered rudder hydraulic system.

McRuer since added to this description the information that the reversible motor that put a yaw damping input in series with pilot's inputs was a turbo-supercharger waste gate servo, as for the B-47. The long cable that runs from the cockpit to the hydraulic servo valves on the clamshell rudders was expected to serve as a backup for the series-installed yaw damper servos. Unfortunately, initial yaw damper actuator motions stretched the cables until the hydraulic servo valve friction was overcome. This created a dead spot until corrected by a reduction in hydraulic valve friction.

McRuer and Richard J. Kulda made the preliminary stability analysis by the method of equivalent stability derivatives, used in the literal approximate factors for the spiral and Dutch roll modes. The detailed design used Bode and Nyquist diagrams, much as in the case of the B-47. The YB-49's yaw damper had no washout to cancel the yaw rate signal in steady turns. Compared with current practice, the five weeks or so that it took to design, round up parts, install, and check out the YB-49's yaw damper is of course quite short.

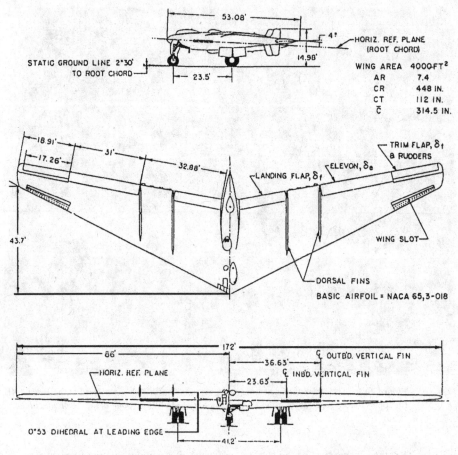

Figure 20.2 The outboard flaps on the Northrop YB-49 are split at the trailing edge to act as rudders. They provide yawing moments for the airplane's series-type yaw damper. The YB-49 and Boeing XB-47 were the first airplanes with series-type yaw dampers. (From Ashkenas and Klyde, NASA CR 181806, 1989)

20.5.3   *The Northrop F-89 Sideslip Stability Augmenter*

The straight-wing twin-jet F-89 was originally flown with a conventional rate gyro yaw damper, with washout for steady turns. The rate gyro was replaced with a sideslip sensor, to reduce adverse (top) rudder angle in lead pursuit courses more than the yaw rate washout could do. The F-89 sideslip stability augmenter improved directional stability as well as Dutch roll damping.

20.6   Root Locus Methods of Analysis

One of the remarkable stories in the stability and control field is the invention of the root locus analysis method by Walter R. Evans. This came relatively late in the game for a fundamental advance in control system analysis. The root locus method first appeared in Evans' Master's degree thesis at the University of California, Los Angeles, followed by a North American Aviation report. The first root locus journal paper (Evans, 1948) was published over objections from referees who thought the work was of little merit. While the method is known simply as the root locus method in the United States, Russian papers quite properly call it the Evans method. The root locus method received wide publicity with Dr. William Bollay's 1950 Wright Brothers Lecture, but even before the 1948 journal publication it had "spread like wildfire," according to Duane McRuer. This happened because John Moore at UCLA and Phillip Whittaker at MIT lectured on the method, using drafts of the North American report.

The essence of the root locus method is the set of rules that Evans discovered for the migration of roots of an open-loop system to the roots of the same system when the loop is closed. Airframe open-loop roots are nothing more than the airframe roots discussed by G. H. Bryan in 1911, the short- and long-period airplane modes of motion, and the aperiodic modes such as the spiral and roll modes. Evans found that the open-loop modes migrate toward the open-loop zeros as closed-loop gain is increased from zero (the open-loop case) to infinity. The open-loop zeros are the roots of the numerator function of each element's transfer function.

Root locus methods survived from the 1950s to the present day as one of the most widespread flight control system analysis and synthesis methods. Modern variants are

z-Plane The z-plane root locus makes use of the z-transform, where $z = e^{Ts}$ and s is the Laplace transform variable (Bollay, 1951). The complex number z is defined only at the switching or sampling times T of a sampled-data or digital control system. Since the states of sampled-data or digital control systems are likewise defined only at sampling times T, the z-plane root locus can be used for stability and performance analysis of these systems. In the z-plane, real parts of the variable z are plotted along the abscissa, while imaginary parts are plotted along the ordinate. The Evans s-plane root locus rules apply as well in the z-plane. Only the region of stability and lines of constant damping ratio differ.

w- and w'-Planes An improved transformation method for dealing with sampled-data or digital systems appeared in the 1950s, called the w-plane. The complex variable w is created by a bilinear transformation on z, or $w = (z-1)/(z+1)$. Richard F. Whitbeck and L. G. Hofmann (1978) describe a scaled version of the w-domain with even better properties. This is the w'-domain, in which $w' = 2w/T$. In contrast to the z-plane, the left or stable half of the w'-plane corresponds to the left or stable half of the s-plane. Powerful analogies exist between the s- and w'-domains, allowing use of conventional root locus and Bode (Bollay, 1951) design tools. As a drawback, w' transfer functions are far more complex algebraically than s transfer functions.

Unified Analysis Using Bode Root Locus The Bode root locus is a hybrid method developed by Duane McRuer (McRuer, Ashkenas, and Graham, 1973) that adds to the conventional Bode plot the amplitude ratios of the various loci in the s-plane. For any given loop gain actual closed-loop roots, all of the conventional frequency response quantities and the sensitivity to gain changes are seen in this plot.

Root Locus Sensitivity Vectors Sensitivity vectors can be drawn from unaugmented airplane poles, such as the Dutch roll complex conjugate pair, giving the directions and magnitude in the complex plane of the migration of those poles for individual feedbacks. The effects on a Dutch roll mode of unconventional feedbacks, such as sideslip and lateral acceleration to the ailerons, can be compared. Root locus sensitivity vectors were first published by Duane McRuer and Robert Stapleford (1963).

20.7   Transfer-Function Numerators

Airplane transfer-function denominator factors, or roots, govern airplane motions following initial disturbances. Stable roots, having negative real parts, lead to subsidence of oscillatory or aperiodic motions. The same is true for the denominators of closed-loop transfer functions, and early root locus work, such as the material in Dr. William Bollay's 1950 Wright Brothers Lecture, dealt with roots, or poles, of the closed-loop denominator. Transfer function numerator factors are called zeros. A response survey to step inputs for a systematic variation in pole–zero combinations (Elgerd and Stephens, 1959) gives striking results, particularly for the case of two real poles and one real zero. Depending on whether the zero is between the poles or to the right, the step response appears either deadbeat or with a large overshoot.

Transfer-function zeros play an important role in the closed-loop responses of stability-augmentation systems. The details are too involved to go into here, but some examples can be touched upon. In the altitude or glide path loop in which errors are corrected by elevator or stabilizer control, a zero called $1/T_{h1}$ can be in the right half of the root locus plane. This occurs on the back side of the power required curve, or at airspeeds below the minimum drag point. Loop closure drives a closed-loop real root into the right half-plane, with consequent divergence. An inner stability augmentation loop can correct this.

Another example is the complex zero associated with bank angle control by the ailerons. The Systems Technology, Inc., symbol for the undamped natural frequency associated with this zero is ω_ϕ. For values of ω_ϕ that exceed the Dutch roll undamped natural frequency ω_d, loop closure excites the Dutch roll and closed-loop stability is degraded. A complete tutorial discussion of this problem, as well as the altitude control zero problem, is given by Duane T. McRuer and Donald E. Johnston (1975).

20.8   Transfer-Function Dipoles

The problem of the complex zero in the bank angle–aileron transfer function happens to be one of a class of transfer-function dipole problems in stability-augmenter design. In many lightly damped airplane modes, the root in question is near a complex zero. The pole–zero pair is called a dipole. Root locus rules generally make sure that, for the pair, the locus that originates at the pole ends at the zero, forming a semicircle along the way.

When the dipole is close to the root locus imaginary axis, the semicircle can pass into the unstable, or right-half, plane. Conversely, by assuring that the semicircle forms to the left, closing the loop increases the stability of that lightly damped mode. This is called phase stabilization. By far the most important application of phase stabilization is to the bending and torsional modes of an elastic airplane with stability augmentation. As in the case of the bank angle transfer function, the modes are phase-stabilized when the dipole zero has a lower undamped natural frequency than the pole, or root.

20.9   Command Augmentation Systems

Command augmentation systems, or CAS, are a relatively recent form of airplane stability augmentation. Pilot control inputs, usually filtered or shaped, are compared with measured airplane motions, with the differences being sent to the control surface actuation servos (Figure 20.3). In early command augmentation applications, such as in the McDonnell Douglas F-4, F-15, and F-18 and the Rockwell B-1, the augmentation system has limited authority. There are parallel direct links from the sticks to the control surface servos.

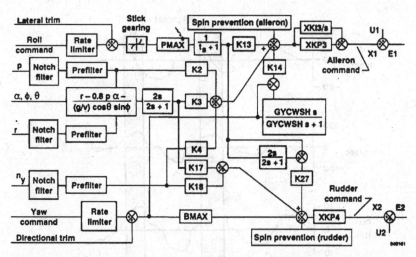

Figure 20.3 Block diagram for the lateral-directional command augmentation system for the X-29A research airplane. Key features are the rate-limiters on roll and yaw commands. These minimize the possibility of rate-limiting the control servos due to large pilot control inputs. (From Clarke, Burken, Bosworth, and Bauer, NASA TM 4598, 1994)

The command augmentation systems of later airplanes such as the fly-by-wire General Dynamics F-16 have full-authority and high-command gains. Full-authority roll command augmentation systems have worked very well, with sharp, rapid, and precise responses to control inputs (Mitchell and Hoh, 1984). Some problems come along with these successes. Oversensitivity to small inputs, overcontrol with large inputs, and the phenomenon called roll-ratcheting can occur.

20.9.1 *Roll-Ratcheting*

Roll-ratcheting bears a resemblance to the aileron buffeting that occurs on sharp-nosed Frise ailerons. The limit cycle oscillations occur at about 3 cycles per second when the ailerons are hard over, and the flight records even look the same (compare Figure 20.4 with Figure 5.6). However, the two phenomena could hardly be more different.

Roll-ratcheting arises from interactions among a variety of mechanisms. These include arm neuromuscular effects, limb and stick mass effective stick bobweights, force-sensing side stick gains, and roll command prefiltering. At 2 to 3 cycles per second, pilot voluntary efforts are not involved, so that roll-ratcheting is not a form of the pilot-induced oscillations discussed in Chapter 21.

A major effort was made to pin down roll-ratcheting parameters, using a fixed based simulator (Johnston and McRuer, 1977). The progress of that investigation, which brought in flight test data from the NT-33 variable-stability research airplane as well as the F-16, is given in fascinating detail by Irving L. Ashkenas in a summary paper (1988). There is a convincing correlation involving the stick sensing force gradient (degrees per second of roll rate per pound of stick force) and roll time constant T_R, in seconds. A single line divides roll command augmentation systems into ratcheting and nonratcheting cases. However, this particular correlation is thought to hold only for nonmoving or force-type side sticks, such as installed in the F-16 airplane.

The role of arm neuromuscular effects as a prime component of roll-ratcheting is questioned by Gibson (1999). In Delft TU studies, a simple assumption of a lateral bobweight

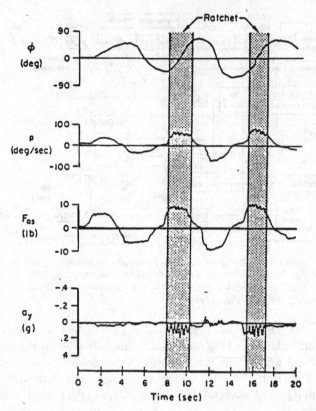

Figure 20.4 Flight record of roll ratcheting during banking maneuvers. (From Mitchell and Hoh, *Jour. of Guidance*, 1984)

loop was found to produce roll-ratcheting. A later paper from DVL, Braunschweig (Koehler, 1999) returns to the neuromuscular model with refinements, adding torso and hip dynamics to that of the arm. The Koehler paper claims good correlation with an F-16 XL roll-ratcheting incident. Gibson notes that a spectacular roll-racheting incident involving the F-18 is described by Klyde (1995). A mild rachet occurred on the BAe FBW (fly-by-wire) Jaguar airplane, which was cured by adding a stick damper and by changes to high-frequency control dynamics.

A prudent design approach to avoid roll-ratcheting might be to adhere initially to the Ashkenas 1988 force gradient/roll time constant criterion, supplementing this with detailed stability analyses that account for both neuromuscular and bobweight effects.

20.10   Superaugmentation, or Augmentation for Unstable Airplanes

The design goal of classical stability augmentation, such as in the XB-47 and YB-49, is to merely restore acceptable flying qualities such as well-damped Dutch roll or pitching oscillations to airplanes that have poor characteristics. The loss is usually due to operation at high altitudes, or at low speeds with airplanes designed for high speeds.

With the success of classical stability augmentation designers have become bold. They now offer the prospect of significant flight performance gains by flying unstable airframes. Superaugmentation makes inherently unstable airplanes have stable flight characteristics while retaining the performance gains. The greatest performance gains appear possible for airplanes inherently unstable in pitch. According to Peter Mangold, maximum lift coefficient increases of 25 percent and trim drag decreases of 20 percent can be obtained with a longitudinally unstable tailless design as compared with its stable counterpart. These large improvements arise from the down-trailing-edge angles required to trim an unstable flying wing, increasing wing camber. A negative camber is required to trim a stable flying wing. More modest gains are available for tailed airplanes, where down-trimming tail loads operate on longer moment arms to the airplane's center of gravity.

According to Duane McRuer, the designers of the YB-49 gave a great deal of thought to flying that airplane with unstable loadings, to take advantage of the performance gains. However, Waldemar O. Breuhaus claims that the first actual application of superaugmentation was made on a North American AT-6 trainer. A Sperry airline-type A-12 automatic pilot flew the AT-6 quite successfully with a negative static margin of 6.7 percent of the wing chord. In current practice, the General Dynamics F-16 was designed to fly with a slight negative static margin. Depending on store loadings, this is the case in practice. To reduce horizontal tailtrim loads during cruise, fuel is pumped into cells in the horizontal tails of the McDonnell Douglas MD-11 and the Boeing 747-400.

There are two general paths to superaugmentation. The most obvious is to artificially increase back to stable levels those derivatives that characterize longitudinally unstable airplanes. These are the derivatives M_α, M_q, and M_u. Feedbacks to the longitudinal control of angle of attack, pitching velocity, and airspeed, respectively, are required. However, there are practical problems with feeding back angle of attack and airspeed signals at high gain. Vertical and horizontal air turbulence components appear as objectionable noise inputs. Atmospheric turbulence noise effects may be partially compensated with complementary filters, which replace the higher frequency atmospheric turbulence signals with their inertially derived equivalents, inertial vertical and horizontal velocity. The airframe thus acts as a filter on the high-frequency turbulence signals.

The alternate path to superaugmentation relies on inertially based signals, such as pitching velocity, derived pitch attitude, and normal acceleration. Inertial signals lend themselves to modern redundant mechanizations. For example, five or six rate gyros in a skewed orientation can provide fail-operational capability for all three airplane axes. Figure 20.5 illustrates

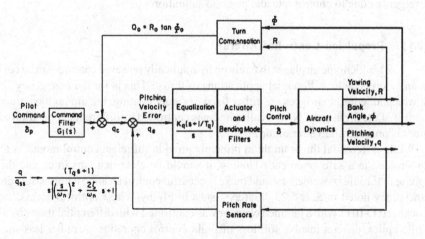

Figure 20.5 Example of a block diagram for an inertially based superaugmentation system for longitudinally unstable airplanes. The denominator s in the Equalization box integrates the pitching velocity error signal q_e, providing a pseudopitch attitude signal. (From Myers, McRuer, and Johnston, NASA CR 170419, 1984)

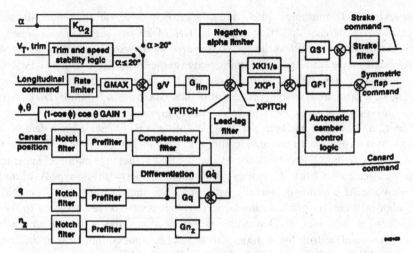

Figure 20.6   Block diagram for the X-29A research airplane superaugmented pitch control system, designed to overcome unstable static margins of up to 35 percent of the wing mean chord. Pitch angular acceleration is synthesized with filtered canard position. Normal acceleration feedback provides proper stick force per g. (From Clarke, Burken, Bosworth, and Bauer, NASA TM 4598, 1994)

superaugmentation by pitch rate feedback, integrated for a derived pitch attitude (McRuer, Johnston, and Myers, 1985). An aperiodic divergence caused by static longitudinal instability is eliminated as system gain is increased. A new longitudinal short-period mode appears with relatively high damping and frequency, making possible precision attitude control, provided that proper attention is paid to flight path response (Gibson, 1995). Large stability margins are attained. This provides robustness against parameter changes. The X-29A pitch control system, illustrated in Figure 20.6, is another inertially based superaugmented case.

Control system rate or position saturation can be particularly deadly for superaugmented airplanes. Once the control surfaces are operating at actuator-limited rates or against the surface stops the design reverts back toward the unaugmented, or unstable, case. In design studies, one must identify the command or disturbance levels that could cause unacceptable divergences due to control rate and position saturation.

20.11   Propulsion-Controlled Aircraft

Multiengine airplanes that rely on hydraulically powered controls can be controlled in an emergency by differential applications of thrust. This is for the emergency situation in which all control surfaces are either fixed or freely floating, but are no longer under the control of the flight crew. NASA calls airplanes controlled by differential operation of thrust propulsion-controlled aircraft, or PCA.

While differential thrust might in principle provide sufficient control moments to guide an airplane to a safe emergency landing, it should be clear that lags in engine thrust response to throttle movements would make successful control an almost impossible task. The emergency noted in Sec. 5.23, "Safety Issues in Fly-by-Wire Control Systems," where a Lockheed L-1011 with a jammed elevator was controlled with differential thrust by a highly skilled pilot, did not involve full loss of flight control operation. A rather less successful outcome of differential thrust control occurred with a DC-10 that had lost all flight control operation (Tucker, 1999).

The difficulty of throttle-only control for emergency control of airplanes with failed control systems led NASA to authorize a research program for propulsion-controlled aircraft that would lead to workable systems. The key concept was the use of stability-augmentation techniques that would overcome the thrust lag problem, without requiring unusual pilot skills. The research was authorized by the then director of NASA's Dryden Flight Research Center, Kenneth J. Szalai, and took place starting in 1990 (Burken and Burcham, 1997).

The NASA PCA program was carried far enough to prove in flight testing that in the absence of surface hardovers or large mistrim conditions, a three-engine commercial jet, the MD-11, could be returned to an airport and landed without the aid of aerodynamic surfaces. The tail engine was used in the tests, although the PCA system is designed primarily for airplanes with two wing engines. The MD-11 engines were modified for PCA operation with full-authority digital controls and special idle settings to avoid large time lags in response to thrust change commands. The longitudinal and lateral PCA control laws are illustrated in Figure 20.7. In conclusion, the NASA PCA effort has provided a viable option, with

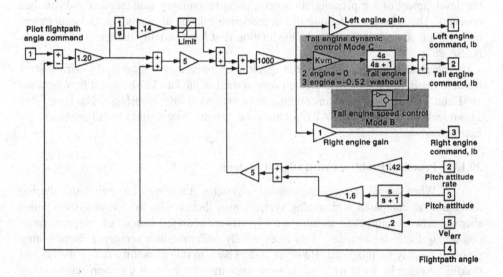

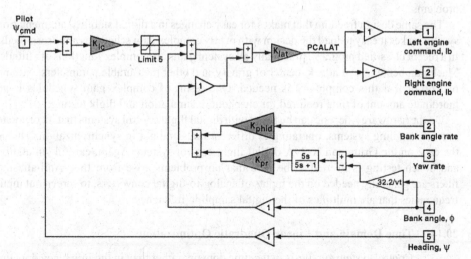

Figure 20.7 Longitudinal (*above*) and lateral (*below*) control laws used in the NASA Propulsion-Controlled Aircraft (PCA) demonstration using a MD-11 jet. (From Burken and Burcham, 1997).

moderate hardware and software costs, for transport designers to consider in the quest for safety.

20.12   The Advent of Digital Stability Augmentation

Airplane digital fly-by-wire flight control systems, which make possible digital stability augmentation, go back to the 1970s. Priority is difficult to establish, since many organizations were doing this work at about the same time. One early application was at the NASA Dryden Flight Research Center, using digital flight hardware from the Apollo program. Although overdesigned in many ways for the airplane application, it made possible an early demonstration of the possibilities of airplane digital augmentation.

That program used a Vought F-8C airplane (Jarvis, 1975). The first step was to fly single-channel digital flight control systems on the F-8C, with backup analog controls in case of failure. The next step was a big one from the standpoint of system complexity: the development of a triplex digital system, using redundancy management and data bus concepts. The subsequent routine use in modern airplanes of redundant, fail-operational digital flight control and stability augmentation is at least partially the result of this early NASA effort.

Another early application was the quadruplex redundant digital fly-by-wire system flown in the BAe FBW Jaguar. Design commenced in the late 1970s, and it flew between 1981 and 1984 in configurations ranging from normal to highly unstable. The BAe FBW Jaguar technology led to the EAP (Experimental Aircraft Programme) and ultimately to the Eurofighter.

20.13   Practical Problems with Digital Systems

When digital stability-augmentation systems first appeared, their most alluring advantage, as compared with analog systems, was their ability to change system gains, shaping networks and even architecture by software changes, instead of requiring time-consuming hardware changes. This is especially attractive in a prototype flight testing program, as may be imagined. However, a drawback to this capability is that the ease of making changes by software modifications encourages a cut and try approach to fixing problems.

The same design freedom that makes for easy changes in a digital stability-augmentation system makes it easy to load the design with overly complex gain schedules and cross-feeds. In a recent classified program, practically all system gains are complex functions of altitude, Mach number, angle of attack, center of gravity, and other measurable parameters, with no real proof that this complexity is needed. One result of complex gain schedules is an inordinate amount of time required for checkout in simulation and flight testing.

On the hardware side, one can be faced with digital flight control systems that incorporate several sampling systems, operating at different rates and not in synchronization. This is the case on the Grumman X-29A digital flight control system. Again, careful simulation and bench testing is needed to be sure that no problems arise from this. Anti-aliasing filters are generally needed on the inputs of analog-to-digital converters, to screen out input frequencies that are multiples of the digital sampling frequency.

20.14   Time Domain and Linear Quadratic Optimization

Control system synthesis in the time domain, rather than in the frequency domain, is often called modern control theory. Optimal controller design is generally involved.

Although one usually thinks of modern control theory in connection with full automatic control, it is applied as well to the design of stability-augmentation systems.

Linear quadratic (LQ) optimization methods have been used for a number of stability-augmentation system designs. These methods have their origins in the work of R. E. Kalman. Airframe and controller equations are cast in the state matrix form discussed in the previous chapter. The optimal controller is a linear feedback law that minimizes an integral cost function J of the form

$$J = \int [x^{\mathrm{T}} Q x + \delta^{\mathrm{T}} R \delta] dt,$$

where x is the system state vector, δ is the control vector, and Q and R are weighting matrices that express the designer's ideas on what constitutes optimal behavior for this case. The optimal control law takes the form of a linear set of feedback gains $\delta = Cx$, where C is the gain matrix of constants. The gain matrix C is computed by a matrix equation called the Riccati equation.

The linear quadratic approach to controller design is attractive because it is an organized method for finding feedback gains. The method produces an optimal set of feedbacks, but only for the arbitrarily chosen weighting matrix values. One can argue that if the weighting matrix values are poorly chosen, the resulting system can be far from ideal. In fact, it is not uncommon for designers using linear quadratic methods to tinker with weighting matrix values until a reasonable-looking system emerges. This puts the optimal design method on all fours with ordinary cut-and-try methods.

The problem of assigning weighting matrix values aside, there have been numerous variants of the linear quadratic approach to controller design and any number of applications in the literature and in practice. A typical application is to the design of a lateral-directional command augmentation system (Atzhorn and Stengel, 1984). The criterion function includes control system rate as a means of limiting high-frequency or rapid control motions. Displacement and rate saturation are significant nonlinearities that cannot be treated with the linear quadratic approach, except by the use of describing functions (Hanson and Stengel, 1984). Other linear quadratic stability-augmentation designs that may be found in the literature include departure-resistant controls, superaugmented (unstable airplane) pitch controls, and multiloop roll–yaw augmentation.

According to Robert Clarke and his associates at the NASA Dryden Flight Research Center, the Grumman X-29A research airplane's flight controls were originally designed using an optimal model-following technique. Simplified computer, actuator, and sensor models were used in the original analysis, leading to an unconservative design. A classical approach was chosen in the end, with lags introduced by the actual hardware compensated for by the addition of lead-lag filters (Clarke, Burken, Bosworth, and Bauer, 1994).

Another interesting linear quadratic stability augmentation design adds feed-forward compensation for nonlinear terms that cannot be included in the linearized design. This is the stability-augmentation system for the Rockwell/Deutsche Aerospace X-31 research airplane. Feed-forward compensation is added for nonlinear engine gyroscopic and inertial coupling effects (Beh and Hofinger, 1994).

20.15 Linear Quadratic Gaussian Controllers

Linear quadratic Gaussian (LQG) controllers add to the linear quadratic (LQ) designs random disturbances and measurement errors. LQG designs are discussed at length in a 1986 text and a 1993 IEEE paper by Professor Robert F. Stengel. The form taken by

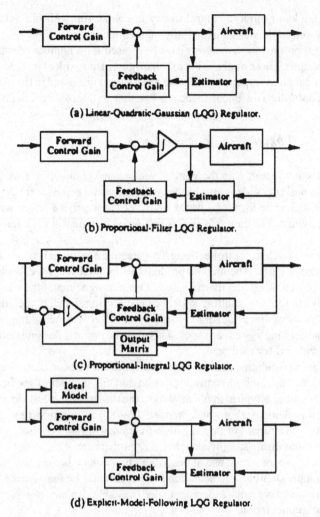

Figure 20.8 Various control system forms that can be represented with the structured linear quadratic regulator (LQG) method. (From Stengel, *IEEE Trans. on Systems, Man, and Cybernetics*, © 1993 IEEE)

the discrete-time LQG optimal controller is

$$u_k = C_F y_k^* - C_B x_k,$$

where y_k^* is the desired value of an output vector and x_k is the Kalman filter state estimate.

The LQG design approach is very flexible because of the number of parameters that can be chosen arbitrarily. At one extreme, a scalar one-input, one-output design can be produced. Measurement and control redundancies can be represented if measurement and control vector sizes exceed that of the state vector. Also, integral compensation and explicit model-following structures can be produced (Figure 20.8).

LQG designs are among the most advanced to be in use by stability-augmentation engineers, as this is written. Even more advanced control concepts continue to pour out of university and other research centers. The same 1993 paper by Stengel cited above provides a good survey of advanced control concepts, including expert systems, neural networks, and intentionally nonlinear controls.

20.16   Failed Applications of Optimal Control

The failure of optimal control methods to produce a satisfactory flight control system for the Grumman X-29A airplane was noted in Sec. 14. This failure is by no means an isolated event. Additional instances can be found in which optimal control methods in the hands of experienced engineers have failed to produce safe and satisfactory flight control systems. What has gone wrong? Several experts who have witnessed these failures discuss the problem:

Phillip R. Chandler and David W. Potts (1983), U.S. Air Force Flight Dynamics Laboratory "[T]he infinite bandwidth constant compensation elements which are required [for LQR] violate the very heart of the feedback problem.... LQR therefore is an elegant mathematical solution to a nonengineering problem.... SVT (Singular Value Theory) [Doyle, 1979] is a very crude method of coping with uncertainty in the LQR or LQG procedure. It makes assumptions that are not valid for flight control.... LQR with all its ramifications and refinements is totally unsuited for the flight control servomechanism problem."

John C. Gibson (2000), formerly with English/Electric/British Aerospace "[Robert J.] Woodcock told me that there have been several missile and aircraft projects in serious trouble due to the use of such [LQG] methods.... While optimization methods are continually being improved, they cannot yet (and may never) guarantee a safe and satisfactory FCS [flight control system] design without the strictest guidance and detailed physical understanding of experienced control and handling qualities engineers. This is true for highly advanced and demanding types of aircraft. Every signal path must be clearly visible and easily related to specific aerodynamic or inertial characteristics of the airframe. In simple aircraft without complexity, there is no advantage over straightforward engineering methods anyway."

Michael V. Cook (1999, 2000), Senior Lecturer, Cranfield University "There exists an enormous wealth of published material describing the application of so-called, 'modern control methods' to the design of flight control systems for piloted aeroplanes. It is also evident, with the exception of a very small number of recent applications, that there is a conspicuous lack of enthusiasm on the part of the airframe manufacturers to adopt this design technology, especially for the design of command and stability augmentation systems for piloted airplanes. Having an industrial background I am well aware of the many reasons why modern control has not been taken onboard seriously by the manufacturers – academic control specialists don't share my view, and in many cases probably don't even understand it! ... I know that my views are shared by the control people in —— who, in private are not at all complimentary about the academic control specialists in the UK. I am also aware that the Boeing view is similar to that of ——. I've seen some appallingly bad control systems design theses (not from Cranfield)."

Steven Osder (2000), Osder Associates, Arizona "We [Osder and Dunstan Graham] used to lament the absurdity of papers [on robustness theory] that were filling the journals and we amused each other by citing specific examples of such departures from reason and logic.... At the [Boeing] helicopter company, we took each of those University of – [robust flight control] designs and tested them against more complete [nonlinear] models of the [Apache] aircraft.

In every case, these robust flight control designs always fell out of the sky. In one case [which used eigenstructure assignment], even testing against a linear model, but with only a 10 percent variation in a single B [control] matrix term, our simulations resulted in a crash."

Duane T. McRuer (2001), Chairman, Systems Technology, Inc. "At STI we have spent an enormous amount of time and effort searching for ways to make optimal control practical – at least 20 major reports and papers, with some tremendously capable folk (e.g., Dick Whitbeck, Greg Hofmann, Bob Stapleford, Peter Thompson, et al.). Our focus has been on finding performance indices, special schemes, etc., to make optimal control solutions jibe with good design practice. . . . We have just never been happy with the results for stability augmentation design."

In the light of the foregoing comments, a design case (Ward, 1996) in which an LQG design for a pitch stability augmentation system was used only as a guideline for a more conventional approach suggests a reasonable use for optimal control techniques. The concept of using LQR optimal control synthesis as a guide or in conjunction with classical methods is also developed by Blight (1996). Blight also comments that LQR methods should be used only on "control problems that actually require modern multivariable methods for their solution." For example, Blight recommends ordinary gain scheduling instead of attempting to design a single robust linear control law for all flight conditions.

20.17   Robust Controllers, Adaptive Systems

Robust flight control systems are designed specifically to perform well in the face of airframe, sensor, and actuator uncertainties and even failures. An early robust flight control system approach was the adaptive control system, a particular research objective of the Honeywell Corporation. This was in the days before airborne digital computers. The modest objective was to identify the airplane's pitch natural frequency by periodic injection of small test pulses of elevator control. Pitch natural frequency variations reflected changes in both center of gravity location and dynamic pressure, or calibrated airspeed. Control system gain was lowered at the higher pitch natural frequencies to maintain system stability.

Modern applications of adaptive control make use of parameter identification, although test signals are still required to keep the parameter identification loop from going unstable. In a 1982 NASA workshop on restructurable controls, reasonably good results were reported for two adaptive schemes (Cunningham, in Montoya, 1983). Horizontal tail effectiveness M_δ was identified on a Vought F-8 sufficiently well for autopilot gain scheduling through the flight envelope. Also, the flutter modes of a wind-tunnel model of a wing with stores (weapons) were identified by maximum likelihood methods.

The same NASA workshop brought a theoretical criticism of all adaptive systems by MIT professor Michael Athans. In his words:

> Over two thousand papers have been written [on adaptive control] and a lot of excitement generated. You may have seen that people are giving courses to industry on how to make adaptive control practical. We have a recent MIT Ph.D. thesis [Rohrs, 1982] finished in November 1982 that Dr. Valvani and I supervised, which proved with a combination of analytical techniques and simulation results that all existing adaptive control algorithms are not worthwhile.
>
> The algorithms may look excellent if you follow their theoretical assumptions, but in the presence of some persistent output disturbances and unmodeled high frequency dynamics all adaptive control algorithms considered become unstable with probability one.

Aside from coping with center-of-gravity and flight condition changes, robustness in control systems already exists in augmentation systems incorporating self-checking redundant digital computers. Robustness against sensor failures has also been demonstrated with redundant inertial sensors in skewed orientations. Failure of one or two sensors leaves the system fully operational. Failure of a single airspeed meter due to icing resulted in the losses of a General Dynamics B-58 Hustler and of an US/German X-31A research airplane. The automatic pilot gain-changing features interpreted the iced meter readings as low airspeed, requiring higher gains (communication from Dr. Peter Hamel).

Robustness against actuator failures, and especially against failures that result in control surfaces that go hard over against a stop and stay there, is another matter. The stirring example of Delta Airlines' pilot McMahan who saved a Lockheed 1011 with one elevator against the up stop is told in Chapter 5, "Managing Control Forces." System concepts for reconfiguring control systems to cope automatically with major failures are still in the early stages.

While waiting for the development of systems that are robust in the face of actuator hardovers, Thomas Cunningham suggests two straightforward aids for the human pilot. The position of each individual control surface should be measured and displayed in the cockpit. Captain McMahan did not know that the 1011 elevator was against its stop. Also, engine controllers should be designed to the higher bandwidths needed for differential thrust control of a crippled airplane.

20.18 Robust Controllers, Singular Value Analysis

The analysis of robust controllers took a different tack from adaptive controls with the work of J. C. Doyle and his associates, starting around 1980. The key to the new approach is a generalization of system gain using the singular values of a matrix. Matrix singular values are another term for the matrix norm, defined as the square root of the sum of the squares of the absolute values of the elements. The matrix norm is the trace of A^*A, where A is the given matrix and A^* is the Hermitian conjugate of A (or the transpose if A is real).

According to the singular value approach, control system robustness against uncertainties in mechanical and aerodynamic properties is assured if the amplitude of the maximum expected uncertainty is less than the minimum system gain at all frequencies.

A simpler, but equally important application of singular value analysis is to system stability margins, without considering uncertainties. Stability margins are guaranteed if the minimum singular values of the system's return difference matrix are all positive (Mukhopadhyay and Newsom, 1984). The system return difference matrix $I + G$ is a matrix generalization of the closed-loop transfer function denominator for a single-input single-output system. This stability margin application of singular value analysis was made for the X-29A research airplane (Clarke et al., 1994).

20.19 Decoupled Controls

Airplane stability augmentation must be rethought when designers choose to add direct normal and side force control surfaces. For example, with direct lift control through a fast-acting wing flap, pitch attitude can be controlled independently of the airplane's flight path, and vice versa. The utility of such decoupled controls for tracking, defensive maneuvers, and for landing approaches is reviewed by David J. Moorhouse (1993).

20.20   Integrated Thrust Modulation and Vectoring

An airplane's propulsion system can be integrated into a stability augmentation system that uses aerodynamic control surfaces. The total system would operate while the airplane remains under the control of the human pilot, qualifying as a stability-augmentation system rather than as an automatic flight control system.

For comparison, the previous coverage of propulsion systems in this book included:

Chapter 4   the effects of conventional, or fixed-configuration, propeller-, jet-, and rocket-propulsion systems on stability and control;

Chapter 10, Sec. 8   thrust vector control to augment aerodynamic surfaces in supermaneuvering;

Chapter 11, Secs. 14 and 15   propulsion effects on modes of motion and at hypersonic speeds;

Chapter 12, Sec. 1   carrier approach power compensation systems, for constant angle of attack approaches;

Chapter 20, Sec. 11   Propulsion-controlled aircraft, designed to be able to return for landing after complete failure of normal (aerodynamically implemented) control systems.

Depending on the number of engines under control, thrust modulation and vectoring systems can supply yawing, pitching, and rolling moments, as well as modulated direct forces along all three axes. Thus, thrust modulation and vectoring integrated into a stability-augmentation system can augment or replace the aerodynamic yawing, pitching, and rolling moments provided by aerodynamic surfaces. The situation is similar to aircraft like the Space Shuttle Orbiter, which carries both aerodynamic and thruster controls. However, in the context of stability augmentation, thrust modulation and vectoring would be used normally at the low airspeeds of approach and landing, rather than in space.

While in principle thrust modulation and vectoring can take the place of aerodynamic control surfaces at the low airspeed where the aerodynamic surfaces are least effective, it is reasonable to ask whether thrust stability-augmentation systems could satisfy flying qualities requirements. In a simulation program at DERA, Bedford (Steer, 2000), integrated thrust vector control was evaluated at low airspeeds on the baseline European Supersonic Commercial Transport (ESCT) design. The nozzles of all four wing-mounted jet engines were given both independent pitch and yaw deflections, providing yawing, pitching and rolling moments. Nozzle deflections were modeled as first-order lags. Conventional pitch rate, pitch attitude, velocity vector roll rate and sideslip command control structures were programmed.

Pitch control by thrust vectoring at approach airspeeds was as good as aerodynamic or elevon control, for a reason peculiar to the very low wing-aspect-ratio ESCT configuration. That is, the airplane has high induced drag at approach angles of attack, requiring large levels of thrust to maintain the glide path, thus making available large pitching moments with thrust deflection. Low airspeed roll and sideslip thrust vector control were positive and suitably damped but did not satisfy MIL-STD-1797A criteria.

20.21   Concluding Remarks

Airplane stability augmentation was born with the B-47 and B-49 systems around 1947. Available analysis methods were at hand in the frequency-based Nyquist and Bode methods developed by electrical engineers. More advanced time-domain analytical concepts appeared about the time an airborne digital flight control computer flew in an F-8C at the NASA Dryden Flight Research Center.

Today's stability and control designer has a remarkably wide choice of modern flight control theoretical concepts to select from and mature digital control hardware to correspond. The expansion of new modern flight control theory shows no signs of tapering off. As always, a moderating factor in selecting any advanced concept is the cost of a thorough validation on an actual project. A recurring theme among design engineers is failure of modern control theory to produce practical flight control system designs. M. V. Cook notes that modern control methods may have more relevance to pilotless aircraft because their performance is more easily defined in purely mathematical terms.

The likelihood is that advanced flight control system concepts will not be adopted in inexpensive personal aircraft, regardless of cost, simply because possible gains in performance and safety will not be there. The same is likely in light commuter transports. However, performance gains with relaxed static stability or superaugmentation become significant for long-range subsonic transport airplanes. Robustness in their flight control systems should help bring about this application and the opportunity for controls designed by optimal, singular-value, neural network, or advanced methods not yet imagined. The case for relaxed or even negative static stability is even stronger for supersonic cruise transports. Two limiting factors are the required pitch control moments to recover from upsets and the interaction with flexibility modes of the high-bandwidth actuators needed to cope with unstable airplanes.

Finally, military airplanes should be where advanced techniques, such as departure resistance, decoupling, real-time vehicle system identification, active flexible wings, integrated thrust vectoring and modulation, and self-repairing controls, will get serious consideration and possible application.

CHAPTER 21

Flying Qualities Research Moves with the Times

Robert R. Gilruth's key flying qualities contribution was to test a significant sample of airplanes for some flight characteristic such as lateral control power and then to separate the satisfactory and unsatisfactory cases by some parameter that could be calculated in an airplane's preliminary design stage. The Gilruth method put design for flying qualities on a rational basis, although Chapter 3 tells of some later backsliding, attempts to specify flying qualities parameters arbitrarily.

Modern times have brought the $100 million and more airplane and development costs for new prototypes into the billions of dollars. This has made for a scarcity of new machines that can be tested in the Gilruth manner and an interest in alternate flying qualities methods. The pilot-in-the-loop method surfaced around 1960 as an alternate way of rationalizing flying qualities and to focus attention on the pilot–aircraft combination as a closed-loop system. Pilot-in-the-loop analysis involves adoption of mathematical models for the human pilot as just another control system element.

The three basic concepts of the pilot-in-the-loop analysis method are (McRuer, 1973):

1. To accomplish guidance and control functions the human pilot sets up a variety of closed loops around the airplane, which, by itself, could not otherwise accomplish such tasks.
2. To be satisfactory, these closed loops must behave in a suitable fashion. As the adaptive means to accomplish this end, the pilot must make up for any dynamic deficiencies by adjustments of his own dynamic properties.
3. There is a cost to this pilot adjustment: in workload stress, in concentration of the pilot's faculties, and in reduced potential for coping with the unexpected. The measure of the cost are pilot commentary and pilot rating, as well as physical and psychological measures.

In this chapter we trace the development of pilot-in-the-loop analysis methods as they apply to airplane flying qualities. Pilot-in-the-loop methods are clearly essential to study closed-loop operations such as tracking, but can they replace or add to the classical Gilruth approach?

21.1 Empirical Approaches to Pilot-Induced Oscillations

Figure 21.1 is a time history of the pilot-induced oscillation that occurred during landing of the Space Shuttle Orbiter Enterprise in 1977. Pilot-induced oscillations (PIO), or airplane–pilot coupling (APC) incidents, in which pilot attempts at control create instability, are a natural subject for pilot-in-the-loop analysis and a major motivating factor for the method's development. However, pilot-induced oscillations appeared long before advanced pilot-in-the-loop methods were in place. Engineers were obliged to improvise solutions empirically, so that airplane programs could proceed.

One cause of pilot-induced oscillations was apparent without much deep study. If control surface rate of movement is restricted for any reason, such as insufficient hydraulic

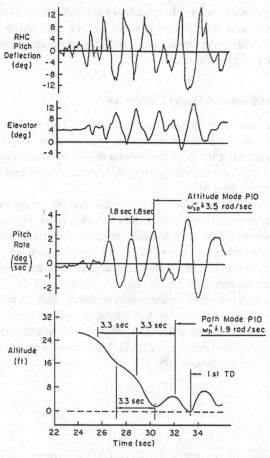

Figure 21.1 Time history of pilot-induced oscillations that occurred during landing of the space shuttle Enterprise, on October 26, 1977. Time lags in the longitudinal control system are considered to have been the primary cause. (From Ashkenas, Hoh, and Teper, AIAA Paper 82-1607-CP, 1982)

fluid flow rate into actuation cylinders, the pilot is unable to reverse control motion quickly enough to stop an airplane motion, once started. A late correction drives the airplane too far in the reverse direction, requiring ever-increasing control motions. Describing function analysis of rate limiting does indeed show destabilizing phase lag. Thus, one empirical design rule for pilot-induced oscillation avoidance is high available control surface rates.

In unpublished correspondence W. H. Phillips comments on other empirical findings on pilot-induced oscillations:

> We found that very light control forces together with sensitive control were very likely to lead to pilot-induced oscillations. Viscous damping on the control stick was not the answer as this put lag in the response to control force as well as the recovery. What was needed was a large force in phase with deflection for rapid stick movements, which could be allowed to wash out quite rapidly. This could be obtained with a spring and dashpot in series. Grumman called this a "sprashpot" and used it successfully in the feel system of the F-11F.... The negative C_{h_α} of flap-type controls causes the control force to fall off after the airplane responds.

An additional empirical approach to solving longitudinal pilot-induced oscillation problems is the double bobweight system described in Chapter 5. An aft bobweight provides heavy stick forces to start a pitch maneuver, by applying pitching acceleration forces to the stick. Stick force falls off as the airplane responds.

21.2   Compensatory Operation and Model Categories

Pilot-in-the-loop analysis methods have had their earliest and most meaningful successes representing compensatory operation. As applied to pilot-in-the-loop operation, in compensatory operation or tracking the pilot operates on displayed or perceived errors to minimize them in a closed-loop fashion. Precognitive pilot operation is essentially open-loop; the pilot is not part of the tracking loop.

Mathematical models for pilot compensatory operation fall into two categories, structural and algorithmic. Structural models reduce the pilot to subsystems such as muscle manipulators and vestibular sensors, each with transfer functions. Structural model pilot transfer functions contain delays, leads, and lags. The overall assemblage must reproduce pilot behavior in an end-to-end fashion. This challenging approach is made possible by careful frequency response measurements on human subjects (McRuer, 1973).

Pilot algorithmic models have grown out of modern optimal control theory. These models include an estimator, such as a Kalman filter, which processes the pilot's observations to provide an estimate of the airplane's state, and a controller, which is a mathematical model for the pilot's regulation and muscular functions (Figure 21.2). Minimization or maximization of a criterion function provides the required results.

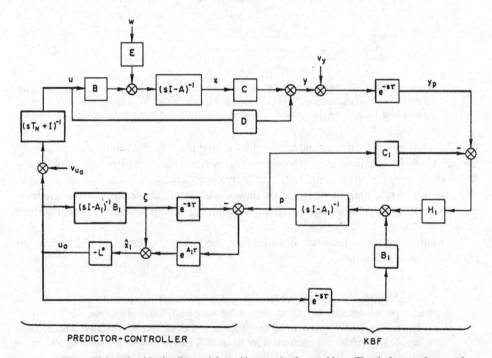

Figure 21.2 Algorithmic pilot model, used in an optimal control loop. The airplane or plant matrices A, B, C, and D, including a noise-shaping filter E, are at the upper left. The airplane's state is estimated by the Kalman-Bucy filter at the lower right. The optimal controller is on the lower left. (From Thompson, AIAA Paper 88-4183, 1988)

It is important to recognize that delay-lead-lag pilot models are needed primarily in analysis of compensatory operation of inner, generally attitude loops. Such loops are closed at high frequencies relative to pilot dynamics. Pure gain models for the pilot are generally adequate for analysis of turn coordination and lower frequency speed and path control loops.

21.3   Crossover Model

The crossover model of compensatory operation grows out of the observation that pilots develop the necessary dynamics to produce in the pilot–airplane combination a particular transfer function in the crossover region of frequencies (McRuer, 1988). The pilot–airplane open-loop transfer function developed has the remarkably simple form of an integrated time delay, or $\omega_c (e^{-\tau s})/s$, where the open-loop gain is ω_c, τ is the delay, and s is the Laplace operator. The open-loop gain ω_c is called the *crossover frequency*, the frequency at which the open-loop amplitude response crosses the 1.0- or 0-db line.

The closed-loop frequency response, or ratio of output to input for the crossover model, is flat at 1.0 at low frequencies, meaning that the output follows exactly the input. As input frequency is raised, the frequency at which the output drops 3 db lower than the input, or to only 70 percent of the input, is considered a cutoff for all practical purposes. This frequency defines the closed-loop system bandwidth. For the crossover model, the frequency that defines closed-loop system bandwidth is also the frequency ω_c for which the open-loop system has a gain of 1.0.

The crossover model time delay τ is actually a low-frequency approximation, valid at crossover frequencies, for numerous pilot and control system delays and higher order lag terms. That part of τ due to the pilot becomes greater as the lead contributed by the pilot increases, a cost of additional pilot effort (McRuer, 1988). This reduces the available crossover frequency for other system lags.

21.4   Pilot Equalization with the Crossover Model

All airplane transfer functions, such as the pitch response to elevator and the roll response to aileron, have first- or second-order denominator functions, arising from mass or inertia. To satisfy the crossover model the pilot must supply a canceling numerator function over the same frequency range. This amounts to lead or anticipation, agreeing with common sense as to what is required for the error elimination in compensatory operation.

The amount of lead or compensation required by the pilot is a direct measure of workload. The pilot lead is reflected in the positive slope of the pilot model amplitude ratio in the Bode diagram, in the vicinity of the crossover frequency. A large positive slope corresponds to excessive lead, high workload, and poor pilot rating. A numerical connection can be made between pilot rating by the Cooper-Harper scale, discussed in Chapter 3, and required lead equalization (Figure 21.3).

21.5   Algorithmic (Linear Optimal Control) Model

The algorithmic or linear optimal control model is partially a structural pilot model in that elements of the optimal controller can be identified with the neuromuscular lag. However, the basic distinction between the algorithmic and structural pilot models is that, except for simple problems, the pilot cannot be represented with a simple transfer function in the algorithmic case. When very simple airplane dynamics (a pure integrator) are postulated in order to be able to generate a pilot transfer function, the linear optimal control pilot

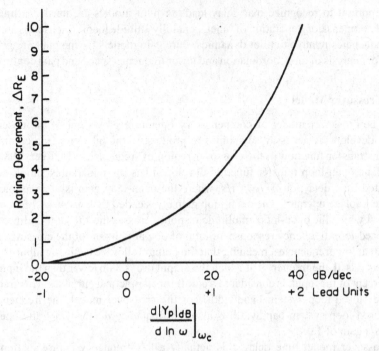

Figure 21.3 Degradations (increases) in pilot rating for tracking tasks associated with degree of pilot lead required. (From McRuer, AGARDograph 188, 1974)

model is found to be of high order, but with characteristics similar to the crossover model (Thompson and McRuer, 1988).

The linear optimal pilot model has been used to advantage in the generation of pilot ratings (Hess, 1976; Anderson and Schmidt, 1987), the analysis of multiaxis problems (McRuer and Schmidt, 1990), and the stability of the pilot–airplane combination in maneuvers (Stengel and Broussard, 1978).

21.6   The Crossover Model and Pilot-Induced Oscillations

The crossover model has proved to be of great value in understanding pilot-induced oscillations. The way has been opened for validating empirical corrections for the phenomenon, such as described by Phillips, and for the development of new concepts in the area and superior flying qualities designs.

Duane McRuer provides a comprehensive survey of pilot-induced oscillations in a report for the Dryden Flight Research Center (McRuer, 1994). Having been experienced by the Wright brothers, pilot-induced oscillations qualify as the senior flying qualities problem. Recent dramatic flight experiences, combined with the availability of advanced analysis methods, have given the subject fresh interest. Between the years 1947 and 1994, there were over 30 very severe reported cases, in airplanes ranging from a NASA paraglider to the space shuttle Orbiter. McRuer proposes three pilot-induced oscillation categories, as follows:

> essentially linear;
> quasi-linear, with surface rate or position limiting;
> essentially nonlinear, including pilot or mode transitions.

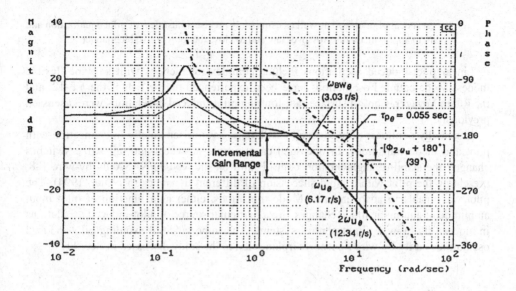

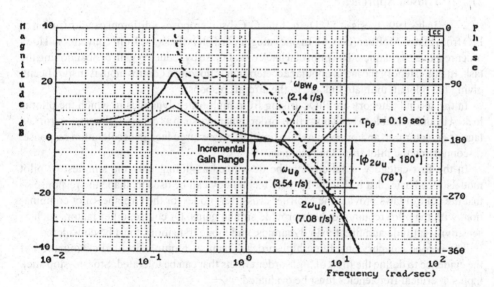

Figure 21.4 Pilot–airplane open-loop frequency responses for two configurations of the USAF/Calspan variable-stability T-33. The upper case, with no pilot-induced oscillations, has the ideal integrator shape in the vicinity of crossover. The lower case, with severe pilot-induced oscillations, has a steeper slope and more phase lag at high frequencies. (From McRuer, STI Technical Rept. 2494-1, 1994)

An important validation of the crossover model approach to the first category was furnished by analysis of fully developed pilot-induced oscillations on the USAF/Calspan variable-stability NT-33 (Bjorkman, 1986). In six severe cases there were large effective open-loop system delays, departing from the ideal integrator-type airframe transfer function in the region of crossover (Figure 21.4). The required pilot dynamics for compensatory operation thus required a great deal of pilot lead as well as exquisitely precise adjustment of pilot equalization and gain to approximate the crossover law and to close the loop in a stable manner.

Linear pilot-induced oscillations include complex interactions with airplane flexible modes. Mode-coupled oscillations have been experienced on the F-111, the YF-12, and the Rutan Voyager. Control surface rate-limiting pilot-induced oscillations were discussed previously.

Essentially nonlinear pilot-induced oscillations have arisen chiefly in connection with pilot and mode transitions. In one such case, weight-on-wheel and tail strike switches changed the stability augmentation control laws on the Vought/NASA fly-by-wire F-8, presenting the pilot with a rapid succession of different dynamics (McRuer, 1994). The pilot was unable to adapt in time. Mode transitions, either as a function of pilot input amplitude or automatic mode changes, are a particular source of pilot-induced oscillations in modern fly-by-wire flight control systems. The importance of avoiding pilot-induced oscillations on fly-by-wire transport airplanes led to the study discussed in Sec. 11.

21.7   Gibson Approach

In his 1999 thesis at TU Delft, John C. Gibson proposes a different categorization of PIO from that of McRuer (Sec. 6). In one category are PIOs that arise from conventional low-order response dynamics. The pilot can back out of these by reducing gain or abandoning the task. In this category the lag in angular acceleration following a control input is insignificant, giving the pilot an intimate linkage to the aircraft response.

In the second category are PIOs arising from high-order dynamics in which the pilot is locked in and is unable to back out. High-order dynamics such as excessive linear control law lags or actuator rate and/or acceleration limiting create large lags in acceleration response, disconnecting the pilot from the response.

In the first category, solutions can be developed assuming only the simplest of pilot models. The basic idea is that fly-by-wire technology can be used to shape the response so that the control laws provide the McRuer crossover model for the airplane–pilot combination, with the pilot required only to provide simple gains. Of course, other factors such as sensitivity, attitude and flight path dynamics, and mode transitions must be considered.

The second category, involving high-order dynamics, requires detailed examination of the evidence to define the limit of high-order effects that can be tolerated. Stop-to-stop stick inputs at critical frequencies must be evaluated.

21.8   Neal-Smith Approach

The connection between excessive lead requirements for control and poor pilot ratings is the basis for the Neal-Smith criterion, dating from 1970. A lead-lag pilot model is assumed, with a fixed time delay of 0.3 second. When this pilot model is combined with the dynamics of the airplane, the model parameters can be adjusted to meet bandwidth and other closed-loop requirements. The resultant pilot model phase lead and closed-loop resonance are compared with pilot opinions to establish acceptable boundaries (Figure 21.5).

The Neal-Smith approach is an important contribution to the rationalization of flying qualities requirements since it makes direct use of the mathematical pilot model. The method has shortcomings in that the required pilot lead is strongly dependent on the required bandwidth, an arbitrary starting point (Moorhouse, 1982).

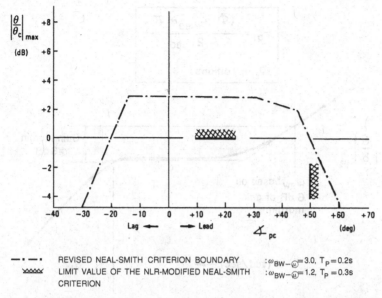

Figure 21.5 Neal-Smith criterion for pitch control. Acceptable short-period behavior occurs below the boundary established by closed-loop peak resonance ratio, the abscissa, and pilot model lead, the ordinate. The hatched boundaries are more restrictive limits proposed for large transports. (From Mooij, AGARD LS-157, 1988)

21.9 Bandwidth-Phase Delay Criteria

The insights furnished by the crossover model for compensatory operation lead to criteria that can be used in control system design, as in the Neal-Smith approach. An important example is the Hoh-Mitchell-Ashkenas bandwidth and phase delay criteria (Hoh, 1988), a combination of two individual metrics, illustrated in Figure 21.6.

The first metric is *aircraft* bandwidth, defined as the frequency at which the phase angle of attitude response to stick force input is −135 degrees. The aircraft bandwidth measures the frequency over which the pilot can control without the need for lead compensation. The second metric is phase delay, defined as the difference in response phase angle at twice the frequency for a −180-degree phase angle and 180 degrees, divided by twice the frequency for a −180-degree phase angle. The phase delay metric approximates the phase characteristics of the effective airplane dynamics, from the region of crossover to that for potential pilot-induced oscillations. Systems with large phase delays are prone to such oscillations.

Boundaries in aircraft bandwidth–phase delay space have been developed using flight and simulator pilot ratings and commentary. Similar boundaries have been especially useful for rotorcraft and special (translatory) modes of control. With these boundaries, designers are able to account for closed-loop pilot–airplane dynamics, using effective airplane dynamics alone. A related airplane-alone criterion based on the crossover model is the Smith-Geddes (1979) criterion frequency. Still another criterion based on airplane-alone dynamics places boundaries in the Nichols plane of the attitude frequency response (Gibson, 1995). The idea is to confine the attitude frequency response within boundaries defined by the best piloted closed-loop flying qualities. All of these boundary methods depend on simple correlation. They should be effective to the extent that new cases resemble those on which the boundaries are based.

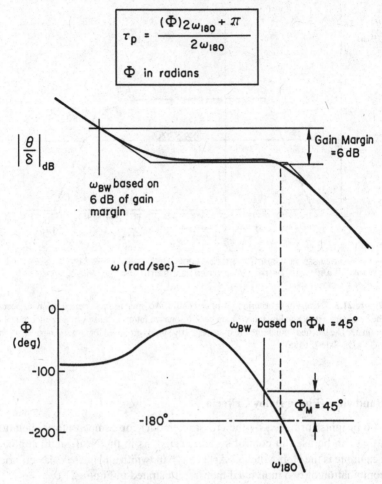

$$\tau_p = \frac{(\Phi)_{2\omega_{180}} + \pi}{2\omega_{180}}$$

Φ in radians

$\left|\frac{\theta}{\delta}\right|_{dB}$

Gain Margin =6 dB

ω_{BW} based on 6 dB of gain margin

ω (rad/sec) →

Φ (deg)

ω_{BW} based on $\Phi_M = 45°$

0

-100

$\Phi_M = 45°$

-180°

-200

ω_{180}

To Obtain Bandwidth:

1. Calculate ω_{BW} based on phase margin

2. Calculate ω_{BW} based on gain margin

3. Use lowest value

Figure 21.6 Definitions of the bandwidth and phase delay criteria. (From MIL-STD-1797A, 1990)

Good design practice suggests using all of these criteria to examine airplane dynamics at issue.

21.10   Landing Approach and Turn Studies

There is a large class of advanced flying qualities studies that has benefitted from modern pilot-in-the-loop technology without requiring delay-lead-lag mathematical models for the pilot. For loop closures made at frequencies of 1.0 radian per second or lower, simple gain models for the pilot appear adequate. Low-frequency closed loops characterize airspeed and path control outer loops.

STOL (short takeoff and landing) flight path control loops are the outer loops around higher frequency pitch attitude inner loops. The path loops can be closed with simple pilot gains, assuming tight control of the inner loop (Ashkenas, 1988). Where some modest pilot lead is required to achieve the ideal integrator-type pilot–airplane transfer function in the crossover region, pilot ratings will be degraded.

Wings-level turn and turn coordination studies fall into the same category of low-frequency closed loops for which simple gain pilot models are adequate. The Ashkenas-Durand reversal parameter and the Heffley closed-loop studies of the naval carrier approach problem (Chapter 12) are yet additional examples of the use of simple pilot gain models.

21.11   Implications for Modern Transport Airplanes

Historically, pilot-induced oscillations (PIO) associated with fly-by-wire technology have occurred in military and experimental aircraft, which usually introduce advanced technologies before they appear on civil transports. This has provided a breathing space for that category of PIO problems to be worked out before exposing the traveling public to new hazards. However, fly-by-wire technology is now standard for new transport airplanes, bringing the possibility of PIO.

A U.S. National Research Council (NRC) report (McRuer, 1997) is intended to alert all interested parties to this hazard and to offer recommendations to avert serious problems in the future. Aside from the evident need to continue research and pilot training in this area, a few striking conclusions and recommendations emerge from the NRC report:

1. Parameters measured by on-board flight recorders, the "black boxes," need to be at higher data rates, to capture PIO events that may have contributed to accidents. Dr. Irving Statler, who is involved in a major part of NASA's Aviation Safety Program, states that the highest data rate found in black box recorders is only 8 samples per second, as compared with the 20–30 samples per second needed to capture PIO events.
2. Highly demanding tasks with known and suspected triggering events for PIO should be included in simulation, flight test, and certification. These tests should use pilots with experience and training in PIO events.
3. Current certification procedures should be revised to incorporate existing techniques for mitigating the risk of PIO.

The warnings that were sounded by the NRC report of potentially dangerous PIOs in commercial aviation should be taken seriously. The recommendations of the experienced group that wrote the report should be put into action.

On the matter of recording PIO events that may have contributed to accidents, an ambitious approach is under study at the Aerospace Corporation and at RTCA (Grey, 2000). This is a satellite-based aircraft monitoring system and data archive that does away with the need for on-board flight recorders. The satellite-based system could provide real-time, high-data-rate information for accident prevention or diagnosis. Such a system is seen as a logical outgrowth of developments in the field of communications.

21.12   Concluding Remarks

Pilot-in-the-loop studies are partially in the realm of traditional scientific research and partially one of the technologies belonging to airplane stability and control. As pure research, one explores the interesting interactions of human beings and machines performing various flying tasks. With pilot and crew error said to be responsible for a large number of accidents, there is a strong motivation for support of this research enterprise.

To date, clear successes of pilot-in-the-loop studies as technology are the applications of the crossover and linear optimal models to explaining and avoiding a large group of pilot-induced oscillations. The crossover model has also been the foundation for metrics used in specifying flying qualities. Other accomplishments, such as better understanding of pilot loop closures in STOL and carrier approaches and in turn coordination, have relied on normal bandwidth-maximizing closed-loop design techniques, with the pilot represented as a simple gain. Pilot-in-the-loop methods have thus appeared as the next step following the classical flying qualities work, which emphasized control power and suitable control forces.

A potential for dangerous pilot–airplane interactions in advanced commercial transport airplanes has been uncovered. A well-considered plan to mitigate these problems should be put into action.

CHAPTER 22

Challenge of Stealth Aerodynamics

The invention of aircraft that are almost invisible to ground or surface-to-air-missile radars promises to be an effective defensive measure for reconnaissance and attack airplanes. This development has taken six paths so far, the first three of which are a distinct challenge to stability and control designers:

Faceted airframes   replace the smooth aerodynamic shapes that produce attached flows and linear aerodynamics. Radar returns from faceted shapes, such as the Lockheed F-117A, are absent except for the instants when a facet faces the radar transmitter.

Parallel-line planforms   have the same sweep angle on wing leading and trailing edges and on surface tips and sharp edges. Parallel-line planforms concentrate radar returns into narrow zones that are easily missed by search radars. This is the Northrop B-2 stealth method, augmented by special materials and buried engines.

Suppressed vertical tails   are either shielded from radar by wing structure or eliminated altogether. The Lockheed F-22 has shielded vertical tails, the B-2 none at all.

Blended aerodynamics   eliminate internal corners such as wing–fuselage intersections. Internal corners can act as radar corner reflectors. The Rockwell B-1 uses this technique to reduce its radar signature.

Buried engines and exhausts   hide compressor fan blades and hot exhaust pipes from radar and infrared seekers.

Radar-absorbent materials   are used, generally nonmetallic. This is a highly classified subject.

The challenges of faceted airframes, parallel-line planforms, and suppressed vertical tails to stability and control engineers are illustrated by current stealth airplanes.

22.1   Faceted Airframe Issues

The Lockheed F-117A's faceted airframe flies in the face of conventional aerodynamic wisdom, which requires smooth surfaces to maintain attached flow under the widest possible ranges of angles of attack, sideslip, and angular velocities (Figure 22.1). On the other hand, the aerodynamic forces and moments of faceted airframes are reasonably linear functions of these variables for sufficiently small ranges.

Large-wing sweepback, 67 1/2 degrees in the case of the F-117A, extends the linear ranges somewhat, making facet edges into side edges instead of breaks normal to the flow direction. Still, the stability and control engineer who is faced with a faceted airframe such as the F-117A must expect to restrict flight parameters in order to avoid nonlinear and unstable aerodynamic moments that exceed available control power. The F-117A was originally called "The Hopeless Diamond" by Lockheed aerodynamicists.

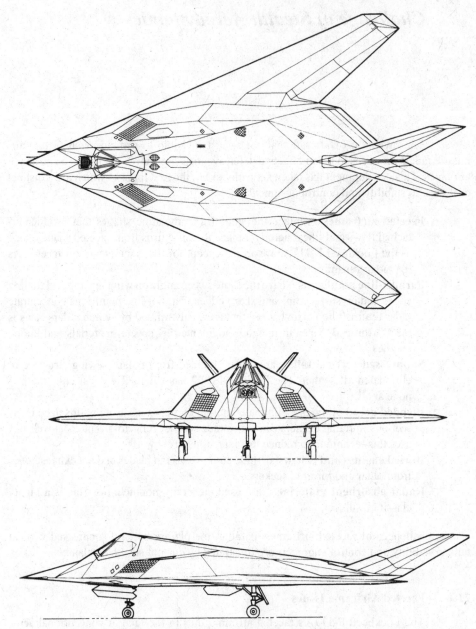

Figure 22.1   Faceted structure of the Lockheed F-117A Stealth Fighter. (From Lockheed Advanced Development Company, J. W. Ragsdale)

On the F-117A, the angle of attack is hard-limited, but sideslip angles are unlimited with the landing gear down for cross-wind landings. With landing gear up, the sideslip angle is nulled by closed-loop control, a normal loop closure. F-117A longitudinal static margins are low or negative within the angle-of-attack limit range, but air combat maneuvers can be made within that range. Severe pitchups and pitchdowns occur outside of the angle-of-attack limit range (Farley and Abrams, 1990). Without augmentation, the airplane is directionally unstable over large parts of its operational envelope.

The four F-117A elevons have relatively large travels of 60 degrees up and down, which are necessary to deal with nonlinear and unstable moments within the angle-of-attack limit range. The two vertical tails are all-moving, for the same reason. The F-117A has quadruple fail-safe fly-by-wire controls, using F-16 technology. An 18-foot-diameter braking parachute doubles as a spin chute, an unusual feature for a service airplane. Nominal landing speed is 160 knots, at an angle of attack of 9 1/2 degrees.

22.2 Parallel-Line Planform Issues

The Northrop B-2 planform strongly distorts from the ideal the additional span load distributions, or the span loadings due to symmetric angle of attack and to rolling. Each planform internal corner, marked "C" in Figure 22.2, produces a sharp local peak in the additional span loading, as do the triangle-shaped wing tips. Premature stalling can be expected in the vicinity of the corners at high angles of attack and at high rolling velocities.

The resultant nonlinearity in the variation of lift with angle of attack at high angles is not in itself a stability and control problem. However, the yawing and rolling moment due to rolling derivatives C_{n_p} and C_{l_p}, normally negative in sign, become positive at combined high angle-of-attack and rolling velocity values. This problem is countered in the B-2 with an angle-of-attack limiter and with artificial stability increments that are tailored to C_{n_p} and C_{l_p} values below the angle-of-attack limits.

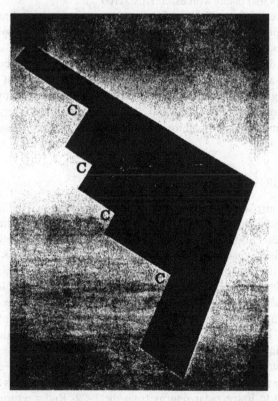

Figure 22.2   Internal sharp corners, marked "C" on the Northrop B-2 bomber planform. These corners, the result of the stealth parallel-line planform, produce sharp local peaks in the wing's additional span-load distribution. Premature stalling can be expected in the vicinity of the corners at high angles of attack and rolling velocities.

A second B-2 stability and control issue is the elimination of vertical tails, requiring split ailerons to supply yawing moments in response to pilot and autopilot inputs. The split ailerons act as differential drag devices. The brake drag, and hence yawing moment, is nonlinear with brake opening, requiring further tailoring to produce predictable control moments.

As with all isolated sweptback wings, the B-2 has an inherent low level of positive static directional stability. However, the sweptback hinge lines of the split ailerons result in directional instability whenever the brakes are opened symmetrically at large angles, as an airspeed-control device. The open brakes themselves act as low-aspect-ratio wings, with lift components that produce destabilizing yawing moments when the whole wing is yawed to the airstream. The speed brakes are opened at landing approach airspeeds, putting an additional requirement on the logic that provides artificial directional stability at low airspeeds.

In addition to a very small level of inherent static directional stability, the B-2's all-wing shape has next to no side force derivative in sideslip, or C_{y_β}. This creates a flight instrument problem, in that the normal ball-bank component of the turn and slip instrument cannot function as an indication of airplane sideslip. The standard ball-bank component is a lateral accelerometer, calibrated to produce one ball width at a tilt angle of 4.5 degrees, or a lateral acceleration of 0.08 g, in level flight. With virtually no side force developed in sideslip, there is no lateral acceleration to displace the ball. This instrumentation problem is aside from flight dynamics problems that occur with essentially zero C_{y_β}.

The Northrop B-2 was the very first parallel-planform configuration to be built. Engineers who were willing to talk about its development concede that there were some unpleasant stability and control surprises. Radar signature considerations probably rule out alleviation of distortions in span load by local wing twist for future applications. Correction for undesirable (positive in sign) values of C_{n_p} by artificial stability increments is of course available with sophisticated digital flight control systems, but the amount of correction available is limited for configurations without vertical tails. This is because the rudder power, or ability to generate yawing moments, is small for split aileron controls as compared with conventional vertical tail surfaces.

22.3 Shielded Vertical Tails and Leading-Edge Flaps

Radar return from an airplane's bottom is an important consideration if the airplane is to operate where hostile ground radars rather than air- or space-borne radars are the major defense. Intersections of vertical tails with wing or fuselage surfaces act as corner reflectors, increasing radar returns. This is the reason why vertical tails are located entirely above wing surfaces on airplanes such as the Lockheed F-117A and F-22 (Figure 22.3).

Vertical surface shielding from radar by above-wing mountings has the undesirable effect of shielding the vertical surfaces from the airflow at large positive angles of attack. Premature departures into uncontrolled flight and spins result. Canting vertical tail tips outward, as on the F-117A and F-22 designs, is intended to put at least the tail tips out in unshielded flow at large positive angles of attack.

The importance of radar returns from an airplane's bottom was brought out by Fulghum (1994). He reported that some reduction in the number of underwing doors, access panels, and drain holes was required to lower radar returns from the F-22. Radar returns from seams or junctures between fixed and movable surfaces are another consideration. Leading-edge flaps in particular are a cause of concern because of the lower seam between the flap and wing. The F-22 is equipped with leading-edge flaps. The F-22's leading-edge flaps

Figure 22.3 The Lockheed F-22, showing its vertical tail intersections with the wings that are shielded from ground radars. The vertical tail tips are canted outward to retain some effectiveness at high angles of attack. (From Lockheed Martin Corporation)

Figure 22.4 NASA 8-Ft. Transonic Wind-Tunnel model of a U.S. Air Force Multirole fighter concept, designed to have no vertical tail. Directional stability and control would be provided by thrust vectoring and low-radar-return control surfaces. (NASA photo 93–01934)

are an important contributor to the airplane's air combat capability, including its reported ability to fly stably at an angle of attack of 60 degrees. Leading- and trailing-edge flaps are programmed with Mach number and angle of attack to maintain lateral and directional stability.

22.4   Fighters Without Vertical Tails

The designers of the B-2 stealth bomber proved that the stability and control requirements for a subsonic-level bomber can be met without vertical tails. What is not clear is whether the more severe fighter stability and control requirements can be met without vertical tails.

All designs in a USAF Wright Laboratory multirole fighter study have either small vertical tails or none at all (Figure 22.4) (Oliveri, 1994). The preferred replacement for normal vertical tails is thrust vectoring and split ailerons. These controls were used successfully on the NASA/Boeing X-36 Fighter Agility Research Aircraft. A 28-percent-scale remotely piloted model was flown in 1997, reaching an angle of attack of 40 degrees.

A thrust vectoring scheme used to replace fighter vertical tails must have a high-bandwidth actuator responding to sideslip signals for directional stability as well as other stability-augmentation system signals and commands from the pilot. Unless split ailerons are used, for safety reasons it would seem necessary for a thrust vectoring sideslip loop to provide directional stability even at idle thrust. Alternately, engine thrust could be diverted left and right when idle thrust is needed and modulated for directional stability and control.

All in all, stability and control engineers should come well prepared at design meetings where stealth is the topic.

CHAPTER 23

Very Large Aircraft

It is not at all certain that the supersonic Concorde will be followed by fleets of new supersonic cruise civil transports. However, the prospect of subsonic commercial jet transports larger and heavier than the Boeing 747-400 is almost a certainty, with some Airbus A380 superjumbo jets already on order. Thus, it is reasonable to review the expected stability and control problems for very large airplanes.

23.1   The Effect of Higher Wing Loadings

Higher wing loadings than on airplanes of the Boeing 747-400 and advanced 777 classes seem inevitable for commercial airplanes of the 1,000-seat category, if these airplanes are to fit into current airport terminals, runways, taxiways, and maintenance facilities that have had reasonable modifications. Folding wings, tandem main wings, or some other radical departures from current technology would get around the necessity of higher wing loadings, but radical innovations are unlikely in airplanes that will be as expensive as superjumbo jets. All-wing superjumbo jets have been studied by several groups, but the Boeing and Airbus designs for superjumbo jets show quite conventional arrangements.

Some of the stability and control consequences of using high wing loadings in very large airplanes can be predicted. Higher wing loadings than current practice imply higher fuel weights relative to the aerodynamic forces generated by the wings and stabilizing surfaces. Dynamic fuel slosh effects, a nonproblem for 747-class airplanes, will require a fresh analytical look.

23.2   The Effect of Folding Wings

Folding wings as a device for fitting very large airplanes into airport ramps, taxiways, and maintenance areas would have stability and control consequences. While fairly long wing sections could fairly readily be folded up, shortening fuselages by folding hardly seems practical. This means that tail length in terms of wing spans would be reduced.

Longitudinal stability and control would be adversely affected by the shorter relative tail length of folding wing designs. Higher downwash and larger trim drag would result. There would be increased yawing moment due to rolling. An extended-span transport airplane would have some of the flight behavior of a high-performance sailplane, with strong demands on rudder power to coordinate rolls. However, the most adverse effect and most difficult to correct would be reduced rudder power for low-speed control with an engine out.

Even without folding wings, the current Airbus A380 layout has 10- to 20-percent reduced tail length in terms of wing span than current designs, such as the Boeing 747-400. The same adverse trends in stability and control as for the folding wing case, noted above, might be expected.

23.3   Altitude Response During Landing Approach

An early pilot-induced oscillation experience with the Space Shuttle Orbiter ALT-FF5 landing approach tests (Powers, 1986) led to some concern about the shuttle's combination of large airplane size and distance of the pilot ahead of the center of gravity, and initially reversed altitude response with elevon control deflections. On the shuttle, more than 2 seconds elapses before altitude starts to increase following a step nose-up elevon deflection (Phillips, 1979). F. A. Cleveland (1970) gives a value of 0.8 second for this delay parameter on the C-5A Galaxy and speculates that glide path control during landing approach would be unsatisfactory for time delays above 2 seconds.

Cleveland's analysis of the effects of airplane size gives a linear growth in the altitude delay parameter with size. Time delay would reach 2 seconds on a C-5A scaled up about 2.4 times, requiring direct lift control or canard surfaces. Digital sampling delays in the pitch control circuit and elevon rate limiting are known to contribute to the Space Shuttle's pilot-induced oscillation tendency. According to Robert J. Woodcock (1988)

> extensive, continual training [for the Space Shuttle Orbiter] has resulted in very good landing performance (smooth landings with small dispersion, mostly on long runways, in smooth air).

The conclusion is that large airplane sizes might indeed lead to landing approach problems related to delayed altitude responses following pitch control inputs and special design features to correct the problem.

An alternate criterion for the control response of very large airplanes is the generic response to abrupt control inputs, as shown in Figure 10.5. Grantham, Smith, Person, Meyer, and Tingas (1987) compared large airplane flying qualities with the generic pitch response requirements of MIL-F-8785C. Maximum effective time delays of 0.10 to 0.12 second are specified in MIL-F-8785C, based on tests of relatively small tactical airplanes. These requirements may be much too stringent for large airplanes in all but very high-gain operations. Satisfactory, or Level 1, flying qualities are obtained in simulation of a Lockheed 1011–class transport for effective time delays of 0.15 second, suggesting even larger acceptable values for superjumbo jets. On the other hand, the effective longitudinal axis dynamics of the Boeing 777 compares with some fighters.

John Gibson (1995) suggests that the observed sluggish response of the Space Shuttle Orbiter is caused by poor control laws as much as airplane size. Increasing pitch rate overshoot in a pullup would improve the observed sluggish flight path angle response. This further suggests that the flexibility of fly-by-wire technology can provide the necessary quickness of response in large airplanes without requiring special features such as direct lift control or canards.

23.4   Longitudinal Dynamics

The flight characteristics of current-day large airplanes such as the Lockheed C-5A Galaxy can offer some clues as to what to expect from the really large machines of the future. For example, the C-5A's longitudinal short-period oscillation approaches in frequency that of the phugoid oscillation (Mueller, 1970). Normally, the short-period oscillation is much higher in frequency than the phugoid oscillation.

This is not surprising since the phugoid oscillation frequency depends only on an airplane's true airspeed and is invariant with airplane size. However, airplane short-period frequency is proportional to the square root of the quantity (linear dimension cubed divided by pitching moment of inertia). That combination of parameters varies roughly as the reciprocal of a linear dimension, such as wing span or fuselage length, implying lower frequency short-period oscillations for large airplanes.

The practical consequence, according to Mueller, is that the C-5A is difficult to trim for a particular airspeed. Pilots report that the airplane wanders about a trim point.

Additional longitudinal dynamics problems could be encountered on very large airplanes as a result of low-frequency flexible modes.

23.5   Roll Response of Large Airplanes

Experience with C-5A roll response shows possible shortcomings of current requirements when applied to really large airplanes. William D. Grantham (1983) reports:

> [C]onsiderable effort and expense were initially expended on the C-5A in an attempt to meet a requirement for rolling to an 8 degree bank angle in one second. It was later determined from flight tests that the handling qualities of the C-5A were totally acceptable with less than one-half such roll capability.

This implies that careful research is needed to establish time-to-bank requirements for superjumbo jets, to avoid overdesign.

23.6   Large Airplanes with Reduced-Static Longitudinal Stability

In a paper on stability augmentation for a large, flexible airplane that appears to be the Boeing 777, Greta Ward (1996) makes an interesting observation on the limitations of reduced-static longitudinal stability when applied to large airplanes. Reduced-static longitudinal stability is an attractive feature for long-range airplanes, reducing cruise flight trim drag and fuel consumption. In any such application, a designer must retain a suitable margin of longitudinal control power to recover from inadvertent upsets, upsets that would be opposed by static stability in airplanes of normal longitudinal stability.

Pitch moment of inertia varies as the fifth power of fuselage length, while the maximum available pitching moment produced by a horizontal tail surface varies only as the third power of fuselage length. This implies an upper limit to fuselage length and airplane size if reduced-static longitudinal stability is to be used.

23.7   Large Supersonic Airplanes

A successor to the supersonic Concorde is likely to be a large airplane, in the sense considered here. The stability and control problems of supersonic flight and low-speed flight of a low-aspect-ratio design would be added to those of large airplanes. A general review of the combined problems (Steer and Cook, 1999) reflects experience with the Concorde. The authors conclude that a foreplane, or canard surface, would be needed in place of the tailplane used in the Boeing/NASA High-Speed Commercial Transport (HCST) design.

23.8   Concluding Remarks

Stability and control problems that may be encountered by very large (1,000-seat) airplanes can be reasonably well anticipated. We assume that compatibility with current ground facilities will constrain airplane size at the same time that weight is increased. The following problems are expected:

> more prominent dynamic fuel slosh effects and more severe flexible mode effects;
> increased adverse yaw in rolls and reduced rudder power for low-speed control with an engine out, because of reduced tail length in terms of wing span;

possible requirement for direct lift control or canards, to overcome delay in altitude
  response to pitch control;

trimming problems as a result of the merging of longitudinal short-period and
  phugoid frequencies;

control overdesign unless appropriate required time-to-bank and acceptable control
  delay criteria are found;

reduced static longitudinal stability for lower cruise fuel consumption may not be
  available because of limited control pitching moments for recovery from upsets;

flight control system complexity to accommodate maneuver load control and gust
  alleviation.

CHAPTER 24

Work Still to Be Done

Stability and control has advanced over the years along two paths, in categories that might be called "fundamental" and "reactive." Fundamental developments, such as Bryan's small-perturbation equations of motion and Soulé's and Gilruth's flying qualities requirements, have arisen out of the genius of their developers. These fundamental developments seem to have come along when the time was ripe, and not to meet current crises.

The second category of stability and control advances, those made in reaction to need, are no less important or praiseworthy. Airplanes grew denser and flew higher and faster as aviation itself advanced. Each new stretch in performance and design, such as transonic flight and sweptback wings, brought fresh stability and control challenges and responses by inspired researchers and designers.

It is easy to predict that the future will bring further airplane performance gains, bringing with them fresh stability and control challenges and reactions. But are there fundamental advances yet to be made? Are there pockets of systematic ignorance that have been bypassed and grand formulations yet to be found? This is almost certainly the case.

One reason to think so is that a spectacular growth in airplane stability and control theory was interrupted at the end of the 1950s, losing for us a generation's work in the field. This happened when the old NACA became what was for all purposes the "space agency" NASA in 1958. One can regretfully speculate on what new fundamental discoveries in airplane stability and control might have been made by people such as Harry Goett, Robert Gilruth, Christopher Kraft, Joseph Loftus, Charles Mathews, and Walter Williams had they not been called away to NASA's space program at Houston, Goddard, and other space centers. Similar losses to the field of course took place at research centers, universities, and factories all over the world, when space programs were staffed at those places.

U.S. Air Force–sponsored aeronautical research, as distinct from space-related work, virtually disappeared in the late 1950s by edict of the general who headed Air Force research and development. This was part of an Air Force attempt to not only carve out a mission in space, but to be the sole manager of the U.S. space program. Air Force support for aeronautical research was not restored until President Eisenhower declared that the U.S. space effort would be run by NASA.

Recent times have brought additional losses in potential aeronautical talent to fields such as biotechnology, molecular biology, material sciences, computers, and communications, all of which have the novelty and glamor once attached to aeronautics. On the other hand, enrollment in aeronautics courses continues strong worldwide, bringing new talent to the field.

A few fundamental airplane stability and control areas that seem to have been bypassed are as follows:

A general theory is needed for vertical path control in manually flown landing approaches. The theory should be able to predict for any size airplane the upper limits of thrust and pitch response delays and the download due to pitch control deflection. The theory should also identify applications in which direct lift control is needed for vertical path control.

An improved predictive method is needed for incipient and developed spins, to sort out the general effects of wing-stalling characteristics, tail contributions, and fuselage shape. NASA researchers disparage the existing TDR/TDPF preliminary design criteria of the 1960s without offering anything better.

Generalized predictive methods are needed for handling during takeoff and landing ground runs, including data on modern tires and struts. Reliable mathematical models could predict the need for stability augmentation during ground rolls and steering difficulties in cross-winds.

The body of knowledge on static aeroelastic effects on stability and control needs to be overhauled and made accessible to ordinary designers. Experts in the aeroelastics field may think the subject is well understood, yet static aeroelasticity remains problematical for most working-level stability and control engineers. There is nothing like the assurance these engineers feel while working with the equations of rigid-body motion or with the design of stability augmenters.

The landing maneuver needs to be reinvented, particularly for small personal airplanes. Even with automatic flight controls reaching new levels of reliability and low cost, future private pilots will probably land their machines by hand. Because really good landings are so difficult to make, the time spent by pilots in flight training and flight currency now goes mainly into takeoffs and landings. New hardware, as radical as the nose-wheel landing gear was in its day, will probably be required, rather than new procedures alone.

Fly-by-wire control system reliability must reach new high levels, now that the general traveling public is carried in fly-by-wire machines. A particular concern is that pilot-induced oscillations may appear under circumstances not yet encountered in specific machines. Stability and control engineers must provide advice to the industry on design criteria, testing, and appropriate training and flight recording methods.

These are unfinished airplane stability and control tasks, fundamental problems left for future researchers, designers, and inventors, in addition to the reactive work that will crop up as the result of advances in other areas of aviation.

Short Biographies of Some Stability and Control Figures

Abzug, Malcolm J. 1920–, b. New York, NY. B.S. (1941) Mass. Inst. of Tech., M.S. (1959), PhD. Engr. (1962), U. of Calif. at Los Angeles. After government laboratory work, he joined Douglas Aircraft, where he was stability and control lead engineer for the A2D-1 and A4D-1. His later industrial experience was at Sperry Gyroscope, TRW Systems, and Northrop on the A-9A, YF-17, and B-2 programs.

A'Harrah, Ralph 1931–, b. Warren, PA. B.S. Aero. (1955), Penn. State U. A'Harrah's career is balanced between North American Aviation, the U.S. Department of Defense, and NASA. He used ground-based fight simulation as a tool in solving flight dynamics problems associated with hazardous flight. On the AGARD Flight Mechanics Panel, he developed V/STOL flying qualities criteria.

Anderson, Seth B. 1918–, b. Los Altos Hills, CA. B.S. (1941), M.S. (1942), Purdue U. Anderson's long career at NACA and NASA dealt with handling quality requirements for conventional and VTOL airplanes. He is the principal author of AGARD Report 577 on V/STOL handling criteria.

Ashkenas, Irving L. 1916–, b. New York, NY. B.S. (1937), M.S.M.E. (1938), Ae.E. (1939), Calif. Inst. of Tech. His stability and control career started in industry, first at North American Aviation, then with the Northrop P-61 spoiler ailerons and design requirements for the XB-35 power controls and artificial-feel systems. He is best known for applying man-in-the-loop theory for flying qualities prediction and as a co-author of *Aircraft Dynamics and Automatic Control*.

Bairstow, Leonard 1880–1963, b. Halifax, Yorkshire, U.K. Royal College of Science, London. Bairstow's major stability and control contributions were the extension of the Bryan equations of motion to the nonsymmetric steady-flight case and development of efficient methods for root extraction, both done in 1914. The 1939 (second) edition of his *Applied Aerodynamics* was a useful stability and control reference for years.

Barnes, Arthur G. 1929–, b. Wigan, U.K. B.S. (1950), Manchester U. RAF and RauxAF pilot. His career in the United Kingdom industry from 1954 to 1990 included research and development for flight controls, flying qualities, and flight simulation. Barnes proposed the original numerical rating scale for pilot opinion on flying qualities. He is a consultant to the Kungl Tekniske Hogskola (KTH) and SAAB in Sweden.

Bihrle, William, Jr. 1925–, b. New York, NY. B.Ae S. (1945), Rensselaer Poly. Inst. Bihrle contributed to the stability and control designs of the Republic F-105 and XF-103 airplanes. He invented the widely used control anticipation parameter for pullups and plays a leading part in developing advanced spin tunnel rotary balance techniques and methods for improving high angle of attack stability and control.

Bowman, James S., Jr. 1924–, b. Burlington, NC. B.S. (1951), N.C. State Coll. As a leading NASA expert on spinning, Bowman consulted with military and commercial designers on spin problems for many years. He is the author or co-author of more than 40 reports on spinning, including NASA TP 2939 on pressure distribution at spinning attitudes.

Bratt, Robert W.   1918–, b. Palisade, MN. B.S. (1941), M.S. (1942), U. of Michigan. Bratt was a stability and control engineer at the El Segundo Division of Douglas. He pioneered in the application of digital computers to maneuvering flight. He solved drop vehicle instability problems involving aeroelasticity and inertial coupling. He later became Chief of Preliminary Design at Northrop.

Breuhaus, Waldemar O.   1918–, b. Lowell, OH. B.S.Ae. (1940), Carnegie Inst. of Tech., M.S. (1961), State U. of New York at Buffalo. Breuhaus was in charge of stability and control at Vought-Sikorsky during World War II. At Cornell Aero. Lab., later Calspan, he was responsible for the development of the B-26 and T-33 variable-stability airplanes, and he used these machines in flying qualities requirement research.

Bryan, George Hartley   1864–1928, b. Cambridge, U.K. Cambridge U. Bryan's monumental contribution to the field was the equations of aircraft motion, developed in 1911 in essentially modern form from a preliminary study (with W. S. Williams) in 1904. He later made contributions to compressible flow theory.

Cantrell, Coy R.   1924–, b. Muskogee, OK. B.S. (1953), M.S. (1954), Calif. Inst. of Tech. Cantrell's long career at Lockheed's Advanced Development Company (Skunk Works) started in 1954. He shared stability and control responsibility for the SR-71, the Have Blue prototype, and the F-117A, whose air data measurement system he designed. He was also involved in the YF-22A Advanced Tactical Fighter prototype.

Cook, Michael V.   1942–, b. Colchester, U.K. B.Sc. (1965), U. of Southampton, M.Sc. (1967) Coll. of Aeronautics, Cranfield. At Elliott Flight Automation, Ltd., Cook was involved with flight control system research and design on the Hovermarine HM2 hovercraft, the Westland Lynx helicopter, the Panavia Tornado, and the Jaguar fly-by-wire. He teaches at Cranfield College and is the author of *Flight Dynamics Principles* (1997).

Cook, William H.   1915–, b. Plainview, TX. B.S.M.E. (1934), Rensselaer Poly. Inst., M.S. (1938), Mass. Inst. of Tech. Cook was a designer of the Boeing High-Speed Wind Tunnel and was involved with the stability and control development of many Boeing designs, including the B-29, XB-47, and 707. He was co-inventor of the B-47 electronic yaw damper, one of the first of its kind.

Cooper, George E.   1916–, b. Burley, ID. B.S. (1940), U. of Calif. Cooper combined NACA/NASA engineering and research test pilot careers to become an important stability and control contributor. He is the Cooper of the Cooper-Harper handling qualities rating system and the author of a NASA Technical Note that is a text for test pilot training schools.

Czinczenheim, Joseph   1919–1994, b. Hungary. La Sorbonne, Centre Superieur de Mécanique, Paris. He worked on stability and control problems of the STOL Breguet 941, the transonic Breguet Taon, and the BAC-Breguet-Dassault Jaguar. Later, he was involved with certification of the Dassault Civil Transport /C and with stability and control of several Israeli prototypes. He was a member of the AGARD Flight Mechanics Panel.

Doetsch, Karl-H.   1910–, b. Kaldenhusen, Germany. Dipl.-Ing. (1934), TH Aachen, Dr. -Ing. (1943), TU Berlin. Professor Doetsch is an aeronautical scientist as well as a 3,000-hour test pilot. His contributions are fly-by-wire control (Avro 707C, Do 27, Pembroke), flight simulation, flight recording, and advanced aircraft flight controls. He chaired the

AGARD Flight Mechanics Panel and has made special efforts to broaden international cooperation in education and research.

Duncan, William Jolly   1894–1960, b. Hillhead, Glasgow. D.Sc. (1930), U. of London. Duncan was co-author of the important textbook *Elementary Matrices* and author of the 1952 book *Control and Stability of Aircraft*. His other contributions were in the theories of aileron reversal, tail buffeting, aerodynamic derivatives, and flap hinge moments.

Dunn, Orville R.   1916–1997, b. Wayne, PA. B.S. (1939), Mass. Inst. of Tech. Dunn was chief of stability and control at the Douglas Aircraft Santa Monica Division during the designs of the DC-4, C-74, DC-6, DC-7, and DC-8 transports. He produced a useful synthesis of methods for control force reduction by various tab systems. As Director of Aerodynamics he saw the DC-10 through certification.

Efremov, Alexander V.   1944–, b. Gorky Cty, U.S.S.R. Ph.D. (1973), D.Sc. (1996), Moscow Aviation Inst. As an expert in flight dynamics and control and in pilot-in-the-loop problems, Dr. Efremov participated in the flight control system designs for the aerospace vehicle Buran, the airship ALA-40, and the TU-204 and IL-96 airplanes. He is a member of the SAE control and guidance systems committee.

Etkin, Bernard   1918–, b. Toronto, Canada. B.A.Sc. (1941), M.A.Sc. (1947), U. of Toronto, D.Eng. (Hon) (1971), Carleton U. Dr. Etkin had a long career at the University of Toronto, becoming University Professor in 1982. He wrote three standard stability and control texts, which have German, Russian, and Chinese editions. Etkin made many contributions to the theory of flight dynamics, including flight in turbulence and dynamic longitudinal stability at high altitude.

Gates, Sidney B.   1893–1973, b. Watton, England. Gates was a brilliant theorist who did remarkable work on analyzing spins and predicting spin recovery with minimal facilities. Gates is responsible for the important flying qualities parameters of static and maneuver margins and stick force per g. With A. V. Stephens, he established the effect of air density on spins. The scope of his stability and control work is truly wide. Gates was the British counterpart of R. R. Gilruth in flying qualities research.

Gee, Brian   1933–, b. Manchester, U.K. B.Sc. (1954), Manchester U. Gee was head of the Flight Control Systems Design Group at British Aerospace, Warton, involved with the Jaguar and Fly-by-Wire Jaguar, the Toronado, EAP, Eurofighter, and the RAE VAAC Harrier. His main contributions were in the areas of component requirements, digital flight control specifications, and system clearance for flight control/structural mode interactions.

Gera, Joseph   1937–, b. Szentes, Hungary. B.Ae. (1961), Auburn U., M. Appl. Mech. (1965), U. of Virginia. At NASA Langley and Dryden Flight Research Facility Gera contributed to understanding the effects of wind gradients on pitch stability. He led efforts at Dryden to integrate simulators into flight research and to measure stability margins "on-line" for such aircraft as the X-29A.

Gibson, John C.   1929–, b. Swatow, China. M.Sc. (1958), Cranfield, Ph.D. (1999), Delft U. of Technology. At English Electric/British Aerospace, 1952–1992, he worked on the flight control systems of the Lightning, TSR-2, and Jaguar and developed new fly-by-wire handling design methods and criteria for the Tornado, BAe FBW Jaguar, the Experimental Aircraft Programme (EAP), the Eurofighter, and the VAAC (vectored-thrust) Harrier. He is responsible for the phase-gain, dropback, and other criteria used to prevent pilot-induced oscillations by design.

Gilruth, Robert R. 1913–2000, b. Nashwauk, MN. M.S., U. of Minnesota. He joined NACA in 1937. His major stability and control contributions were design methods for static longitudinal stability and roll performance and an early complete set of flying qualities requirements. He later was Director of the NASA Manned Spacecraft Center. He retired in 1973 and was a consultant to NASA from 1974 to 1983.

Glauert, Hermann 1892–1934, b. Sheffield, U.K. B.S. (1915), Trinity Coll., Cambridge. Glauert's notable work was in unsteady lift, airfoil theory, control surface effectiveness, and propeller theory. He originated the lag in downwash theory that explained damping discrepancies in the longitudinal short-period mode. He made the first nondimensionalization of the equations of airplane motion.

Goett, Harry J. 1910–, b. New York, NY. B.S. (1931), Holy Cross, B.S. (1933), New York U. Goett's important contribution to stability and control came at NACA, on methods of predicting flying qualities from wind-tunnel tests. In charge of large NASA Ames wind tunnels, he directed high-lift and stability research on swept wings. He later became the Director of NASA's Goddard Research Center.

Goto, Norohiro 1943–, b. Sasebo, Nagasaki, Japan. B. Eng. (1966), D. Eng. (1972), U. of Tokyo. Dr. Goto developed methods to identify pilot-control behavior in practical multi-input and multi-output aircraft control systems. At Kyushu University he is developing an autonomous flight control system for an unmanned observation blimp. He had been an NRC research associate at NASA Ames Research Center and a Fulbright Scholar at M.I.T.

Graham, F. Dunstan 1922–1992, b. Princeton, NJ. B.S.E. (1943), M.S.E. (1947), Princeton U. As an aerodynamicist at Boeing in 1947 and 1948, Graham made an early analysis of inertial coupling on a pilotless aircraft. At Lear, Inc., he was in charge of automatic controls development for the KC-135 and other jet aircraft. However, he is best known as the co-author with McRuer and Ashkenas of *Aircraft Dynamics and Automatic Control* and the co-author with McRuer of *Analysis of Nonlinear Control Systems*.

Hamel, Peter G. 1936–, b. Hamburg, Germany. Dipl.-Ing. (1963), Dr. -Ing. (1968), Tech U. Braunschweig (TUBS), S.M. (1965), Mass. Inst. of Technology. Dr. Hamel had a long career as the director of the Institute of Flight Research of the German Aerospace Research Center (DLR) and as a professor at TUBS. He is recognized internationally for the development and use of in-flight simulators. He is a leader in European vehicle system identification and in handling qualities research.

Harper, Robert P., Jr. 1926–, b. Gallipolis, OH. S.B. Ae. (1952), S.M.Ae. (1953), Mass. Inst. of Tech. Harper was a Calspan engineer and test pilot who is noted for his part in developing the Cooper-Harper flying qualities rating. He was project engineer on the F-94 and NT-33A variable-stability airplanes during simulation of reentry vehicles and the X-15, as well as during basic flying qualities research.

Harris, Thomas Aubrey 1903–1987, b. Whites, VA. B.S. (1929), William and Mary. Harris designed the NASA Langley Atmospheric and 7 by 10-foot wind tunnels during a long career at Langley. He was an expert on flaps and tabs, and he contributed to numerous wind-tunnel studies of control surfaces.

Haus, Frederic Charles 1896–1993, b. St. Gilles, Belgium. Brussels U. (1922). In a long, productive career, Professor Haus headed the famous aeronautical laboratory of Rhode-St.-Genèse, published a 1930 book (in French) on airplane stability and control, served as professor at both Ghent and Liège Universities, and was a member of AGARD panels on flight mechanics, guidance, and control.

Heald, Ervin R. 1917–, b. Sultan, WA. B.S.A.E. (1940), U. of Michigan. Heald headed stability and control at the El Segundo division of Douglas Aircraft during the years when that division produced new airplanes on the average of one every two years. He took part in the stability and control work on the U.S. Navy's XSB2D-1, XBT2D-1, AD-1, XA2D-1, D-558-1, D-558-2, F3D-1, F4D-1, XF5D-1, A3D-1, and A4D-1. Later, Heald was Chief Engineer for the U.S. Air Force's C-17 transport.

Heppe, R. Richard 1923–, b. Kansas City, MO. B.A. (1944), M.S. (1945), Stanford U., A.E. (1946), Calif. Inst. of Tech. At Lockheed Aircraft, Heppe made significant contributions to understanding the inertial coupling problems of the F-104 and other USAF fighters, and helped find corrections for those problems. He contributed in the unlimited angle-of-attack maneuvering areas of the YF-22A prototypes. He became president of the Lockheed-California Company.

Hodgkinson, John 1943–, b. Ilseworth, U.K. B.Sc. (1965), U. of Southampton, M.S. (1971), St. Louis U. After training at British Aerospace, Warton, he joined McDonnell and then led controls R&D at Northrop. He later was at Eidetics and McDonnell Douglas (Boeing). Hodgkinson's stability and control contributions are in equivalent systems, agility, and safety. He is the author of *Aircraft Handling Qualities.*

Hunsaker, Jerome C. 1886–1969, b. Creston, IA. B.S. (1908), Annapolis, M.S. (1912), Mass. Inst. of Tech., D.Sc. (1914), Williams Coll. Dr. Hunsaker was the author of NACA Technical Report No. 1 on inherent dynamic stability, 1915. He taught airplane stability and control at MIT, starting in 1914, and headed the Department of Aeronautical Engineering at MIT for many years.

Jex, Henry R. 1929–, b. Baltimore, MD. S.B. (1951), Mass. Inst. of Tech., M.S. (1958), Calif. Inst. of Tech. Jex developed analytical models of operator–vehicle control and applied them to handling qualities, landing displays, and workload studies. He is the principal developer of the critical-instability tracking task, used for detecting impaired pilots and drivers. Jex designed the control system for the first autostabilized-while-flapping ornithopter, the Q-N pterodactyl replica.

Johnston, Donald E. 1924–1995, b. Huron, SD. B.S. Eng. (1952)., U. of California, Los Angeles. Johnston's contributions have been in the fields of man/machine control analysis, synthesis, simulation, and full-scale flight test. He was a vice president of Systems Technology, Inc., where he was assigned to the most critical investigations. He conducted studies into control problems of the F-4, F-111, F-14, F-16, and F-18 airplanes and designed control laws for the McDonnell Douglas C-17 cargo airplane.

Jones, Bennett Melvill 1887–1975, b. Birkenhead, England. B.S. (1909), Emmanuel Coll., Cambridge. He joined the National Physical Laboratory in 1910. He contributed the "Dynamics of the Airplane" division in W. F. Durand's *Aerodynamic Theory*, published in 1934. This is a key reference, the first complete derivation of aircraft equations of motion, in modern form. His research at Cambridge was on stalling. Jones was a pilot and a decorated gunner in World War I.

Jones, Robert T. 1910–1999, b. Macon, MO. U. of Missouri, 1928, Catholic U. of America, 1933. After working as an airplane designer for the Nicholas Beasley Company, Jones joined NACA in 1934. His long career there produced notable stability and control contributions in lateral control, in the theory of two-control flight, in all-movable controls, and in a very early (1936) application of operator theory to the solution of the equations of aircraft motion.

Kalviste, Juri 1935–1996, b. Tartu, Estonia. B.S. (1957), M.S.E.E. (1960), U. of Washington. He worked on the flight control designs of the Boeing X-20 and Northrop YF-17 airplanes and on the ATF proposal. Kalviste made innovative formulations of the large-amplitude equations of airplane motion to develop departure parameters and methods of combining rotary balance and oscillatory aerodynamic data.

Katayanagi, Ryoji 1946–, b. Gumma Prefecture, Japan. B.S.M.E. Waseda U., M.S. (1972), Ph.D. (2000), U. of Tokyo. At Mitsubishi Heavy Industries, Katayanagi analyzed flying qualities and flight controls of the T-2 trainer. He designed flight control laws for the T-2CCV research airplane, the QF-104 drone, and the F-2 fighter. His research interests are multiloop flight controls and PIOs. He leads the engineering team for the NAL scaled supersonic research airplane.

Koppen, Otto C. 1901–1991, B.S. (1924), Mass. Inst. of Tech. Koppen's career went back to the design of the Ford "Flying Flivver," a contemporary of the Ford Model A. He joined MIT in 1929 and taught airplane stability and control and airplane design courses there until his retirement in 1965. Koppen did early work on the effects of closing loops on stability. He designed one of the first two-control airplanes, the Skyfarer, as well as the famous STOL Helioplane. Koppen test-flew the Helioplane prototype in 1949 and continued to fly at a ripe age, getting an FAA instrument rating at age 80.

Larrabee, E. Eugene 1920–, b. Marlboro, MA. B.S.Me. (1942), Worcester Poly. Inst., M.S.Ae. (1948), Mass. Inst. of Tech. Larrabee did stability and control design work on the Curtiss C-46, XF15C-1, and XP-87 airplanes. He developed stability derivative extraction methods using time vector analysis. He taught airplane stability and control at MIT and Northrop University for many years. He is a recognized expert on propeller design.

Lecomte, Pierre 1925–, b. France. Ecole Polytechnique, ENSAE. Lecomte was Professor of Flight Mechanics at ENSAE and author of the book *Mécanique du Vol*. He initiated a new handling qualities approach based on normal and peripheral flight envelopes and a theoretical explanation of wing drop. He was a test pilot in the French Flight Test Center, a Concorde evaluator at Aerospatiale, and chairman of the AGARD Flight Mechanics Panel.

McDonnell, John D. 1937–, b. Hollywood, CA. B.S. (1960), M.S. (1965), U. of Calif. at Los Angeles. At Systems Technology, Inc., he contributed to the analysis and evaluation of flying qualities. At McDonnell Douglas he contributed to the design and evaluation of avionics and control systems for the DC-10, MD-80, T-45, C-17, MD-11, and the space shuttle (HUD). He was the chief avionics engineer and chief avionics FAA DER at McDonnell Douglas, Long Beach.

McRuer, Duane T. 1925–, b. Bakersfield, CA. B.S. (1945), M.S.E.E. (1948), Calif. Inst. of Tech. McRuer is perhaps best known to stability and control engineers as the senior author of "The Green Book," whose real title is *Aircraft Dynamics and Automatic Control*. His enormous personal contributions to the field include mathematical models for human control and information processing in closed-loop systems and a well-tested theory of vehicle handling qualities.

McWha, James 1939–, b. Millisle, N. Ireland. B.S., Queens U., Belfast. McWha was chief engineer of flight systems at Boeing Commercial Group throughout the development of the fly-by-wire 777 transport. Prior to a 30-year employment at Boeing, he worked at Shorts Brothers, N. Ireland. He is vice chairman of an SAE control and guidance subcommittee and a member of a NASA Flight Controls and Guidance Panel.

Milliken, William F., Jr. 1911–, b. Old Town, ME. B.S. Ae. and Math. (1934), Mass. Inst. of Tech. At Cornell Aeronautical Laboratory, later Calspan, he was a leader in the application of servomechanism techniques to airplane stability and control, including the determination of airplane stability derivatives and transfer functions from flight-test frequency-response measurements.

Mueller, Robert K. 1909–1994, b. Waterbury, CT. B.S. (1932), M.S. (1934), ScD. (1936), Mass. Inst. of Tech. Mueller produced one of the first electronic analog computers while doing his Sc.D. thesis. He also developed time vector analysis of airplane dynamics at that time. He invented the Microsyn transducer, used in servomechanism systems.

Mulder, Jan A. (Bob) 1943–, b. The Hague, The Netherlands. MSc. Aero. (1968), Ph.D. (1986), Delft U., a student of Professor Otto Gerlach. Dr. Mulder is head of the division of Control and Simulation, Aerospace Engineering, Delft U. of Technology and an active captain on the Boeing B-757. His current research interests are in intelligent flight control and dynamic flight-test techniques.

Neumark, Stefan 1897–1967, b. Lodz, Poland. Dipl. Ing. and Sc.D., Tech. U. of Warsaw. Dr. Neumark was a most versatile engineer. In stability and control, he was noted for the atmospheric density change effect on the phugoid mode and for the theory of airplane stability under constraints. He also contributed to the theories of dynamic stability with rudder free and of gust effects on automatic control.

Nguyen, Luat T. 1947–, b. Vietnam. B.S. (1968), M.S. (1970), E.A.A. (1970), Mass. Inst. of Tech. Nguyen is a NASA expert on aircraft flight dynamics at high angles of attack. He has contributed to control system design for enhanced maneuverability and departure resistance.

Osder, Stephen 1925–, b. New York, NY. B.E.E. (1946), City Coll. of New York, M.S. (1951), Johns Hopkins U. He pioneered in the design of digital flight control, fly-by-wire systems, and redundancy management. He was Director of R and D at Sperry and Chief Scientist, Flight Controls and Avionics at McDonnell Douglas Helicopters. His flight control design experience included the QF-104 drone, MD-80, DC-10, NASA CV-990, AH-64 helicopter, various reentry bodies, NASA STOLAND and Space Shuttle Autoland, and recently the Boeing Canard Rotor Wing aircraft.

Perkins, Courtland D. 1912–, b. Philadelphia, PA. B.S. (1935), Swarthmore Coll., M.S. (1941), Mass. Inst. of Tech. Perkins helped launch the stability and control function at Wright Field in World War II. He wrote the stability and control portion of the important text *Airplane Performance, Stability and Control*. He taught the subject at Princeton University and later became Chief Scientist of the U. S. Air Force.

Phillips, W. Hewitt 1919–, b. Port Sunlight, Merseyside. S.B. (1939), S.M. (1940), Mass. Inst. of Tech. Phillips was a well-known model aircraft builder before joining NACA in 1940. His achievements in stability and control are many, but they are perhaps topped by his discovery of the roll or inertia-coupling phenonemon in 1947. Phillips also made important theoretical contributions to the design of spring tabs, the landing approach problem, gust alleviation, and pilot–airplane interactions that cause instability.

Pinsker, Werner J. G. 1918–, b. Mannheim, Germany. BEUTH (1939), Berlin College. Pinsker was the preeminent expert in inertial coupling at the British Royal Aircraft Establishment. He also contributed to the theories of landing large airplanes and of nose-slice departures. As a consultant, he helped solve lateral-directional problems of the multi-national Tornado airplane.

Poisson-Quinton, Phillipe 1919–, b. Loches, France. La Sorbonne, ENSAE (1945). He was a professor/lecturer on aerodynamics, flying qualities, and control systems and a visiting professor at Princeton U. in 1975. At ONERA, he initiated transonic research on flying qualities and on optimized shapes for aircraft from V/STOL to hypersonic types. He delivered the 1967 AIAA Wright Brothers Lecture and was a member of the AGARD Flight Mechanics Panel.

Reid, Lloyd D. 1942–, b. North Bay, Canada. B.A.Sc. (1964), M.A.Sc. (1965), Ph.D. (1969), U. of Toronto. He is Associate Director of the University of Toronto Institute for Aerospace Studies. He is co-author of two standard stability and control textbooks. Dr. Reid's research contributions include the study of aircraft response to the planetary boundary layer and the development and operation of a facility for flight simulation of the effects on stability and control of pilot–aircraft interactions.

Relf, Ernest Frederick 1888–1970, b. Beckenham, Kent. A.R.C.Sc. Relf combined mastery of aerodynamic theory with extraordinary talents as an experimentalist. He devised methods for the testing of autorating wings, yawed propellers, and apparent mass. In 1922, he built a small, powerful electric motor for use in powered wind-tunnel models.

Ribner, Herbert S. 1913–, b. Seattle, WA. B.S. (1935), Calif. Inst. of Tech., M.S. and Ph.D. (1937, 1939), Washington U. Dr. Ribner's contributions to stability and control are in propeller and slipstream theory and in gust response. While at NACA, he solved the problem of the forces on yawed propellers, an important factor in static stability.

Rodden, William P. 1927–, b. San Francisco, CA. B.S. (1947), M.S. (1948), U. of Calif., Ph.D. (1958), U. of Calif. at Los Angeles. Dr. Rodden made important stability and control contributions as a co-developer of the Doublet Lattice method for oscillating lifting surfaces and of the first correct equations of motion for quasi-steady aircraft utilizing restrained aeroelastic derivatives. He is a co-author of the *MSC/NASTRAN Aeroelastic Analysis User's Guide*.

Root, L. Eugene 1911–1992, b. Lewiston, ID. B.S., U. of the Pacific, M.S., Cal. Inst. of Tech. As chief of aerodynamics at the El Segundo plant of the Douglas Aircraft Company, Root led the team that developed in a systematic way excellent flying qualities for the U.S. Navy SBD Dauntless and AD Skyraider aircraft. Root went on to become one of the founders of the RAND Corporation and later president of the Lockheed Missile and Space Company.

Roskam, Jan 1930–, b. The Hague, The Netherlands. M.S.A.E. (1964), Delft U. of Technology, Ph.D. (1965), U. of Washington. Dr. Roskam worked for Cessna (1957–1959) and Boeing (1959–1967) on a variety of airplane projects. He is a major stability and control influence through his teaching at the University of Kansas, his consulting work, and his papers and textbooks.

Ross, A. Jean 1931–, b. Sussex, U. K. B.Sc. (1953), Ph.D. (1956), U. of Southampton. At the RAE and DERA, Farnborough, Ross specialized in modeling and analysis of aircraft responses to nonlinear dynamics and aerodynamic effects. She contributed to wing rock theory and to model testing of spin prevention and maneuver limitation systems. She participated in experimental wind-tunnel and free-flight work, culminating in the active control of forebody vortices.

Schairer, George S. 1913–, b. Pittsburgh, PA. B.S. (1934), Swarthmore Coll., M.S. (1935), Mass. Inst. of Tech. During Schairer's long career at the Boeing Company, he was responsible for the stability and control designs of many of their airplanes. He redesigned the Stratoliner vertical tail to include one of the first dorsal fins and was responsible for sweeping the B-47's wing. He became Boeing's Corporate Vice President for Research.

Shaw, David E. 1932–, b. Bradford, U.K. B.Sc. Aeronautics (1954), Queen Mary Coll., London. He worked in all areas of aerodynamics at AV Roe Weapons Division and at BAe Warton. Shaw's career highlights were clearance of the Lightning for rapid rolling, Tornado wing design, and Aerodynamics Team Leader for the Experimental Aircraft Programme (EAP), from design to full-flight clearance.

Smith, Terry D. 1947–, b. Norwich, U.K. B.Sc.Eng. (1968), Imperial Coll., London. As a flight-test engineer specializing in flight controls, he was deeply involved in stability and control testing and flight control system development on the Jaguar, Tornado, and Eurofighter Typhoon. He led the flight-test teams for both the fly-by-wire Jaguar and the Experimental Aircraft Programme (EAP) digital flight control systems.

Soulé, Hartley A. 1904–, b. New York, NY. B.S.Ae. (1927), New York U. Soulé started at NACA in 1927. He pioneered in spin research and made the first comprehensive measurements of airplane flying qualities. He was a co-inventor of the NACA Stability Wind Tunnel. Soulé wrote a set of flying qualities requirements that eventually led to civil standards and military specifications.

Stengel, Robert F. 1938–, b. East Orange, NJ. S.B. (1960), Mass. Inst. of Tech., M.S.E. (1965), M.A. (1966), Ph.D. (1968), Princeton U. At the Draper Laboratory, Stengel followed airplane flying qualities principles in designing the manual attitude-control system for the Project Apollo Lunar Module. At Princeton, he converted the Navion variable-stability research airplane to digital control and conducted flying qualities and control system research.

Szalai, Kenneth J. 1942–, b. Milwaukee, WI. B.S.E.E. (1964), U. of Wisconsin, M.S.M.E. (1970), U. of Southern California. Mr. Szalai was principal investigator for the NASA Dryden F-8 Digital Fly-by-Wire program, the first of its type. He led the development of the U.S.–Russian Tu-144 supersonic flying laboratory. As director of the NASA Dryden Flight Research Center, he supervised research in thrust vectoring, high-angle-of-attack aerodynamics, and advanced flight controls. The X-29, X-31, X-36, and X-38 experimental programs were under his direction.

Thomas, H. H. B. M. (Beaumont) 1917–2000, b. Llanelli, U. K. B.Sc. (1939), D.Sc. (1980), U. of Wales, OBE 1979. In the Aerodynamics Department of the RAE, Farnborough, his expertise was in stability and control at the edges of the flight envelope. He contributed to control surface aerodynamics during World War II and to the dynamic stability of slender aircraft during the basic research that led to the Concorde. He also contributed to spin entry and recovery testing and analysis.

Toll, Thomas A. 1914–, b. Bridgewater, SD. B.S. (1941), U. of Calif. Toll made a wide range of stability and control contributions, including control surface aerodynamic balance, swept wings, and variable geometry. He is perhaps best known for two valuable summary reports, on lateral control research and on the supersonic transport.

Tonon, Aldo   1957–, b. Caracas, Venezuela. Politecnico of Turin (1982). At Alenia of Turin (formerly Aeritalia), Tonon's major activity was in combat aircraft controls development. He was on the AMX program and then on the Eurofighter 2000 control law design, starting with the EAP technology demonstrator.

Wanner, Jean-Claude L.   1930–, b. Brest, France. Ing. (1950), École Polytechnique, Ing. (1955), ENSAE. Dr. Wanner's career in airplane stability and control includes serving as a military pilot, flight test engineer, and as professor in a number of institutions, including the ENSAE. He is author of the French text *Mécanique du Vol*. He pioneered in using computer methods in the teaching of stability and control.

Washizu, Kyuichiro   1921–1981, b. Ichinomiya, Aichi, Japan. B.Eng. (1942), Imperial U. of Tokyo, Dr.Eng. (1957), U. of Tokyo. Dr. Washizu's important contribution to airplane stability and control was to train a generation of Japanese engineers in the field, having spent time in the United States to study the educational system. He did research on human controllability limits and finite-element methods and is the principal author of the stability and control section in Japan's *Handbook of Aerospace Engineering* (1974).

Weick, Fred E.   1899–1993, b. Chicago, IL. B.S. (1922), U. of Illinois. Weick was at NACA's Langley Aeronautical Laboratory from 1925 until 1936, contributing to lateral control research. He developed the W-1 pusher airplane, incorporating important stability and control innovations. The W-1 was a two-control airplane that had limited up-elevator travel and a tricycle landing gear. He later became known as the designer of the Ercoupe, the first agricultural airplane, the Ag-1, and a series of Piper aircraft.

Westbrook, Charles B.   1918–2001, b. Port Jervis, NY. M.S. (1946), Mass. Inst. of Tech. Westbrook joined the USAF Flight Dynamics Laboratory in 1945 as head of stability and control. He oversaw the development of post-war flying qualities specifications and the *USAF Stability and Control Handbook*. Westbrook managed for the Air Force much flying qualities research, including work on variable-stability airplanes.

White, Roland J.   1910–2001, b. Missoula, MT. B.S. (1933), U. of Calif., M.S.M.E. (1934), M.S.A.E. (1935), Calif. Inst. of Tech. His long stability and control career started at Curtiss-Wright, St. Louis, where he incorporated a springy or "vee" tab to the C-46 Commando, adding to its allowable aft cg travel. White designed a mechanical yaw damper for the Boeing B-52 and made one of the first servo analyses of electronic yaw dampers, for the B-47.

Wykes, John H.   1925–1988. B.S. (1949), Mass. Inst. of Tech., M.S., U. of So. Calif. Wykes was a leading stability, control, and aeroelastics engineer at the Rockwell International Aircraft Division from 1949 to 1986, where he contributed to the designs of the F-86, F-100, F-107, B-70, and B-1 airplanes. He joined Northrop in 1987 to work on their YF-23A airplane. In addition to innovative work on stability augmentation, he also was responsible for the design of the B-1 gust alleviation system.

Zimmerman, Charles H.   1907–1995, b. Olathe, KS. B.S. (1929), U. of Kansas, M.S.Ae. (1954), U. of Virginia. Zimmerman started at NACA in 1929. He produced the classical NACA dynamic longitudinal and lateral stability analyses in 1935 and 1937, complete with stability boundary design charts. This was a considerable accomplishment for those times and the main design source for dynamic stability for years afterwards. He was instrumental in developing the Langley 20-foot spin and free-flight tunnels.

References and Core Bibliography

References are grouped by chapters. Additional citations have been made, to make the reference list serve also as a core bibliography. As a core bibliography, only the most significant papers and reports on airplane stability and control are included.

Stability and control historical surveys are listed in the Preface. Stability and control textbooks are listed at the end of Chapter 2.

Abbreviations used in this section and in other places in the book are as follows:

AFFDL	Air Force Flight Dynamics Laboratory
AFWAL	Air Force Wright Aeronautical Laboratories
AGARD	Advisory Group for Aerospace Research and Development
AIAA	American Institute of Aeronautics and Astronautics (formerly IAS, Institute of the Aeronautical Sciences)
ASME	American Society of Mechanical Engineers International
AWST	Aviation Week and Space Technology
CAA	U.K. Civil Aviation Authority
DERA	Defense Evaluation and Research Agency (U.K.) (formerly Royal Aircraft Establishment)
DLR	German Aerospace Center (formerly DFVLR)
FAA	U.S. Federal Aviation Administraion
FAR	U.S. Federal Aviation Regulation
IEEE	Institute of Electrical and Electronics Engineers (formerly AIEE, American Institute of Electrical Engineers)
JAR	European Joint Aviation Requirement
NAE	National Aeronautical Establishment (Canada)
NAL	National Aerospace Laboratory (Japan)
NASA	National Aeronautics and Space Administration (formerly NACA, National Advisory Committee for Aeronautics)
NASDA	National Aerospace Development Agency (Japan)
NATC	Naval Air Test Center
NRC	National Research Council (Canada)
RAeS	Royal Aeronautical Society
R & M	Reports and Memoranda, British A.R.C.
RTO	Research and Technology Organization (of NATO)
SAE	Society of Automotive Engineers
STI	Systems Technology, Inc.

Chapter 1. Early Developments in Stability and Control

Bryan, George H. 1911. *Stability in Aviation*. London: MacMillan.

Gibson, John C. 2000. Unpublished comments on the first edition of *Airplane Stability and Control*, Cambridge U. Press, 1997.

Jex, Henry R. and Fred. E. C. Culick 1985. Flight Control Dynamics of the 1903 Wright Flyer, *Proc. of the 12th Atmospheric Flight Control Conf.*, New York: AIAA, pp. 534–548.

Jones, B. Melvill 1934. Dynamics of the Airplane, in *Aerodynamic Theory*, Vol. V, ed. W. F. Durand, Berlin: Springer, pp. 1–222.

Lednicer, David 2001. Unpublished comments in letter dated February 2, referring to NACA Report 254 by A. J. Fairbanks on pressure distribution tests of a model of the Fokker D VII.

Chapter 2. Teachers and Texts

Milliken, William F., Jr. 1947. Progress in Dynamic Stability and Control Research, *Jour. of the Aeronautical Sciences*, Vol. 14, No. 9, pp. 493–518.

Chapter 3. Flying Qualities Become a Science

Anderson, Seth B., Hervey C. Quigley, and Robert C. Innis 1965. Stability and Control Considerations for STOL Aircraft, AGARD Rept. 504.

Anon. 1980. Military Specification. Flying Qualities of Piloted Airplanes, MIL-F-8785C.

Anon. 1983. *Design and Airworthiness Requirements for Service Aircraft*, Defense Standard 00-970/Issue 1, Vol. 1, Book 2, Part 6 – Aerodynamics, Flying Qualities and Performance, Ministry of Defense, UK.

Anon. 1987. Flying Qualities of Piloted Vehicles, MIL-STD-1797, USAF.

Anon. 1994. Joint Aviation Requirements – JAR 25-Large Airplanes, Section 1 – Requirements, Subpart B – Flight, Joint Aviation Authority.

Ashkenas, Irving L. 1985. Collected Flight and Simulation Comparisons and Considerations, in AGARD CP 408. (64 references).

Ashkenas, Irving L. et al. 1973. Recommended Revisions to Selected Portions of MIL-F-8785(ASG) and Background Data, AFFDL-TR-73-76.

Assadourian, A. and John A. Harper 1953. Determination of the Flying Qualities of the Douglas DC-3 Airplane, NACA TN 3088.

Barnes, Arthur G. 1988. The Role of Simulation in Flying Qualities and Flight Control System Related Development, in AGARD LS 157.

Belsley, Steven E. 1963. Man-Machine System Simulation for Flight Vehicles, *IEEE Transactions on Human Factors in Electronics*, Vol. HFE-4, No. 1, pp. 4–14.

Breuhaus, Waldemar O. 1991. The Variable Stability Airplane, from a Historical Perspective, *American Aviation Historical Society Jour.*, Vol. 36, No. 1, pp. 30–55.

Bryan, George H. 1911. *Stability in Aviation*, London: MacMillan.

Chalk, Charles R., T. P. Neal, T. M. Harris, F. E. Pritchard, and Robert J. Woodcock 1969. Flying Qualities of Piloted Airplanes, *Background Information and Users Guide for MIL-F-8785(ASG)*, AFFDL-TR-69-72.

Cook, M. V. 1994. The Theory of the Longitudinal Static Stability of the Hang Glider, *The Aeronautical Jour.*, Vol. 98, No. 978, pp. 292–304.

Cook, M. V. 1997 *Flight Dynamics Principles*, London: Arnold, pp. 212–214.

Cooper, George E. and Robert P. Harper, Jr. 1969. The Use of Pilot Rating in the Evaluation of Aircraft Handling Qualities, NASA TN D-5153.

Coyle, Shawn 1996. *The Art and Science of Flying Helicopters*, Ames: Iowa State University Press, pp. 91–92.

Craig, Samuel J. and Robert K. Heffley 1973, Factors Governing Control in a STOL Landing Approach, *Jour. of Aircraft*, Vol. 10, No. 8, pp. 495–502.

Donlan, Charles J. 1944. An Interim Report on the Stability and Control of Tailless Airplanes, NACA Rept. 796.

Fielding, C. and M. Lodge 2000. Stability and Control of STVOL Aircraft: The Design of Longitudinal Flight Control Laws, *The Aeronautical Jour.*, Vol. 104, No. 1038, pp. 383–389.

Gawron, V. J. and Reynolds, P. A. 1995. When In-Flight Simulation Is Necessary, *Jour. of Aircraft*, Vol. 32, No. 2, pp. 441–415.

Gilruth, Robert R. 1943. Requirements for Satisfactory Flying Qualities of Airplanes, NACA Rept. 755.

Gilruth, Robert R. and M. D. White 1941. Analysis and Prediction of Longitudinal Stability of Airplanes, NACA Rept. 711.

Gilruth, Robert R. and W. N. Turner 1941. Lateral Control Required for Satisfactory Flying Qualities Based on Flight Tests of Numerous Airplanes, NACA Rept. 715.

Glauert, H. 1934. Airplane Propellers, Divsion L of *Aerodynamic Theory*, Vol. IV, ed. by W. F. Durand, California: Durand Reprinting Committee, pp. 348–351.

Goett, Harry J., Robert P. Jackson, and Steven E. Belsley 1944. Wind-Tunnel Procedure for Determination of Critical Stability and Control Characteristics of Airplanes, NACA Rept. 781.

Hansen, James R. 1987. *Engineer in Charge*, NASA SP-4305.

Harris, D, J. Gautrey, K. Payne, and R. Bailey 2000. The Cranfield Aircraft Handling Qualities Rating Scale: A Multidimensional Approach to the Assessment of Aircraft Handling Qualities, *The Aeronautical Jour.*, Vol. 104, No. 1034, pp. 191–198.

Hoh, Roger H. and M. B. Tischler 1983. Status of the Development of Handling Criteria for VSTOL Transition, AIAA Paper 83-2103.

Hoh, Roger H. 1981. Development of Handling Quality Criteria for Aircraft with Independent Control of Six Degrees of Freedom, AFWAL-TR-81-3027.

Hoh, Roger, H. 1983. Bring Cohesion to Handling-Qualities Engineering, *Astronautics and Aeronautics*, June, pp. 64–69.

Hoh, Roger H., David G. Mitchell, and Steven R. Sturmer 1987. Handling Qualities Criteria for STOL Landings, STI Paper 407.

Kandalaft, R. N. 1971. Validation of the Flying Qualities Requirements of MIL-F-8785B(ASG), AFFDL-TR-71-134.

Kayten, Gerald G. 1945. Analysis of Stability and Control in Terms of Flying Qualities of Full-Scale Airplanes, NACA Rept. 825.

Mitchell, D. G. and R. H. Hoh 1983. Handling Qualities Criteria for STOL Flight Path Control for Approach and Landing, AIAA Paper 83-2106.

Mooij, H. A. 1985. *Criteria for Low-Speed Longitudinal Handling Qualities of Transport Aircraft with Closed-Loop Flight Control Systems*. Dordrecht: Martinus Nijhoff.

Moorhouse, David J. and Robert J. Woodcock 1982. Military Specification – Flying Qualities of Piloted Airplanes, AFWAL-TR-81-3109. *Background Information and User Guide for MIL-F-8785C*, (120 references).

Myers, Thomas T., Donald E. Johnston, and Duane T. McRuer 1987. Space Shuttle Flying Qualities and Criteria Assessment, NASA CR 4049.

Phillips, William H. 1948. Appreciation and Prediction of Flying Qualities, NACA Rept. 927.

Phillips, William H. 1994. Effects of Model Scale on Flight Characteristics and Design Parameters, *Jour. of Aircraft*, Vol. 31, No. 2, pp. 454–457.

Rolfe, J. M. and K. J. Staples (eds.) 1986. *Flight Simulation*, London: Cambridge U. Press.

Shafer, Mary F. 1993. In-Flight Simulation at the NASA Dryden Flight Research Facility, *American Aviation Historical Society*, Vol. 38, No. 4, pp. 261–277.

Shanks, G. T., S. L. Gale, C. Fielding, and D. V. Griffith 1996. Flight Control and Handling Research with the VAAC Harrier Aircraft, in *Advances in Aircraft Flight Control*, ed. M. B. Tischler, London: Taylor and Francis, pp.159–186.

Soulé, Hartley A. 1940. Preliminary Investigation of the Flying Qualities of Airplanes, NACA Rept. 700.

Stapleford, R. L., D. T. McRuer, R. H. Hoh, D. E. Johnston, and R. K. Heffley 1970. Outsmarting MIL-F-8785B(ASG), Systems Technology Inc. TR-190-1.

Stengel, Robert F. 1979. In-Flight Simulation with Pilot-Center of Gravity Offset and Velocity Mismatch, *Jour. of Guidance and Control*, Vol. 2, No. 6, pp. 538–540.

Stinton, Darrol 1996. *Flying Qualities and Flight Testing of the Airplane*, Reston, VA: AIAA, pp. 5, 6.

Thomas, H. H. B. M. and D. Küchemann 1974. Sidney Barrington Gates 1893–1973, *Biog. Memoirs of Fellows of the Royal Soc.*, Vol. 20, pp. 181–212.

Tischler, M. B. and R. H. Hoh 1982. Handling Qualities Criterion for Flight Path Control of V/STOL Aircraft, AIAA Paper 82-1292.

Warner, Edward P. and Frederick H. Norton 1970. Preliminary Report on Free Flight Tests, NACA Rept. 70.

Westbrook, Charles B. and Duane T. McRuer 1979. Handling Qualities and Pilot Dynamics, *Aero/Space Engineering*, Vol. 18, No. 5, pp. 26–32.

Woodcock, Robert J. and J. T. Browne 1986. The Mil-Prime Standard for Aircraft Flying Qualities, *Proc. of the AIAA Atmospheric Flight Mechanics Conf.*, New York: AIAA, pp. 232–238.

Chapter 4. Power Effects on Stability and Control

Gilruth, Robert R. and M. D. White 1941. Analysis and Prediction of Longitudinal Stability of Airplanes, NACA Rept. 711.

Goett, Harry J. and Noel Delany 1944. Effect of Tilt of the Propeller Axis on the Longitudinal Stability Characteristics of Single-Engine Airplanes, NACA Rept. 774.

Lee, John G. 1984. *It Should Fly Wednesday.* Connecticut: Mystic Publications.

McKinney, Marion O., Jr., Richard E. Kuhn, and John P. Reeder 1964. Aerodynamics and Flying Qualities of Jet V/STOL Airplanes, SAE Paper 864A.

Millikan, Clark B. 1940. The Influence of Running Propellers on Airplane Characteristics, *Jour. of the Aeronautical Sciences*, Vol. 7, No. 3, pp. 85–103.

Phillips, William H., H. L. Crane, and Paul A. Hunter 1944. Effect of Lateral Shift of the Center of Gravity on Rudder Deflection Required for Trim, NACA WR L-92.

Relf, E. F. 1922. An Electric Motor of Small Diameter for Use Inside Aeroplane Models, British R & M 778.

Ribner, Herbert S. 1944. Notes on the Propeller and Slipstream in Relation to Stability, NACA WR L-25.

Ribner, Herbert S. 1945. Formulas for Propellers in Yaw and Charts of the Side-Force Derivative, NACA Rept. 819.

Ribner, Herbert S. 1945. Propellers in Yaw, NACA Rept. 820.

Ribner, Herbert S. 1946. Field of Flow About a Jet and Effect of Jets on Stability of Jet-Propelled Airplanes, NACA ACR L6C13.

Smelt, Ronald and H. Davies 1937. Estimation of Increase in Lift Due to Slipstream, British R & M 1788.

Squire, H. B. and J. Trouncer 1944. Round Jets in a General Stream, British R & M 1974.

Vetter, Hans C. 1953. Effect of a Turbojet Engine on the Dynamic Stability of an Aircraft, *Jour. of the Aeronautical Sciences*, Vol. 20, No.11, pp. 797, 798.

Chapter 5. Managing Control Forces

Ames, Milton B., Jr. and Richard I. Sears 1941. Determination of Control-Surface Characteristics from NACA Plain-Flap and Tab Data, NACA Rept. 721.

Baumgarten, G. and W. Heine 1996. A New Reconfiguration Concept for Flight Control Systems in Case of Actuator and Control Surface Failures. *Proc. of the 20th Congress of the ICAS, Sorrento, Italy, Sept 8–13, 1996.*

Brown, W. S. 1941. Spring Tab Controls, British R & M 1979.

Bryant, L. W. and R. W. G. Gandy 1939. An Investigation of the Lateral Stability of Aeroplanes with Rudder Free, NPL Rept. S&C 1097.

Bureau of Aeronautics 1953. The Hydraulics System, Rept. AE-61-4 IV.

Chambers, Joseph R. 2000. *Partners in Freedom*, Monographs in Aerospace History No. 19, NASA SP-2000-4519, pp. 154, 155.

Choi, Seong-Wook, Keun-Shik Chang, and Honam Ok 2001. Parametric Study of Transient Spoiler Aerodynamics with Two-Equation Turbulence Models, *Jour. of Aircraft*, Vol. 38, No. 5, pp. 888–894.

Dunn, Orville R. 1949. Aerodynamically Boosted Surface Controls and Their Application to the DC-6 Transport, *I.A.S.-R.Ae.S. Proc.*, pp. 503–533.

Gates, Sidney B. 1940. Note on Differential Gearing as a Means of Aileron Balance, British R & M 2526.

Gates, Sidney B. 1941. Notes on the Spring Tab, Rept. B.A. 1665, British R.A.E.

Glauert, Hermann 1927. The Theoretical Relationships for an Aerofoil with Hinged Flap, British R & M 1095.

Glenn, John E. 1963. Manual Flight Control System Functional Characteristics, *IEEE Transactions on Human Factors in Electronics*, Vol HFE-4, No. 1, pp. 29–38.

Goranson, R. Fabian 1945. Flight Tests of Experimental Beveled Trailing Edge Frise Ailerons on a Fighter Airplane, NACA TN 1085.

Gough, Melvin N. and A. P. Beard 1936. Limitations of the Pilot in Applying Forces to Airplane Controls, NACA TN 550.

Graham, Dunstan and Duane T. McRuer 1991. Retrospective Essay on Nonlinearities in Aircraft Flight Control, *Jour. of Guidance, Control, and Dynamics*, Vol. 14, No. 6, pp. 1089–1099.

Greenberg, Harry 1944. Calculation of Stick Forces for an Elevator with a Spring Tab, NACA WR L-139.

Greenberg, Harry and Leonard Sternfield 1943. A Theoretical Investigation of the Lateral Oscillations of an Airplane with Free Rudder with Special Reference to the Effect of Friction, NACA Rept. 762.

Harschburger, H. E. 1983. Development of Redundant Flight Control Actuation Systems for the F/A-18 Strike Fighter, SAE Paper 831484.

Hess, R. A., W. Siwakosit, and J. Chung 2000. Accommodating a Class of Actuator Failures in Flight Control Systems, *Jour. of Guidance, Control, and Dynamics*, Vol. 23, No. 3, pp. 412–419.

Howard, R. W. 2000. Planning for Super Safety: The Fail-Safe Dimension, *The Aeronautical Jour.*, Vol. 104, No. 1041, pp. 517–555.

Jiang, Jin and Qing Zhao 2000. Design of Reliable Control Systems Possessing Actuator Redundancies, *Jour. of Guidance, Control, and Dynamics*, Vol. 23, No. 4, pp. 709–718.

Jones, Robert T. and Milton B. Ames 1942. Wind Tunnel Investigation of Control-Surface Characteristics. V. – The Use of a Beveled Trailing Edge to Reduce the Hinge Moment of a Control Surface, NACA ARR.

Jones, Robert T. and Doris Cohen 1941. An Analysis of the Stability of an Airplane with Free Controls, NACA Rept. 709.

Jones, Robert T. and Harold F. Kleckner 1943. Theory and Preliminary Flight Tests of an All-Movable Vertical Tail Surface, NACA WR L-496.

Jones, Robert T. and Albert I. Nerkin 1936. The Reduction of Aileron Operating Force by Differential Linkage, NACA TN 586.

Lyle, Bruce S. 1983. Development of Control Surface Actuation Systems on Various Configurations of the F-16, SAE Paper 831483.

Maskrey, Robert H. and W. J. Thayer 1978. A Brief History of Electrohydraulic Servomechanisms, *Jour. of Dynamic Systems, Measurement and Control* (ASME), Vol. 100, No. 2, pp. 110–116.

Mathews, Charles W. 1944 An Analytical Investigation of the Effect of Elevator Fabric Distortion on the Longitudinal Stability and Control of an Airplane, NACA ACR L4E30.

McAvoy, William H. 1937. Maximum Forces Applied by Pilots to Wheel-Type Controls, NACA TN 623.

McLean, D. 1999. Aircraft Flight Control Systems, *The Aeronautical Jour.*, Vol. 103, No. 1021, pp. 159–165.

McMahan, Jack 1983. Appendix A of Restructurable Controls, NASA CP 2277.

McRuer, Duane T. (chair) 1997. *Report of the Committee on the Effects of Aircraft-Pilot Coupling on Flight Safety*, Washington, DC: National Academy Press, pp. 26, 55–73.

Osder, Stephen 1999. Practical View of Redundancy Management – Application and Theory, *Jour. of Guidance, Control, and Dynamics*, Vol. 22, No. 1, pp. 12–21.

Perrin, W. G. 1928. The Theoretical Relationships for an Aerofoil With a Multiply Hinged Flap System, British R & M 1171.

Phillips, William H., B. P. Brown, and J. L. Matthews, Jr. 1953. Review and Investigation of Unsatisfactory Control Characteristics Involving Instability of Pilot-Airplane Combination and Methods of Predicting These Difficulties from Ground Tests, NACA RM L53F17a.

Phillips, William H. 1944. Application of Spring Tabs to Elevator Controls, NACA Rept. 797.

Rogallo, Francis M. 1944. Collection of Balanced Aileron Test Data, NACA WR (ACR) 4A11.

Root, L. Eugene 1939. Empennage Design with Single and Multiple Vertical Surfaces, *Jour. of the Aeronautical Sciences*, Vol. 6, No. 9, pp. 353–360.

Schaefer, W. S., L. J. Inderhees, and John F. Moynes 1991. Flight Control Actuation System for the B-2 Advanced Technology Bomber, *Proc. of the SAE Aerospace Atlantic Conf.*, SAE Paper 911112, 14pp.

Schmitt, Vernon R., James W. Morris, and Gavin D. Jenney 1998. *Fly-By-Wire – A Historical and Design Perspective*, SAE IBSN 0-7680-0218-4.

Sears, William R. 1987 (July) Flying Wing Could Stealthily Reappear, *Aerospace America*, Vol. 25, No. 7, pp. 16–19.

Silverstein, Abe and S. Katzoff 1940. Aerodynamic Characteristics of Horizontal Tail Surfaces, NACA Rept. 688.

Toll, Thomas A. 1947. A Summary of Lateral Control Research, NACA Rept. 868.

Tomayko, James E. 2000. Computers Take Flight: A History of NASA's Pioneering Digital Fly-by-Wire Project, NASA SP-200-4224.

Waterman, A. W. 1983, The Boeing 767 Hydraulic System, SAE Paper 831488.

Weick, Fred C. 1987. *From the Ground Up*, Washington, DC: Smithsonian Inst.

White, Roland J. 1950. Investigation of Lateral Dynamic Stability in the XB-47 Airplane, *Jour. of the Aeronautical Sciences*, Vol. 17, No. 3, pp. 133–148.

Yeung, W. W. H., C. Xu, and W. Gu 1997. Reduction of Transient Adverse Effects of Spoilers, *Jour. of Aircraft*, Vol. 34, No., 4, pp. 479–484.

Chapter 6. Stability and Control at the Design Stage

Abbott, Ira H., A. E. Von Doenhoff, and L. S. Stivers, Jr. 1945. Summary of Airfoil Data, NACA Rept. 824.

Anon. 1974. Royal Aeronautical Society (RAeS) Data Sheets, Engineering Sciences Data Item No. 74011, Aerodynamics Sub-series.

Campbell, John P. and Marion O. McKinney 1952. Summary of Methods for Calculating Dynamic Lateral Stability and Response and for Estimating Lateral Stability Derivatives, NACA Rept. 1098.

Bloy, A. W. and K. A. Lea 1995. Directional Stability of a Large Receiver Aircraft in Air-to-Air Refueling, *Jour. of Aircraft*, Vol. 32, No. 2, pp. 453–455.

DeYoung, John 1948. Theoretical Symmetric Span Loading at Subsonic Speeds for Wings Having Arbitrary Plan Form, NACA Rept. 921.

DeYoung, John 1951. Theoretical Antisymmetric Span Loading for Wings of Arbitrary Plan Form at Subsonic Speeds, NACA Rept. 1056.

DeYoung, John 1952. Theoretical Symmetric Span Loading due to Flap Deflection for Wings of Arbitrary Plan Form at Subsonic Speeds, NACA Rept. 1071.

De Young, John 1976. Vortex Lattice Utilization, NASA SP-405.

Diederich, Franklin W. 1951. Charts and Tables for Use in Calculations of Downwash of Wings of Arbitrary Plan Form, NACA TN 2353.

Falkner, V. M. 1943. The Calculation of Aerodynamic Loading on Surfaces of Any Shape, British A.R.C.R. and M. No. 1910.

Hoak, Donald E. et al. 1976 (rev.). USAF Stability and Control DATCOM, U.S. Air Force Flight Dynamics Lab.

House, Rufus O. and Arthur R. Wallace 1941. Wind-Tunnel Investigation of Effect of Interference on Lateral-Stability Characteristics of Four NACA 23012 Wings, an Elliptical and Circular Fuselage, and Vertical Fins, NACA Rept. 705.

Jameson, A., W. Schmidt, and E. Turkel 1981. Numerical Solution of the Euler Equation by Finite Volume Methods Using Runge-Kutta Time-Stepping Schemes, AIAA Paper 81-1259.

Jones, Robert T. 1946. Properties of Low-Aspect Ratio Pointed Wings at Speeds Below and Above the Speed of Sound, NACA Rept. 835.

Kaminer, Isaac I., Richard M. Howard, and Carey S. Buttrill 1997. Development of Closed-Loop Tail-Sizing Criteria for a High Speed Civil Transport, *Jour. of Aircraft*, Vol. 34, No. 5, pp. 638–664.

Kayten, Gerald G. and William Koven 1945. Comparison of Wind-Tunnel and Flight Measurements of Stability and Control Characteristics of a Douglas A-26 Airplane, NACA Rept. 816.

Multhopp, Hans 1941. Aerodynamics of the Fuselage, NACA TM 1036.

Munk, Max M. 1923. The Aerodynamic Forces on Airship Hulls, NACA Rept. 184.

Pearson, Henry A. and Robert T. Jones 1938. Theoretical Stability and Control Characteristics of Wings with Various Amounts of Taper and Twist, NACA Rept. 635.

Pulliam, T. H. and J. L. Steger 1989. Implicit Finite Difference Simulations of Three-Dimensional Compressible Flow, *AIAA Jour.*, Vol. 18, No. 2, pp. 159–167.

Silverstein, Abe and S. Katzoff 1939. Design Charts for Predicting Downwash Angles and Wake Characteristics behind Plain and Flapped Wings, NACA Rept. 648.

Smith, A. M. O. 1962. Incompressible Flow About Bodies of Arbitrary Shape, IAS Paper No. 62–143, presented at the IAS National Sciences Meeting, Los Angeles, CA, June.

Chapter 7. The Jets at an Awkward Age

Abzug, M. J. 1956. Application of Matrix Operators to the Kinematics of Airplane Motion, *Jour. of the Aeronautical Sciences*, Vol. 23, No. 7, pp. 679–684.

Gunston, Bill 1973. *Bombers of the West*, New York: Scribners.

Chapter 8. The Discovery of Inertial Coupling

Abzug, M. J. 1954. Effects of Certain Steady Motions on Small-Disturbance Airplane Motions, *Jour. of the Aeronautical Sciences*, Vol. 21, pp. 749–762.

Bergrun, Norman and Paul Nickel 1953. A Flight Investigation of the Effect of Steady Rolling on the Natural Frequencies of a Body-Tail Combination, NACA TN 2985.

Gates, O. B. and K. Minka 1959. Note on a Criterion for Severity of Roll-Induced Instability, *Jour. of the Aero/Space Sciences*, Vol. 26, No. 6, pp. 287–290.

Phillips, William H. 1948. Effect of Steady Rolling on Longitudinal and Directional Stability, NACA TN 1627.

Phillips, William H. 1992. Recollections of Langley in the Forties, *Jour. of the Amer. Aviation Historical Soc.*, Vol. 37, No. 2, pp. 116–127.

Pinsker, W. J. G. 1957. Critical Flight Conditions and Loads Resulting from Inertia Cross-Coupling and Aerodynamic Stability Deficiencies, RAE Tech Note Aero 2502.

Rhoads, D. W. and John M. Schuler 1957. A Theoretical and Experimental Study of Airplane Dynamics in Large-Disturbance Maneuvers, *Jour. of the Aeronautical Sciences*, Vol. 24, No. 7, pp. 507–526, 532.

Schy, Albert A. and M. E. Hannah 1977. Prediction of Jump Phenomena in Roll Coupled Maneuvers of Airplanes, *Jour. of Aircraft*, Vol. 14, pp. 375–382.

Seckel, E. 1964. *Stability and Control of Airplanes and Helicopters*, New York: Academic Press.

Stengel, Robert F. 1975. Effect of Combined Roll Rate and Sideslip Angle on Aircraft Flight Stability, *Jour. of Aircraft*, Vol. 12, No. 8, pp. 683–685.

Thomas, H. H. B. M. and P. Price 1960. A Contribution to the Theory of Aircraft Response in Rolling Manoeuvres Including Inertia Cross-Coupling Effects, British R & M 3349.

Weil, Joseph and Richard E. Day 1956. An Analog Study of the Relative Importance of Various Factors Affecting Roll Coupling, NACA RM H56A06.

Westbrook, Charles B. (ed.) 1956. *Transactions of the Wright Air Development Center Conference on Inertia Coupling of Aircraft*, 56WCLC-1041.

Young, J. W., A. A. Schy, and K. G. Johnson 1978. Prediction of Jump Phenomena in Aircraft Maneuvers, Including Nonlinear Aerodynamic Effects, *Jour. of Guidance and Control*, Vol. 1, No. 1, pp. 26–31.

Chapter 9. Spinning and Recovery

Abzug, M. J. 1977. Spin and Recovery Characteristics of the Rockwell Model 680F, ACA Systems Rept. R-143.

Adams, W. M. 1972. Analytical Prediction of Airplane Equilibrium Spin Characteristics, NASA TN D-6926.

Anderson, Seth B., Einar K. Enevoldson, and Luat T. Nguyen 1983. Pilot Human Factors in Stall/Spin Accidents of Supersonic Fighter Aircraft, in AGARD CP-347.

Arena, Andrew S., Jr., R. C. Nelson, and L. B. Schiff 1990. An Experimental Study of the Nonlinear Dynamic Phenomenon Known as Wing Rock, AIAA Paper 90-2812-CP, pp. 173–183.

Beaurain, L. 1977. General Study of Light Plane Spin, Aft Fuselage Geometry, Part 1, NASA TTF-17-446.

Beyers, Martin E. 1995. Interpretation of Experimemtal High-Alpha Aerodynamics – Implications for Flight Prediction, *Jour. of Aircraft*, Vol. 32, No. 2, pp. 247–261 (80 references).

Bihrle, William, Jr. 1957. Analytic Investigations of an Unconventional Airplane Spin, in *Trans. of the Wright Air Development Center Airplane Spin Symposium*, ed. C. B. Westbrook and H. K. Doetsch, 57 WCLC-1688 and -1774.

Bihrle, William, Jr. 1981. Influence of Wing, Fuselage, and Tail Design on Rotational Flow Aerodynamics Beyond Maximum Lift, *Jour. of Aircraft*, Vol. 18, No. 11, pp. 920–925.

Bihrle, William, Jr., and Billy Barnhart 1978. Design Charts and Boundaries for Identifying Departure Resistant Fighter Configurations, Naval Air Devel. Center Rept. 76154-30.

Bihrle, William, Jr. and Billy Barnhart 1983. Spin Prediction Techniques, *Jour. of Aircraft*, Vol. 20, No. 2, pp. 97–101.

Bowman, James S., Jr. 1971. Summary of Spin Technology as Related to Light General Aviation Airplanes, NASA TN D-6575.

Bowman, James S., Jr. 1989. Measurements of Pressures on the Tail and Aft Fuselage of an Airplane Model During Rotary Motions at Spin Attitudes, NASA TP-2939.

Burk, Sanger M., Jr., James S. Bowman, Jr., and William L. White 1977. Spin-Tunnel Investigation of the Spinning Characteristics of Typical Single-Engine General Aviation Designs. I – Low-Wing Model A: Effects of Tail Configurations, NASA TP-1009.

Chambers, Joseph R. 2000. *Partners in Freedom*, Monographs in Aerospace History No. 19, NASA SP-2000-4519, pp. 91–94, 102–105, 223.

Chambers, Joseph R. and H. Paul Stough III 1986. Summary of NASA Stall/Spin Research for General Aviation Configurations, AIAA Paper 86-2597.

Ericsson, Lars E. 1993. Slender Wing Rock Revisited, *Jour. of Aircraft*, Vol. 30, No. 3, pp. 352–356.

Fremaux, C. M. 1995. Estimation of the Moment Coefficients for Dynamically Scaled, Free-Spinning Wind-Tunnel Models, *Jour. of Aircraft*, Vol. 32, No. 6, pp. 1407–1409.

Gates, Sidney B. and L. W. Bryant 1926. The Spinning of Aeroplanes, British R & M 1001.

Goman, M. and A. Khrabov 1994. State-Space Representation of Aerodynamic Characteristics of an Aircraft at High Angles of Attack, *Jour. of Aircraft*, Vol. 31, No. 5, pp. 1109–1115.

Heller, Michael, Robert J. Niewoehner, and Kenneth P. Lawson 2001. F/A-18E/F Super Hornet High-Angle-of-Attack Control Law Development and Testing, *Jour. of Aircraft*, Vol. 38, No. 5, pp. 841–847.

Jahnke, C. C. and Fred E. C. Culik 1994. Application of Bifurcation Theory to the High-Angle-of-Attack Dynamics of the F-14, *Jour. of Aircraft*, Vol. 31, No. 1, pp. 26–34.

Jaramillo, Paul T. and M. G. Nagati 1995. Multipoint Approach for Aerodynamic Modeling in Complex Flowfields, *Jour. of Aircraft*, Vol. 32, No. 6, pp. 1335–1341.

Johnson, Joseph L., William A. Newsom, and Dale R. Satran 1980. Full-Scale Wind-Tunnel Investigation of the Effects of Wing Leading-Edge Modifications on the High Angle of Attack Aerodynamic Characteristics of a Low-Wing General Aviation Airplane, AIAA Paper 80-1844.

Johnston, Donald E. and Jeffrey R. Hogge 1976. Nonsymmetric Flight Influence on High Angle of Attack Handling and Departure, *Jour. of Aircraft*, Vol. 13, No. 2, pp. 112–118.

Kalviste, Juri 1978. Aircraft Stability Characteristics at High Angles of Attack, in AGARD CP 235.

Kalviste, Juri 1982. Use of Rotary Balance and Forced Oscillation Test Data in a Six-Degree-of-Freedom Simulation, AIAA Paper 82-1364.

Kalviste, Juri and Bob Eller 1989. Coupled Static and Dynamic Stability Parameters, AIAA Paper 89-3362.

Lee, Dong-Chan and M. G. Nagati 1999. Improved Coupled Force and Moment Parameter Estimation for Aircraft, AIAA Paper 99-4174, in *Collection of the AIAA Atmospheric Flight Mechanics Conference Technical Papers*, Portland, OR, 9–11 Aug. 1999, pp. 484–492.

Lee, Dong-Chan and M. G. Nagati 2000. Angular Momentum Control in Nonlinear Flight, *Jour. of Aircraft*, Vol. 37, No. 3, pp. 448–453.

Levin, Daniel and Joseph Katz 1992. Self-Induced Roll Oscillations of Low-Aspect-Ratio Rectangular Wings, *Jour. of Aircraft*, Vol. 29, No. 4, pp. 698–702.

Lowenberg, M. H. and Y. Patel 2000. Use of Bifurcation Diagrams in Piloted Test Procedures, *The Aeronautical Jour.*, Vol. 104, No. 1035, pp. 225–235.

Lutze, F. H., W. C. Durham, and W. H. Mason 1996. Unified Development of Lateral-Directional Departure Criteria, *Jour. of Guidance, Control, and Dynamics*, Vol. 19, No. 2, pp. 489–493.

Mangold, Peter 1991. Transformation of Flightmechanical Design Requirements for Modern Fighters into Aerodynamic Characteristics, in *Manoeuvering Aerodynamics*, AGARD CP-497.

Mitchell, David G. and Donald E. Johnston 1980. Investigation of High-Angle-of-Attack Maneuvering-Limiting Factors, AFWAL-TR-80-3141, Part II.

Moul, Martin T. and John W. Paulson 1958. Dynamic Lateral Behavior of High Performance Aircraft, NACA RM L58E16.

Neihouse, Anshal I., Jacob H. Lichtenstein, and Philip W. Pepoon 1946. Tail Design Requirements for Satisfactory Spin Recovery, NACA TN 1045.

Neihouse, Anshal I., Walter J. Klinar, and Stanley H. Sher 1960. Status of Spin Research for Recent Airplane Designs, NASA R-57.

Nelson, R. C. and A. S. Arena, Jr. 1992. An Experimental Investigation of Wing Rock of Slender Wings and Airplane Configurations, *Proc. of the Fluid Dynamics at High Angles of Attack Conference*, Tokyo.

Pinsker, W. J. G. 1967. Directional Stability in Flight with Bank Angle Constraint as a Condition Defining a Minimum Acceptable Value for n_v, RAE Technical Report 67-127.

Polhamus, Edward C. 1966. A Concept of the Vortex Lift of Sharp-Edge Delta Wings Based on a Leading Edge Suction Analogy, NASA TN D-3767.

Relf, E. F. and T. Lavemder 1918. Auto-Rotation of Stalled Aerofoils and Its relation to the Spinning Speed of Aeroplanes, British R & M 549.

Relf, E. F. and T. Lavender 1922, 1925. British R & M's 828 and 936.

Ross, A. Jean and Luat T. Nguyen 1988. Some Observations Regarding Wing Rock Oscillations at High Angles of Attack, AIAA Paper 88-4371-CP.

Ross, Holly M. and John N. Perkins 1994. Tailoring Stall Characteristics Using Leading-Edge Droop Modifications, *Jour. of Aircraft*, Vol. 31, No. 4, p. 767.

Scher, Stanley H. 1954. An Analytical Investigation of Airplane Spin Recovery Motion by Use of Rotary Balance Aerodynamic Data, NACA TN 3188.

Scher, Stanley H. 1955. Pilot's Loss of Orientation in Inverted Spins, NACA TN 3531.

Sitz, David M., David C. Nelson, and Mark R. Carpenter 1997. KEEP EAGLE F-15E High Angle of Attack Flight Test Program, *Jour. of Aircraft*, Vol. 34, No. 3, pp. 265–270.

Skow, Andrew M. and G. E. Erickson 1982. Modern Fighter Aircraft Design for High Angle of Attack Maneuvering, in *High Angle of Attack Aerodynamics*, in AGARD LS-121.

Soulé, Hartley A. and N. A. Scudder 1931. A Method of Flight Measurement of Spins, NACA Rept. 377.

Stephens, A. V. 1931. Free-Flight Spinning Investigations with Several Models, British R & M 1404.

Stephens, A. V. 1966. Some British Contributions to Aerodynamics, *Jour. of the Royal Aeronautical Soc.*, Vol. 70, No. 661, pp. 71–78.

Stough, H. Paul III, James M. Patton, Jr., and Steven M. Sliwa 1987. Flight Investigation of Tail Configuration on Stall, Spin, and Recovery Characteristics of a Low-Wing General Aviation Research Airplane, NASA TP-2644.

Tobak, Murray, G. T. Chapman, and L. B. Schiff 1984. Mathematical Modeling of the Aerodynamic Characteristics in Flight Mechanics, NASA TM-85880.

Tobak, Murray and L. B. Schiff 1976. On the Formulation of the Aerodynamic Characteristics in Aircraft Dynamics, NASA TR R-456.

Tristrant, D. and O. Renier 1985. AGARD CP 386, Paper No. 22.

Weissman, R. 1974. Status of Design Criteria for Predicting Departure Characteristics, AIAA Paper 74-791.

Westbrook, Charles B. and H. K. Doetsch (eds.) 1957. *Trans. of the Wright Air Development Center Airplane Spin Symposium*, 57WCLC-1688 and -1744.

Yip, Long P., Holly M. Ross, and David R. Robelen 1992. Model Flight Tests of a Spin-Resistant Trainer Configuration, *Jour. of Aircraft*, Vol. 29, No. 5, pp. 799–805.

Zimmerman, Charles H. 1936. Preliminary Tests in the NACA Free-Spinning Wind Tunnel, NACA Rept. 557.

Chapter 10. Tactical Airplane Maneuverability

Anon. 1980. Military Specification, Flying Qualities of Piloted Airplanes, MIL-F-8785C.

Anon. 1987. Flying Qualities of Piloted Vehicles, MIL-STD-1797, U.S. Air Force.

Arena, A. S., Jr., R. C. Nelson, and L. B. Schiff 1995. Directional Control at High Angles of Attack Using Blowing Through a Chined Forebody, *Jour. of Aircraft*, Vol. 32, No. 3, pp. 596–602.

Barham, Robert W. 1994. Thrust Vector Aided Maneuvering of the YF-22 Advanced Tactical Fighter Prototype, in AGARD CP 548.

Bihrle, William, Jr. 1966. A Handling Qualities Theory for Precise Flight-Path Control, AFFDL-TR-65-198.

Blight, James D., R. Lane Dailey, and Dagfinn Gangsass 1996. Practical Control Law Design Using Multivariable Techniques, in *Advances in Aircraft Flight Control*, ed. M. B. Tischler, London: Taylor and Francis, pp. 231–267.

Chambers, Joseph R. 2000. *Partners in Freedom*, Monographs in Aerospace History No. 19, NASA SP-2000-4519, p. 41.

Ericsson, Lars E. and Martin E. Byers 1997. Conceptual Fluid/Motion Coupling in the Herbst Supermaneuver, *Jour. of Aircraft*, Vol. 34, No. 3, pp. 271–277.

Field, Edmund J. and Ken F. Rossitto 1999. Approach and Landing Longitudinal Flying Qualities for Large Transports Based on In-Flight Results, AIAA Paper AIAA-99-4095.

Gautrey, J. E. and M. V. Cook 1998. A Generic Control Anticipation Parameter for Aircraft Handling Qualities Evaluation, *Aeronautical Jour.*, Vol. 102, No. 1013, pp. 151–159.

Gibson, John C. 1995. The Definition, Understanding and Design of Aircraft Handling Qualities, TU Delft Report LR-756.

Gibson, John C. 2000. Unpublished comments in letter, dated 10 December.

Greenwell, Douglas I. 1998. Frequency Effects on Dynamic Stability Derivatives Obtained from Small-Amplitude Oscillatory Testing, *Jour. of Aircraft*, Vol. 35, No. 5, pp. 776–783.

Hodgkinson, J., W. J. LaManna, and J. L. Hyde 1976. Handling Qualities of Aircraft with Stability and Control Augmentation Systems – A Fundamental Approach, *Aeronautical Jour.*, Vol. 80, No. 782, pp. 75–81.

Hoh, Roger H. and Irving L. Ashkenas 1977. Handling Quality Criterion for Heading Control, *Jour. of Aircraft*, Vol. 14, No. 2, pp. 142–150.

Hoh, Roger H. and David G Mitchell 1996. Handling-Qualities Specification – A Functional Requirement for the Flight Control System, in *Advances in Aircraft Flight Control*, ed. M. B. Tischler, London: Taylor and Francis, pp. 3–33.

Mangold, Peter 1991. Transformation of Flightmechanical Design Requirements for Modern Fighters into Aerodynamic Characteristics, in AGARD CP-497.

Mitchell, David G. and Roger H. Hoh 1982. Low-Order Approaches to High-Order Systems: Problems and Promises, *Jour. of Guidance*, Vol. 5, No. 5, pp. 482–488.

Myers, Thomas T., D. T. McRuer, and D. E. Johnston 1987. Flying Qualities Analysis for Nonlinear Large Amplitude Maneuvers, AIAA Paper 87-2904.

Nguyen, Luat T. and John V. Foster 1990. Development of a Preliminary High Angle of Attack Nose-Down Pitch Control Requirement for High-Performance Aircraft, NASA TM 101684.

Pedreiro, Nelson, Stephen M. Rock, Zeki Z. Celik, and Leonard Roberts 1998. Roll-Yaw Control at High Angle of Attack by Forebody Tangential Blowing, *Jour. of Aircraft*, Vol. 35, No. 1, pp. 69–77.

Ward, Greta N. and Uy-Loi Ly 1996. Stability Augmentation Design of a Large Flexible Transport Using Nonlinear Parameter Optimization, *Jour. of Guidance, Control, and Dynamics*, Vol. 19, No. 2, pp. 469–474.

Zagainov, G. 1993. High Maneuverability, Theory and Practice, AIAA Paper 93-4737 (1993 Wright Brothers Lecture).

Chapter 11. High Mach Number Difficulties

Anderson, Seth B. and Richard S. Bray 1955. A Flight Evaluation of the Longitudinal Stability Characteristics Associated with the Pitch-Up of a Swept-Wing Airplane in Maneuvering Flight at Transonic Speeds, NACA Rept. 1237.

Bilstein, Roger E. 1989, *Orders of Magnitude*, NASA SP-4406.

Chilstrom, Ken and Penn Leary 1993. *Test Flying at Old Wright Field*, Omaha: Westchester House.

Cook, William H. 1991. *The Road to the 707*, Bellevue, WA: TYC Publ. Co.

Ericson, Albert L. 1942. Investigation of Diving Moments of a Pursuit Airplane in the Ames 16-Foot High Speed Wind Tunnel, NACA WR A-65.

Erickson, Albert L. 1943. Wind Tunnel Investigation of Devices for Improving the Diving Characteristics of Airplanes, NACA WR A-66.

Furlong, Chester G. and James G. McHugh 1957. Summary and Analysis of the Low-Speed Longitudinal Characteristics of Swept Wings at High Reynolds Number, NACA Rept. 1339.

Gilyard, Glenn B. and John W. Smith 1978. Flight Experience with Altitude Hold and Mach Hold Autopilots on the YF-12 at Mach 3, in NASA CP 2054, Vol. I.

Hallion, Richard P. 1981. *Test Pilots: The Frontiersmen of Flight*, New York: Doubleday.

Hallion, Richard P. 1984. *On the Frontier: Flight Research at Dryden, 1946–1981*, NASA SP-4303.

Hood, Manley J. and Julian H. Allen 1943. The Problem of Longitudinal Stability and Control at High Speeds, NACA Rept. 767.

Jones, Robert T. 1946. Properties of Low-Aspect-Ratio Pointed Wings at Speeds Below and Above the Speed of Sound, NACA Rept. 835.

Loftin, Laurence K., Jr. 1985. *Quest for Performance*, NASA SP-468.

McRuer, Duane T. et al. 1992. Assessment of Flying-Quality Criteria for Air-Breathing Aerospacecraft, NASA CR 4442.

Myers, Thomas T., David H. Klyde, Duane T. McRuer, and Greg Larson 1993. Hypersonic Flying Qualities, Wright Laboratory Rept. WL-TR-93-3050.

Perkins, Courtland D. 1970. Development of Airplane Stability and Control Technology, *Jour. of Aircraft*, Vol. 7, No. 4, pp. 290–301.

Phillips, Edward H. 1994 (Feb. 21). NTSB: Pilots Need Training for High Altitude Stalls, AWST.

Sachs, Gottfried 1990. Effects of Thrust/Speed Dependence on Long-Period Dynamics in Supersonic Flight, *Jour. of Guidance, Control, and Dynamics*, Vol. 13, No. 6, pp. 1163–1186.

Schuebel, F. N. 1942. The Effect of Density Gradient on the Longitudinal Motion of an Aircraft, *Luftfahrtforschung*, Vol. 19, No. 4, R.T.P Translation 1739.

Shevell, Richard S. 1992. Aerodynamic Bugs: Can CFD Spray Them Away?, *Aerodynamic Analysis and Design*, AIAA Professional Studies Series, Palo Alto, CA: AIAA.

Shortal, Joseph A. and Bernard Maggin 1946. Effect of Sweepback and Aspect Ratio on Longitudinal Stability Characteristics of Wings at Low Speeds, NACA TN 1093.

Stengel, Robert F. 1970. Altitude Stability in Supersonic Cruising Flight, *Jour. of Aircraft*, Vol. 7, No. 5, pp. 464–473.

Sternfield, Leonard 1947. Some Considerations of the Lateral Stability of High-Speed Aircraft, NACA TN 1282.

Chapter 12. Naval Aircraft Problems

Ashkenas, Irving L. and Tulvio S. Durand 1963. Simulation and Analytical Studies of Fundamental Longitudinal Control Problems in Carrier Approach, *AIAA Simulation for Aerospace Conf.*, Columbus, OH.

Bezanson, A, F. 1961. Effects of Pilot Technique on Minimum Approach Speed, NATC Rept. PTR AD-3089.

Craig, Samuel, Robert Ringland, and Irving L. Ashkenas 1971. An Analysis of Navy Approach Power Compensation Problems and Requirements, STI Rept. 197-1.

Drinkwater, Fred J. III and George E. Cooper 1958. A Flight Evaluation of the Factors Which Influence the Selection of Landing Approach Speeds, NASA Memo 10-6-58A.

Heffley, Robert K. 1990. Outer-Loop Control Factors for Carrier Aircraft, Robert Heffley Engineering Rept. RHE-NAV-90-TR-1 (Limited Distrib.).

Neumark, S. 1953. Problems of Longitudinal Stability Below Minimum Drag Speed, and Theory of Stability Under Constraint, RAE Rept. Aero. 2504.

North, David M. 1993 (Aug. 30) Long Development Phase Nearly Over for Goshawk, AWST.

Shields, E.R. and D. J. Phelan 1953. The Minimum Landing Approach Speed of High Performance Aircraft, McDonnell Rept. 3232.

White, Maurice D., Bernard A. Schlaff, and Fred J. Drinkwater III 1957. A Comparison of Flight-Measured Carrier-Approach Speeds with Values Predicted by Several Different Criteria for 41 Fighter-Type Airplane Configurations, NACA RM A57L11.

Wilson, George C. 1992. *Flying the Edge*, Annapolis: Naval Institute Press.

Chapter 13. Ultralight and Human-Powered Airplanes

Anderson, Seth B. and Robert A. Ormiston 1994. A Look at Handling Qualities of High Performance Hang Gliders, AIAA Paper 94-3492 CP.

Brooks, W. G. 1998. Flight Testing of Flexwings, *Aerogram*, Cranfield College of Aeronautics, Vol. 9, No. 2, pp. 20–28.

Cook, Michael V. and Elizabeth A. Kilkenny 1987. An Experimental Investigation into Methods for Quantifying Hang Glider Airworthiness Parameters, Cranfield Institute of Technology, College of Aeronautics Report 8705.

Gracey, William 1941. The Additional-Mass Effects of Plates as Determined by Experiments, NACA Rept. 707.

Grosser, Morton 1981. *Gossamer Odyssey*, Boston: Houghton Mifflin.

Jex, Henry R. 1979. Gossamer Condor Dynamic Stability and Control Analysis, Systems Technology, Inc., Paper 240A.

Jex, Henry R. and David G. Mitchell 1982. Stability and Control of the Gossamer Condor Human-Powered Aircraft by Analysis and Flight Test, NASA CR 3627.

Mitchell, David G. and Henry R. Jex 1983. Flight Testing the Gossamer Albatross Human-Powered Aircraft, AIAA Paper 83-2699.

Roderick, W. E. B. 1986. Wind Tunnel Evaluation of Chinook WT-11 Ultra Light, NAE Aeronautical Note NAE-AN-35, Ottawa.

Rogallo, F. M., John G. Lowry, D. R. Croom, and R.T. Taylor 1960. Preliminary Investigation of a Paraglider, NASA TN D-443.

Chapter 14. Fuel Slosh, Deep Stall, and More

Abzug, M. J. 1959. Effects of Fuel Slosh on Stability and Control, Douglas Rept. ES 29551.

Abzug, M. J. 1999. Directional Stability and Control During Landing Rollout, *Jour. of Aircraft*, Vol. 36, No. 3, pp. 584–590.

Anderson, Seth B., Einar K. Enevoldson, and Luat T. Nguyen 1983. Pilot Human Factors in Stall/Spin Accidents of Supersonic Fighter Aircraft, in AGARD CP 347.

Anon. 1986. Beech Airframe Failure Report (Results of Flight Tests by Prof. Ronald O. Stearman), *The 1986 Pilot's Yearbook*, pp. 1–7.

Archer, Donald D. and Charles L. Gandy, Jr. 1957. T-37A Phase IV Performance and Stability and Control, ATTTC-TR-56-37.

Bollay, William 1937. A Theory for Rectangular Wings of Small Aspect Ratio, *Jour. of the Aeronautical Sciences*, Vol. 4, No. 7, pp. 294–296.

Chambers, Joseph R. 2000. *Partners in Freedom*, Monographs in Aerospace History No. 19, NASA SP-2000-4519, pp. 55, 56, 153.

Covert, Eugene E. 1993. Aerodynamic Hysteresis of Two-Dimensional Airfoils and Wings at Low Speed; An Empiricist's View, *Advances in Aerospace Sciences*, eds. P. Hajela and S. C. McIntosh, Stanford, CA: Stanford U., pp. 81–92.

Crawford, Charles C. and Jones P. Seigler 1958. KC-135A Stability and Control Test, AFFTC-TR-58-13.

Fischenberg, D. and R. V. Jategaonkar 1999. Identification of Aircraft Stall from Flight Test Data, *RTO MP-11*, pp. 17-1–17-8.

Hamel, Peter G. and Ravindra V. Jategaonkar 1996. Evolution of Flight Vehicle System Identification, *Jour. of Aircraft*, Vol. 33, No. 3, pp. 9–28 (with 183 references).

Hamel, Peter G. and R. V. Jategaonkar 1999. The Role of System Identification for Flight Vehicle Applications – Revisited, *RTO MP-11*, pp. 2-1–2-12.

Harper, John A. 1950. DC-3 Handling Qualities Tests, NACA, Paper presented to the Society of Experimental Test Pilots.

Heffley, Robert K. and Wayne F. Jewell 1972. Aircraft Handling Qualities Data, NASA CR 2144.

Iliff, Kenneth W. 1989. Parameter Estimation for Flight Vehicles, *Jour. of Guidance*, Vol. 12, No. 5, pp. 609–622.

Iliff, Kenneth W. and Richard E. Maine 1986. A Bibliography for Aircraft Parameter Estimation, NASA TM 86804.

Katzoff, S. and Harold H. Sweberg 1942. Ground Effect on Downwash and Wake Location, NACA Rept. 738.

Koehler, W. and K. Wilhelm 1977. Auslegung von Eingangssignalen für die Kennwertermittlung, *DFVLR-IB 154-77/40*, December.

Luskin, Harold and Ellis Lapin 1952. An Analytical Approach to the Fuel Sloshing and Buffeting Problems of Aircraft, *Jour. of the Aeronautical Sciences*, Vol. 19, No. 4, pp. 217–228.

Purser, Paul E. and John P. Campbell 1945. Experimental Verification of a Simplified Vee-Tail Theory and Analysis of Available Data on Complete Models with Vee Tails, NACA Rept. 823.

Reed, R. Dale 1997. *Wingless Flight: The Lifting Body Story*, NASA SP 4220, NASA Historical Series.

Schairer, George S. 1941. Directional Stability and Vertical Surface Stalling, *Jour. of the Aeronautical Sciences*, Vol. 8, No. 7, pp. 270–275.

Schy, Albert A. 1952. A Theoretical Analysis of the Effects of Fuel Motion on Airplane Dynamics, NACA Rept. 1080.

Soderman, Paul T. and Thomas N. Aiken 1971. Full-Scale Wind-Tunnel Tests of a Small Unpowered Jet Aircraft with a T-Tail, NASA TN D-6573.

Taylor, Robert T. and Edward J. Ray 1985. Deep-Stall Aerodynamic Characteristics of T-Tail Aircraft, in *Conference on Aircraft Operating Problems*, NASA SP-63.

Weick, Fred C. 1936. Everyman's Airplane – A Move Toward Simpler Flying, *S.A.E. Jour. (Transactions)*, Vol. 38, No. 5, pp. 176–189.

Weiss, Suzanne, Holger Friehmelt, Ermin Plaetschke, and Detlef Rohlf 1996. X-31 System Identification Using Single-Surface Excitation at High Angles of Attack, *Jour. of Aircraft*, Vol. 33, No. 3, pp. 485–490.

Wetmore, J. W. and L. I. Turner, Jr. 1940. Determination of Ground Effect from Tests of a Glider in Towed Flight, NACA Rept. 695.

Chapter 15. Safe Personal Airplanes

Anon. 1994. Used Aircraft Guide, Piper PA-31T Series Cheyennes, *The Aviation Consumer*, Vol. 24, No. 9, pp. 4–11.

Anon. 1999. *Aeronautical Information Manual (AIM)*, ASA-99-FAR-AM-BK, Aviation Supplies and Academics, Newcastle, WA.

Barber, M. R., C. K. Jones, T. R. Sisk, and F. W. Haise 1966, An Evaluation of the Handling Qualities of Seven General-Aviation Aircraft, NASA TN D-3726.

Bar-Gill, Aharon and Robert F. Stengel 1986. Longitudinal Flying Qualities Criteria for Single-Pilot Instrument Flight Operation, *Jour. of Aircraft*, Vol. 23, No. 2, pp. 111–117.

Campbell, John P., Paul A. Hunter, Donald E. Hewes, and James B. Whitten 1952. Flight Investigation of Control Centering Springs on the Apparent Spiral Stability of a Personal-Owner Airplane, NACA Rept. 1092.

Ferree, William M. 1994. Lightening Up on Design, *Technology Review*, MIT, Jan., p. 6.

Goode, M. W. et al. 1976. Landing Practices of General Aviation Pilots in Single-Engine Light Airplanes, NASA TN D-8283.

Greer, H. Douglas et al. 1973. Wind Tunnel Investigation of Static Longitudinal and Lateral Characteristics of a Full-Scale Mockup of a Light Single-Engine High-Wing Airplane, NASA TN D-7149.

Loschke, Paul C. et al. 1974. Flight Evaluation of Advanced Control Systems and Displays on a General Aviation Airplane, NASA TN D-7703.

Pendray, G. Edward 1964. *The Guggenheim Medalists 1929–1963*, New York: The Guggenheim Board of Award of the United Engr. Trustees, Inc.

Phillips, W. H. 1998. *Journey in Aeronautical Research*, Monographs in Aerospace History, Number 12, Washington, DC: NASA.

Phillips, W. H., Helmut A. Kuehnel, and James B. Whitten 1957. Flight Investigation of the Effectiveness of an Automatic Aileron Trim Control Device for Personal Airplanes, NACA Rept. 1304.

Picon, Gary 1994. Autopilot Buyer's Guide, *The Aviation Consumer*, July 1.

Regis, Edward 1995. Spratt, Schmittle, and Freewing, *Air & Space*, Vol. 9, No. 5, pp. 58–65.

Upson, Ralph H. 1942. New Developments in Simplified Control, *Jour. of the Aeronautical Sciences*, Vol. 9, No. 14, pp. 515–520, 548.

Chapter 16. Stability and Control Issues with Variable Sweep

Kroo, Ilan 1992. *The Aerodynamic Design of Oblique Wing Aircraft*, AIAA Professional Studies Series, Palo Alto, CA: AIAA.

Loftin, Laurence H., Jr. 1985. *Quest for Performance*, NASA SP 468.

Nelms, W. 1976. Applications of Oblique Wing Technology, AIAA Paper 76-943.

Nguyen, Luat T. et al. 1980. Application of High Angle of Attack Control System Concepts to a Variable-Sweep Fighter Airplane, AIAA Paper 80-1582-CP.

Polhamus, Edward C. and Thomas A. Toll 1981. Research Related to Variable Sweep Aircraft Development, NASA TM 83121.

Chapter 17. Modern Canard Configurations

Agnew, J. W., G. W. Lyerla, and Sue B. Grafton 1980. Linear and Nonlinear Aerodynamics of 3-Surface Aircraft Concepts, AIAA Paper 80-1581-CP.

Chambers, Joseph R. and Long P. Yip 1984. Aerodynamic Characteristics of Two General Aviation Canard Configurations at High Angles of Attack, AIAA Paper 84-2198.

Jones, B. Melvill 1934. Dynamics of the Airplane, in *Aerodynamic Theory*, Vol. V, ed. W. F. Durand, Berlin: Springer, pp. 208–214.

Lorincz, D. J. 1980. Flow Visualization of the HiMat RPV, NASA CR-163094.

McCormick, Barnes W. 1979. *Aerodynamics, Aeronautics, and Flight Mechanics*, New York: Wiley, pp. 613–617.

Yeager, Jeana, Dick Rutan, and Phil Patton 1987. *Voyager*, New York: Knopf.

Yip, Long P. 1985. Wind-Tunnel Investigation of a Full-Scale Canard-Configured General Aviation Airplane, NASA TP-2382.

Chapter 18. Evolution of the Equations of Motion

Abzug, M. J. 1980. Hinged Vehicle Equations of Motion, AIAA Paper 80-0364.

Abzug, M. J. 1998. *Computational Flight Dynamics*, AIAA Education Series, Reston, VA: AIAA, pp. 105, 106.

Abzug, M. J. and W. P. Rodden 1993. The Centroidal Siren and Computational Flight Mechanics, in *Advances in Aerospace Sciences*, Stanford, CA: Stanford Univ.

Anon. 1996. *AIAA Guide to Reference and Standard Atmospheric Models*, G-003A-1996, Reston,: VA AIAA.

Ashkenas, Irving L. and Duane T. McRuer 1958. Approximate Airframe Transfer Functions and Application to Single-Sensor Control Systems, WADC Technical Report 58-82, Wright-Patterson Air Force Base, Ohio.

Bairstow, Leonard 1920. *Applied Aerodynamics*, London: Longmans, Green and Co.

Bairstow, Leonard, J. L. Nayler, and R. Jones 1914. Investigation of the Stability of an Aeroplane When in Circling Flight, British R & M 154.

Beam, Benjamin H. 1956. A Wind-Tunnel Test Technique for Measuring the Dynamic Rotary Stability Derivatives at Subsonic and Supersonic Speeds, NASA Rept. 1258.

Bernstein, L. 1998. On the Equations of Motion for an Aircraft with an Internal Moving Load Which Is Then Dropped, *Aeronautical Jour.*, Vol. 102, No. 1011, pp. 9–24.

Bray, Richard S. 1984. A Method for Three-Dimensional Modeling of Wind-Shear Environments for Flight Simulator Applications, NASA TM 85969.

Bryan, G. H. and W. E. Williams 1903. The Longitudinal Stability of Aerial Gliders, *Proc. of the Royal Soc.*, Ser. A, 73, No. 489.

Bryant, L. W. and S. B. Gates 1937. Nomenclature for Stability Coefficients, British R & M 1801.

Bryant, L. W., I. M. W. Jones, and G. L. Pawsey 1932. The Lateral Stability of an Aeroplane Beyond the Stall, British R & M 1519.

Chen, Robert T. N. 1983. Efficient Algorithms for Computing Trim and Small-Disturbance Equations of Motion of Aircraft in Coordinated and Uncoordinated, Steady, Steep Turns, NASA TM 84324.

Cook, M. V., J. M. Lipscombe, and F. Goineau 2000. Analysis of the Stability Modes of the Non-Rigid Airship, *Aeronautical Jour.*, Vol. Vol. 104, No. 1036, pp. 279–290.

Frazer, R. A., W. J. Duncan, and A. R. Collar 1950. *Elementary Matrices*, London: Cambridge U. Press.

Gates, Sidney B. 1927. A Survey of Longitudinal Stability Beyond the Stall, with an Abstract for Designer's Use, British R & M 1118.

Glauert, H. 1927. A Nondimensional Form of the Stability Equations of an Aeroplane, British R & M 1093.

Heffley, Robert K. and Wayne F. Jewell 1972. Aircraft Handling Qualities Data, NASA CR 2144.

Hopkin, H. R. 1966. A Scheme of Notation and Nomenclature for Aircraft Dynamics and Associated Aerodynamics, RAE Tech. Rept. 66200.

ICAO, 1955. Standard Atmosphere – Tables and Data for Altitudes to 65,800 Feet, NACA Rept. 1235.

ICAO, 1962. U.S. Standard Atmosphere, 1962, NASA, USAF, U.S. Weather Bureau, Washington, DC.

Johnson, Walter A., Gary L. Teper, and Herman A. Rediess 1974. Study of Control System Effectiveness in Alleviating Vortex Wake Upsets, *Jour. of Aircraft*, Vol. 11, No. 3, pp. 148–154.

Jones, B. M. 1934. Dynamics of the Aeroplane, Divsion N of *Aerodynamic Theory*, Vol. V, ed. by W. F. Durand, California, Durand Reprinting Committee.

Jones, B. M. and A. Trevelyan 1925. Step-by-Step Calculations Upon the Asymmetric Movements of Stalled Airplanes, British R & M 999.

Jones, Robert T. 1936. A Simplified Application of the Method of Operators to the Calculation of the Disturbed Motion of an Airplane, NACA TR 560.

Kamesh, S. and S. Pradeep 1999. Phugoid Approximation Revisited, *Jour. of Aircraft*, Vol. 36, No. 2, pp. 465–467.

Lehman, John M., Robert K. Heffley, and Warren F. Clement 1977. Simulation and Analysis of Wind Shear Hazard, FAA-RD-78-7.

McMinn, John D, and John D. Shaughnessy 1991. Atmospheric Disturbance Model for NASP Applications, in NACA TM 4331.

McRuer, Duane, Irving L. Ashkenas, and Dunstan Graham 1973. *Aircraft Dynamics and Automatic Control*, Princeton, NJ: Princeton U. Press, pp. 296–316, 353–380.

Melsa, James L. and Stephen K. Jones 1973. *Computer Programs for Computational Assistance in the Study of Linear Control Theory*, New York: McGraw-Hill.

Milne-Thomson, M. 1958. *Theoretical Aerodynamics*, London: Macmillan.

Mokrzycki, G. A. 1950. Application of the Laplace Transformation to the Solution of the Lateral and Longitudinal Stability Equations, NACA TN 2002.

Mueller, Robert K. 1937. The Graphical Solution of Stability Problems, *Jour. of the Aeronautical Sciences*, Vol. 4, No. 8, pp. 324–331.

Mulkens, Marc J. M. and Albert O. Ormerod 1993. Measurements of Aerodynamic Rotary Stability Derivatives Using a Whirling Arm Facility, *Jour. of Aircraft*, Vol. 30, No. 2, p. 178.

Myers, Thomas T., David H. Klyde, Duane T. McRuer, and Greg Larson 1993. Hypersonic Flying Qualities, WL-TR-93-3050.

Neumark, S. 1957. Problems of Longitudinal Stability Below Minimum Drag Speed and Theory of Stability Under Constraint, British R & M 2983.

Newell, F. D. 1965. Ground Simulator Evaluations of Coupled Roll-Spiral Mode Effects on Aircraft Handling Qualities, AFFDL-TR-65-39.

Phillips, W. F. 2000. Improved Closed-Form Approximation for Dutch Roll, *Jour. of Aircraft*, Vol. 37, No. 3, pp. 484–490.

Phillips, W. F., C. E. Hailey, and G. A. Gebert 2001. Review of Attitude Representations Used for Airplane Kinematics, *Jour. of Aircraft*, Vol. 38, No. 4, pp. 718–737, (168 references).

Pinsker, W. J. G. 1967. Directional Stability in Flight with Bank Angle Constraint as a Condition Defining a Minimum Acceptable Value for n_v, RAE Tech. Rept. 67-127.

Powers, Bruce G. and Lawrence J. Schilling 1980, 1985. Subroutine DERIVC [of the SIM2 program], NASA Dryden Research Center.

Regan, Frank J. and Satya M. Anandakrishnan 1993. *Dynamics of Atmospheric Re-Entry*, Washington, DC: AIAA, pp. 389–398.

Ribner, Herbert S. 1956. Spectral Theory of Buffeting and Gust Response: Unification and Extension, *Jour. of the Aeronautical Sciences*, Vol. 23, No. 12, pp. 1075–1077, 1118.

Robinson, A. C. 1957. On the Use of Quaternions in Simulation of Rigid Body Motion, Wright Air Development Center Rept. TR-58-17.

Schilling, Lawrence J., Marlin D. Pickett, and David M. Aubertin 1993. The NASP Integrated Atmospheric Model, *1993 National Aero-Space Plane Technology Review*, Monterey, CA., NASA Publication GWP-ZAA.

Spilman, Darin R. and Robert F. Stengel 1995. Jet Transport Response to a Horizontal Wind Vortex, *Jour. of Aircraft*, Vol. 32, No. 3, pp. 480–485.

Sternfield, Leonard 1947. Some Considerations of the Lateral Stability of High-Speed Aircraft, NACA TN 1282.

Stevens, Brian L. and Frank L. Lewis 1992. *Aircraft Control and Simulation*, New York: Wiley, pp. 132–139.

Strumpf, Albert 1979. Stability and Control, *Hydroballistics Design Handbook, Vol. 1*, SEAHAC TR 79-1.

Teper, Gary L. 1969. Aircraft Stability and Control Data, NASA CR 96008.

Thelander, J. A. 1965. Aircraft Motion Analysis, FDL-TR-64-70.

Workman, F. 1924. Analysis of the Motion of an SE 5 Aeroplane by Step-by-Step Integration, Unpublished Report of the British A.R.C. (T 1918).

Zimmerman, Charles H. 1935. An Analysis of Longitudinal Stability in Power-Off Flight with Charts for Use in Design, NACA Rept. 521.

Zimmerman, Charles H. 1937. An Analysis of Lateral Stability in Power-Off Flight with Charts for Use in Design, NACA Rept. 589.

Zipfel, Peter H. 2000. *Modeling and Simulation of Aerospace Vehicles*, AIAA Education Series, Reston, VA: AIAA.

Chapter 19. The Elastic Airplane

Abzug, M. J. 1974. Equations of Motion for an Aircraft with a Semi-Rigid Fuselage, Northrop Rept. NOR 74-112.

Anderson, L. R. 1993. Order Reduction of Aeroelastic Models Through LK Transformation and Riccati Iteration, AIAA Paper 93-3795.

Ashkenas, I. L., R. E. Magdaleno, and D. T. McRuer 1983. Flight Control and Analysis Methods for Studying Flying and Ride Qualities of Flexible Transport Aircraft, NASA CR-172201.

Bisplinghoff, Raymond L. and Holt Ashley 1962. *Principles of Aeroelasticity*, New York: Wiley.

Bisplinghoff, Raymond L., Holt Ashley, and R. L. Halfman 1955. *Aeroelasticity*, Cambridge: Addison-Wesley.

Britt, Robert T., Steven B. Jacobson, and Thomas D. Arthurs 2000. Aeroservoelastic Analysis of the B-2 Bomber, *Jour. of Aircraft*, Vol. 37, No. 5, pp. 745–752.

Buttrill, Carey S. 1989. Results of Including Geometric Nonlinearities in an Aeroelastic Model of an F/A-18, in NASA CP 3031, Part 2.

Buttrill, Carey S., Thomas A. Zeiler, and P. Douglas Arbuckle 1987. Nonlinear Simulation of a Flexible Aircraft in Maneuvering Flight, AIAA Paper 87-2501-CP.

Cavin, R. K. and A. R. Dusto 1977. Hamilton's Principle: Finite-Element Methods and Flexible Body Dynamics, *AIAA Jour.*, Vol. 15, No. 12, pp. 1684–1690.

Cole, Henry A., Jr., Stuart C. Brown, and Euclid C. Holleman 1957. Experimental and Predicted Longitudinal and Lateral-Directional Response Characteristics of a Large Flexible 35° Swept-Wing Airplane at an Altitude of 35,000 Feet, NACA Rept. 1330.

Collar, A. R. and F. Grinsted 1942. The Effects of Structural Flexibility of Tailplane, Elevator, and Fuselage on Longitudinal Stability and Control, British R & M 2010.

Cox, H. R. and A. G. Pugsley 1932. Theory of Loss of Lateral Control Due to Wing Twisting, British R & M 1056.

Diederich, Franklin W. and Kenneth A. Foss 1953. Charts and Approximate Formulas for the Estimation of Aeroelastic Effects on the Loading of Swept and Unswept Wings, NACA Rept. 1140.

Dowell, Earl H., H. C. Curtiss, Jr., R. H. Scanlan, and F. Sisto 1990. *A Modern Course in Aeroelasticity*, 2nd ed., Alphen aan der Rijn: Sijthoff and Noordhoff.

Duncan, W. J. 1943. The Representation of Aircraft Wings, Tails and Fuselages by Semi-Rigid Structures in Dynamic and Static Problems, British R & M 1904.

Dusto, Arthur R. et al. 1974. A Method for Predicting the Stability Derivatives of an Elastic Airplane; Vol. I – FLEXSTAB Theoretical Description, NASA CR 114712.

Dykman, John R. and William P. Rodden 2000. Structural Dynamics and Quasistatic Aeroelastic Equations of Motion, *Jour. of Aircraft*, Vol. 37, No. 3, pp. 538–542.

Etkin, Bernard 1972. *Dynamics of Atmospheric Flight*, New York: Wiley.

Foss, Kenneth A. and Franklin W. Diederich 1953. Charts and Approximate Formulas for the Estimation of Aeroelastic Effects on the Lateral Control of Swept and Unswept Wings, NACA Rept. 1139.

Fung, Y. C. 1955. *An Introduction to the Theory of Aeroelasticity*, New York: Wiley.

Haug, E. J. 1989. *Computer-Aided Kinematics and Dynamics of Mechanical Systems*, Boston: Allyn and Bacon.

Lovell, P. M., Jr. 1948. The Effect of Wing Bending Deflection on the Rolling Moment Due to Sideslip, NACA TN 1541.

Milne, R. D. 1964. Dynamics of the Deformable Airplane, British R & M 3345.

Milne, R. D. 1968. Some Remarks on the Dynamics of Deformable Bodies, *AIAA Jour.*, Vol. 6, No. 3, pp. 556–558.

Newman, Brett and David K. Schmidt 1994. Truncation and Residualization with Weighted Balanced Coordinates, *Jour. of Guidance, Control, and Dynamics*, Vol. 17, No. 6, pp. 1299–1307.

Pai, S. I. and William R. Sears 1949. Some Aeroelastic Properties of Swept Wings, *Jour. of the Aeronautical Sciences*, Vol. 16, No. 2, pp. 105–115, 119.

Phillips, W. Hewitt 1998. *Journey in Aeronautical Research*, NASA Monographs in Aerospace History, Number 12, pp. 145–150.

Rodden, William P. 1955. A Simplified Expression for the Dihedral Effect of a Flexible Wing, *Jour. of the Aeronautical Sciences*, Vol. 22, No. 8, p. 579.

Rodden, William P. 1965. Dihedral Effect of a Flexible Wing, *Jour. of Aircraft*, Vol. 2, No. 5, pp. 368–373.

Rodden, William P. and Erwin H. Johnson 1994. *MSC/NASTRAN Aeroelastic Analysis User's Guide*, Costa Mesa, CA: MacNeal-Schwendler Corp.

Rodden, William P. and J. Richard Love 1985. Equations of Motion of a Quasistatic Flight Vehicle Utilizing Restrained Static Aeroelastic Characteristics, *Jour. of Aircraft*, Vol. 22, No. 9, pp. 802–809.

Schmidt, David K. and David L. Raney 2001. Modeling and Simulation of Flexible Flight Vehicles, *Jour. of Guidance, Control, and Dynamics*, Vol. 24, No. 3, pp. 539–546.

Skoog, Richard B. 1957. An Analysis of the Effects of Aeroelasticity on the Static Longitudinal Stability and Control of a Swept-Wing Airplane, NACA Rept. 1298.

Ward, Greta and Uy-Loi Ly 1996. Stability Augmentation Design of a Large Flexible Transport Using Nonlinear Parameter Optimization, *Jour. of Guidance, Control and Dynamics*, Vol. 19, No. 2, pp. 469–474.

Chapter 20. Stability Augmentation

Ashkenas, Irving L. 1988. Pilot Modeling Applications, Chapter 3 in *Advances in Flying Qualities*, AGARD LS 157.

Ashkenas, Irving L. and David H. Klyde 1989. Tailless Aircraft Performance Improvements with Relaxed Static Stability, NASA CR 181806.

Atzhorn, David and Robert F. Stengel 1984. Design and Flight Test of a Lateral-Directional Command Augmentation System. *Jour. of Guidance*, Vol. 7, No. 3, pp. 361–368.

Beh, H. and G. Hofinger 1994. X-31A Control Law Design, in AGARD CP-548, *Technologies for Highly Manoeuvrable Aircraft*.

Blight, James D., R. Lane Dailey, and Dagfinn Gangsass 1996. Practical Control Law Design Using Multivariable Techniques, in *Advances in Aircraft Flight Control*, ed. By M. B. Tischler, London: Taylor and Francis, pp. 231–267.

Bollay, William 1951. Aerodynamic Stability and Automatic Control, *Jour. of the Aeronautical Sciences*, Vol. 18, No. 9, pp. 569–624.

Bryson, Arthur E., Jr. 1994. *Control of Spacecraft and Aircraft*, Princeton, NJ: Princeton U. Press, pp. 328–342.

Burken, John J. and Frank W. Burcham, Jr. 1997. Flight-Test Results of Propulsion-Only Emergency Control System on MD-11 Airplane, *Jour. of Guidance, Control, and Dynamics*, Vol. 20, No. 5, pp. 980–987.

Chandler, Phillip R. and David W. Potts 1983. Shortcomings of Modern Control as Applied to Fighter Flight Control Design, in *Proc. of the 22nd IEEE Conference on Decision and Control, Vol. 3,* San Antonio, Texas, Dec.16.

Clarke, Robert, John J. Burken, John T. Bosworth, and Jeffery E. Bauer 1994. X-29 Flight Control System: Lessons Learned, NASA TM 4598.

Cook, M. V. 1999. On the Design of Command and Stability Augmentation Systems for Advanced Technology Aeroplanes, *Trans Inst. Measurement and Control*, Vol. 21, No. 2/3, pp. 85–98.

Cook, M. V. 2000. Private correspondence.

Doyle, John C. and Gunter Stein 1981. Multivariable Feedback Design: Concepts for a Classical/Modern Synthesis, *IEEE Transactions on Automatic Control*, Vol. AC-26, No. 1, pp. 4–16.

Elgerd, Olle I. and William C. Stephens 1959. Effect of Closed-Loop Transfer Function Pole and Zero Locations on the Transient Response of Linear Systems, *AIEE Applications and Industry*, No. 42, pp. 121–127.

Evans, Walter R. 1948. Graphical Analysis of Control Systems, *AIEE Trans.*, Vol. 67, pp. 547–551.

Gibson, John 1995. The Definition, Understanding and Design of Aircraft Handling Qualities, TU Delft Report LR-756.

Gibson, John 1999. *Development of a Methodology for Excellence in Handling Qualities Design for Fly by Wire Aircraft*, Series 03 Control and Simulation 06, Delft: Delft U.

Gibson, John 2000. Private correspondence.

Graham, Dunstan and D.T. McRuer 1991. Retrospective Essay on Nonlinearities in Aircraft Flight Control, *Jour. of Guidance, Control, and Dynamics*, Vol. 14, No. 6, pp. 1089–1099.

Hanson, Gregory D. and Robert F. Stengel 1984. Effects of Displacement and Rate Saturation on the Control of Statically Unstable Aircraft, *Jour. of Guidance*, Vol. 7, No. 2, pp. 197–205.

Hoh, Roger H. and David G. Mitchell 1982. Flying Qualities of Relaxed Static Stability Aircraft, DOT/FAA/CT-82/130-I.

Imlay, Frederick H. 1940. A Theoretical Study of Lateral Stability with an Automatic Pilot, NACA Rept. 693.

Jarvis, Calvin R. 1975. An Overview of NASA's Digital Fly-By-Wire Technology Development Program, TN D-7843.

Johnston, Donald J. and Duane T. McRuer 1977. Investigation of Limb-Sidestick Dynamic Interaction with Roll Control, *Jour. of Aircraft*, Vol. 10, No. 2, pp. 178–186.

Klyde, David, Duane McRuer, and Thomas Myers 1995. Unified Pilot-Induced Oscillation Theory. Vol. I: PIO Analysis with Linear and Nonlinear Effective Vehicle Characteristics, Including Rate Limiting, Report No. WL-TR 96-3208, Wright-Patterson A. F. Ohio, Wright Laboratory.

Koehler, R. 1999. Unified Approach for Roll Ratcheting Analysis, *Jour. Of Guidance, Control, and Dynamics*, Vol. 22, No. 5, pp. 718–720.

McRuer, Duane T. 1950. The Flying Wing's Electronic Tail, Honeywell *Flight Lines*, Vol. 1, No. 2, pp. 6, 7.

McRuer, Duane T. 2001. Private correspondence.

McRuer, Duane T. and Donald E. Johnston 1975. Flight Control System Properties and Problems, Vol. 1, NASA CR 2500.

McRuer, Duane T. and Thomas T. Myers 1988. Advanced Piloted Aircraft Flight Control System Design Methodology, Vol. I: Knowledge Base, NASA CR 181726.

McRuer, D. and R. L. Stapleford 1963. Sensitivity and Modal Response for Single Loop and Multiloop Systems, ASD-TR-62-812.

McRuer, Duane, Irving L. Ashkenas, and Dunstan Graham 1973. *Aircraft Dynamics and Automatic Control*, Princeton, NJ: Princeton U. Press, pp. 135–153.

McRuer, Duane T., Donald E. Johnston, and Thomas T. Myers 1985. A Perspective on Superaugmented Flight Control: Advantages and Problems, *Active Control Systems – Review, Evaluation and Projections*, AGARD CP 384.

McRuer, Duane T., Thomas T. Myers, and Peter M. Thompson 1989. Literal Singular-Value-Based Flight Control System Design Techniques, *Jour. of Guidance*, Vol. 12, No. 6, pp. 913–919.

Milliken, William F., Jr. 1947. Progress in Dynamic Stability and Control Research, *Jour. of the Aeronautical Sciences*, Vol. 14, No. 9, pp. 493–519.

Mitchell, David G. and Roger H. Hoh 1984. Influence of Roll Command Augmentation Systems on Flying Qualities of Fighter Aircraft, *Jour. of Guidance*, Vol. 7, No. 1, pp. 99–103.

Montoya, R. J. et al. 1983. Restructurable Controls, NASA Conference Publication 2277.

Moorhouse, David J. 1993. Decoupling of Aircraft Responses, *Stability in Aerospace Systems*, AGARD Rept. 789.

Morgan, H. B. 1947. Control in Low-Speed Flight, *Aeroplane*, Vol. 73, No. 1891, pp. 281–284.

Mukhopadhyay, V. and J. R. Newsom 1984. A Multiloop System Stability Margin Study Using Matrix Singular Values, *Jour. of Guidance*, Vol. 7, No. 5, pp. 582–587.

Myers, Thomas, Duane McRuer, and Donald E. Johnston 1984. Flying Qualities and Control System Characteristics for Superaugmented Aircraft, NASA CR 170419.

Osder, Stephen 1999. Practical View of Redundancy Management – Application and Theory, *Jour. of Guidance, Control, and Dynamics*, Vol. 22, No. 1, pp. 12–21.

Osder, Stephen 2000. Private correspondence.

Phillips, William H. 1989. Flying Qualities from Early Airplanes to the Space Shuttle, *Jour. of Guidance, Control, and Dynamics*, Vol. 12, No. 4, pp. 449–459.

Phillips, W. Hewitt 1998. *Journey in Aeronautical Research*, Monographs in Aerospace History, Number 12, Chapter 13, NASA History Office, NASA Hq., Washington, DC, 20546.

Safanov, M. G., A. J. Laub, and G. L. Hartmann 1981. Feedback Properties of Multivariable Systems: The Role and Use of the Return Difference Matrix, *IEEE Transactions on Automatic Control*, Vol. AC-26, No. 1, pp. 47–65.

Steer, A. J. 2000. Low Speed Control of a Second Generation Supersonic Transport Aircraft Using Integrated Thrust Vectoring, *Aeronautical Journal*, Vol. 104, No. 1035, pp. 237–245.

Stengel, Robert F. 1986. *Stochastic Optimal Control*, New York: Wiley.

Stengel, Robert F. 1993. Toward Intelligent Flight Control, *IEEE Transactions on Systems, Man, and Cybernetics*, Vol. SMC-23, No. 6, pp. 1699–1717.

Stengel, Robert F. and Paul W. Berry 1977. Stability and Control of Maneuvering High-Performance Aircraft, *Jour. of Aircraft*, Vol. 14, No. 8, pp. 787–794.

Ward, Greta N. and Uy-Loi Ly 1996. Stability Augmentation Design of a Large Flexible Transport using Nonlinear Parameter Optimization, *Jour. of Guidance, Control, and Dynamics*, Vol. 19, No. 2, pp. 469–474.

Whitbeck, Richard F. 1968. A Frequency Domain Approach to Linear Optimal Control, *Jour. of Aircraft*, Vol. 5, No. 4. pp. 395–401.

Whitbeck, Richard F. and Dennis G. J. Didaleusky 1980. Multi-Rate Digital Control Systems with Simulation Applications, AFWAL-TR-80-3101, Wright Air Force Base, Wright-Patterson, OH.

Whitbeck, R. F. and L. G. Hofmann 1978. Digital Law Synthesis in the w' Domain, *Jour. of Guidance and Control*, Vol. 1, No. 5, pp. 319–326.

White, Roland J. 1950. Investigation of Lateral Dynamic Stability in the XB-47 Airplane, *Jour. of the Aeronautical Sciences*, Vol. 17, No. 3, pp. 133–148.

Chapter 21. Flying Qualities Research Moves with the Times

Anderson, M. R. and D. K. Schmidt 1987. Closed-Loop Pilot Vehicle Analysis of the Approach and Landing Task, *Jour. of Guidance, Control, and Dynamics*, Vol. 10, No. 2, pp. 187–194.

Ashkenas, Irving L. 1988. Pilot Modeling Applications, in *Advances in Flying Qualities*, AGARD Lecture Series LS-157.

Ashkenas, Irving L., Henry R. Jex, and Gary L. Teper 1984. Analysis of Shuttle Orbiter Approach and Landing, *Jour. of Guidance, Control, and Dynamics*, Vol. 7, No. 1, pp. 106–112.

Dornheim, Michael A. 1993 (May 4). Report Pinpoints Factors Leading to YF-22 Crash, AWST, pp. 52–54.

Gibson, John 1999. *Development of a Methodology for Excellence in Handling Qualities Design for Fly by Wire Aircraft*, Series 03 Control and Simulation 06, Delft: Delft U.

Gibson, John C. 2000. Unpublished comments in letter, dated 10 December.

Grey, Jerry 2000. Monitoring Aircraft in Real Time, *Aerospace America*, Vol. 38, No. 11, pp. 36–40.

Hess, Ronald A. 1976. A Method for Generating Numerical Pilot Opinion Ratings Using the Optimal Pilot Model, NASA TM X-73101.

Hess, Ronald A. 1990. Methodology for the Analytical Assessment of Aircraft Handling Qualities, *Control and Dynamic Systems*, Vol. 33, No. 3, pp. 129–149.

Hoh, Roger H. 1988. Advances in Flying Qualities, in *Advances in Flying Qualities*, AGARD Lecture Series 157.

Kleinman, D. L., S. Baron, and W. H. Levinson 1970. An Optimal Control Model of Human Response, *Automatica*, Vol. 6, No. 3, pp. 357–383.

McRuer, Duane T. 1973. Human Operator System and Subsystem Dynamic Characteristics, in *Regulation and Control in Physiological Systems*, Pittsburgh, PA: Instr. Soc. of Amer., pp. 230–235.

McRuer, Duane T. 1973. Development of Pilot-in-the-Loop Analysis, *Jour. of Aircraft*, Vol. 10, No. 9, pp. 515–524.

McRuer, Duane T. 1988. Pilot Modeling, in *Advances in Flying Qualities*, AGARD Lecture Series LS-157.

McRuer, Duane T. 1990. Pilot-Vehicle Analysis of Multiaxis Tasks, *Jour. of Guidance*, Vol. 13, No. 2, pp. 348–355.

McRuer, Duane T. 1992. Human Dynamics and Pilot-Induced Oscillations, 22nd Minta Martin Lecture, MIT.

McRuer, Duane T. 1994. Pilot-Induced Oscillations and Human Dynamic Behavior, NASA CR 4683.

McRuer, Duane T. (chair) 1997. *Report of the Committee on the Effects of Aircraft-Pilot Coupling on Flight Safety*, Washington, DC: National Academy Press.

McRuer, Duane T. and E. S. Krendel 1974. Mathematical Models of Human Pilot Behavior, AGARDograph. No. 188.

McRuer, Duane T. and David K. Schmidt 1990. Pilot-Vehicle Analysis of Multi-Axis Task, *Jour. of Guidance, Control, and Dynamics*, Vol. 13, No. 2, pp. 348–355.

McRuer, Duane T. et al. 1990. Pilot Modeling for Flying Qualities Applications, *Minimum Flying Qualities, Vol II*, WRDC-TR-89-3125.

Moorhouse, David J. and Robert J. Woodcock 1982. Present Status of Flying Qualities Criteria for Conventional Aircraft, in *Criteria for Handling Qualities of Military Aircraft*, AGARD CP 333.

Neil, T. Peter and Rogers E. Smith 1970. An In-Flight Investigation to Develop Control System Design Criteria for Fighter Airplanes, AFFDL-TR-70-74, Vol. I.

Smith, Ralph H. and Norman D. Geddes 1979. Handling Quality Requirements for Advanced Aircraft Design: Longitudinal Mode, AFFDL-TR-78-154.

Stengel, Robert F and John R. Broussard 1978. Prediction of Pilot-Aircraft Stability Boundaries and Performance Contours, *IEEE Transactions on Systems, Man, and Cybernetics*, Vol. SMC-8, No. 5, pp. 349–356.

Thompson, Peter M. and Duane T. McRuer 1988. Comparison of the Human Optimal Control and Crossover Models, AIAA Paper 88-4183, *AIAA Guidance, Navigation and Control Conf.*, Minneapolis.

Chapter 22. Challenge of Stealth Aerodynamics

Farley, Harold C., Jr. and Richard Abrams 1990. F-117A Flight Test Program, in *Thirty-Fourth Symposium Proceedings*, Beverly Hills, CA: Soc. of Experimental Test Pilots, pp. 141–167.

Fulghum, David A. 1994 (Mar. 14). F-22 Signature Problem Inflicts Weight Penalty, AWST, pp. 30, 31.

Oliveri, Frank 1994 (Mar.). Fundamental Features for Future Fighters, *Air Force Magazine*, pp. 36–40.

Pace, Steve 1992. *The F-117A – The Stealth Fighter*, Blue Ridge Summit, PA: TAB/Aero Books.

Chapter 23. Very Large Aircraft

Cleveland, F. A. 1970. Size Effects in Conventional Aircraft Design, *Jour. of Aircraft*, Vol. 17, No. 6, pp. 483–512.

Condit, P. M., L. G. Kimbrel, and R. G. Root 1966. In-Flight and Ground-Based Simulation of Handling Qualities of Very Large Airplanes in Landing Approach, NASA CR 635.

Gibson, John 1995. The Definition, Understanding and Design of Aircraft Handling Qualities, TU Delft Report LR-756.

Grantham, William D. 1983. Large Aircraft Handling Qualities, *First Annual NASA Aircraft Controls Workshop*, Hampton, VA.

Grantham, William D., Paul M. Smith, Perry L. Deal, and William R. Neely, Jr. 1984. Simulator Study of Several Large, Dissimilar, Cargo Transport Airplanes During Approach and Landing, NASA TP 2357.

Grantham, William D., Paul M. Smith, Lee H. Person, Jr., Robert T. Meyer, and Stephen A Tingas 1987. Piloted Simulator Study of Allowable Time Delays in Large-Airplane Response, NASA TP 2652.

Holleman, Euclid C. and Bruce G. Powers 1972. Flight Investigation of the Roll Requirements for Transport Airplanes in the Landing Approach, NASA TN D-7062.

Mueller, Lee J. 1970. Pilot and Aircraft Augmentation, *Jour. of Aircraft*, Vol. 7, No. 6, pp. 553–556.

Phillips, William H. 1979. Altitude Response of Several Airplanes During Landing Approach, NASA TM 80186.

Pinsker, W. J. G. 1969. The Landing Flare of Large Transport Aircraft, British R & M 3602.

Powers, Bruce G. 1986. Space Shuttle Longitudinal Landing Flying Qualities, *Jour. of Guidance*, Vol. 9, No. 5.

Proctor, Paul 1994 (Feb. 21). Super-Jumbos Pose Design Challenges, *AWST*.

Steer, A. J. and M. V. Cook 1999. Control and Handling Qualities Considerations for an Advanced Supersonic Transport Aircraft, *Aeronautical Jour.*, Vol. 103, No. 1024, pp. 265–271.

Ward, Greta and Uy-Loi Ly 1996. Stability Augmentation Design of a Large Flexible Transport Using Nonlinear Parameter Optimization, *Jour. of Guidance, Control, and Dynamics*, Vol. 19, No. 2, pp. 469–474.

Woodcock, Robert J. 1988. A Second Look at MIL Prime Flying Qualities Requirements, *Advances in Flying Qualities*, AGARD Lecture Series LS-157.

Index